AN INTRODUCTION TO INTEGRAL TRANSFORMS AND THEIR APPLICATIONS

AN INTRODUCTION TO INTEGRAL TRANSFORMS AND THEIR APPLICATIONS

Edited by:

Olga Moreira

ARCLER
PRESS

www.arclerpress.com

An Introduction to Integral Transforms and Their Applications

Olga Moreira

Arcler Press
2010 Winston Park Drive,
2nd Floor
Oakville, ON L6H 5R7
Canada
www.arclerpress.com
Tel: 001-289-291-7705
001-905-616-2116
Fax: 001-289-291-7601
Email: orders@arclereducation.com

ISBN: 978-1-77361-554-7 (Hardcover)

DECLARATION

ABOUT THE EDITOR

Olga Moreira obtained her Ph.D. in Astrophysics from the University of Liege (Belgium) in 2010, her BSc. in Physics and Applied Mathematics from the University of Porto (Portugal). Her post-graduate travels and international collaborations with the European Space Agency (ESA) and European Southern Observatory (ESO) led to great personal and professional growth as a scientist. Currently, she is working as an independent researcher, technical writer, and editor in the fields of Mathematics, Physics, Astronomy and Astrophysics.

TABLE OF CONTENTS

LIST OF CONTRIBUTORS

Jasson Vindas
Mathematics Department, Louisiana State University, Baton Rouge, USA

Ricardo Estrada
Mathematics Department, Louisiana State University, Baton Rouge, USA

Hwajoon Kim
Faculty of General Education, Kyungdong University, Yangju, Gyeonggi 11458, Republic of Korea

Shanoja R. Naik
Department of Mathematics and Statistics, University of Regina, Regina, SK S4S 0A2, Canada

Hans J. Haubold
Office for Outer Space Affairs, United Nations, Vienna International Centre, Vienna 1400, Austria

Darren Crowdy
Department of Mathematics, Imperial College London, 180 Queen's Gate, London SW7 2AZ, UK

Pingrun Li
School of Mathematical Science, Qufu Normal University, Qufu, P.R. China

Oleg Yaremko
Penza State University, Krasnaya Street 40, 440038 Penza, Russia

Natalia Yaremko
Penza State University, Krasnaya Street 40, 440038 Penza, Russia

Norazrizal Aswad Abdul Rahman
Institute of Engineering Mathematics, Universiti Malaysia Perlis, Pauh Putra Main Campus, 02600 Arau, Perlis, Malaysia

Muhammad Zaini Ahmad
Institute of Engineering Mathematics, Universiti Malaysia Perlis, Pauh Putra Main Campus, 02600 Arau, Perlis, Malaysia

Reiner Lenz
Department of Electrical Engineering, Linköping University, SE-581 83 Linköping, Sweden
Department of Science and Technology, Linköping University, SE-60174 Norrköping, Sweden

Kai Liu
School of Electrical Engineering, Southeast University, Nanjing 210096, China

Ren-Jie Hu
School of Electrical Engineering, Southeast University, Nanjing 210096, China

Carlo Cattani
Department of Mathematics, University of Salerno, Via Ponte don Melillo, Fisciano, 84084 Salerno, Italy

Gong-Nan Xie
School of Mechanical Engineering, Northwestern Polytechnical University, Xi'an, Shaanxi 710048, China

Xiao-Jun Yang
Department of Mathematics and Mechanics, China University of Mining and Technology, Xuzhou, Jiangsu 221008, China

Yang Zhao
Department of Electronic and Information Technology, Jiangmen Polytechnic, Jiangmen 529090, China

Leonid P. Yaroslavsky
Department of Physical Electronics, School of Electrical Engineering, Tel Aviv University, 69978 Tel Aviv, Israel

Ali Sahin
Department of Mathematics, Fatih University, Buyukcekmece, 34500 Istanbul, Turkey

Chen An
Institute for Ocean Engineering, China University of Petroleum-Beijing, Beijing 102249, China

Jian Su
Nuclear Engineering Program, COPPE, Universidade Federal do Rio de Janeiro, CP 68509, 21941-972 Rio de Janeiro, RJ, Brazil

LIST OF ABBREVIATIONS

ALF	Associated Legendre Functions
DCT	Discrete Cosine Transform
DFrT	Discrete Fresnel Transform
DFT	Discrete Fourier Transform
FDEs	Fuzzy Differential Equations
FFDEs	Fuzzy Fractional Differential Equations
FG	Functionally Graded
FGM	Functionally Graded Material
FLT	Fuzzy Laplace Transform
FST	Fuzzy Sumudu Transform
GITT	Generalized Integral Transform Technique
LDCT	Lebesgue's dominated convergence theorem
MFT	Mehler-Fock-transform
MRI	Magnetic Resonance Imaging
MSE	Mean Square Error
PSNR	Peak Signal-to-Noise Ratio
RMSE	Root Mean Square Approximation Error
RMSE	Root Mean Square Error
ROC	Region of Convergence
SNR	Signal-to-Noise Ratio

PREFACE

The Fourier, Laplace, Mellin, Hilbert, and Hankel transforms are classic examples of integral equations with numerous applications in several fields of science and engineering. When mapping a function from one domain into another, integral transforms provide an elegant solution for many mathematical problems that are algebraically difficult to solve in their original domain but of simpler solution in the target domain. The Fourier transform, for example, maps an original signal (represented by a function of time) into its corresponding power spectrum in the frequency domain. Therefore, the Fourier transform plays an essential role in data analysis of periodic signals, decomposing the incoming signal into the frequencies that it consists of. The Laplace transform converts a convolution in the original domain into a simple multiplication in the target domain. This transform has extensive applications in the fields of engineering and physics research involving the analysis of time-invariant systems in which the output signal is calculated as the convolution between the impulse response and the input signal. Many applications of probability theory also rely on integral transforms, such as statistics that are based on kernel functions.

This book is a collection of contemporary open access articles which highlight the importance of integral transforms and their applications. In particular, three topics are discussed:

- Analysis of generalized functions for integral transform and their properties;
- Methods for solving the Cauchy problem (initial value or boundary problems) for space-time partial differential equations;
- Applications of integral transform, including fractional methods, to solving systems of differential equations in physics, signal processing, quantum mechanics and mechanical engineering.

The generalization of global estimates for various integral transforms is discussed by Vindas & Estrada (2006). The properties kernels of Laplace-typed integral transform and q-Laplace transforms were studied by Kim (2017) and Naik & Haubold (2016), respectively. Transform methods for convex polygons are derived by Crowdy (2015); the analysis of singular integral equations

(Li, 2017); matrix integral transforms (Yaremko & Yaremko, 2016), Sumudu transforms (Rahman & Ahmad, 2015) are also included in this book.

The remaining content of this book focuses on applications of the integral transform as a method for solving various differential equations. For instance, image transform processing and compression (Yaroslavsky, 2014); the modeling of thermal distributions around a barrier at the interface of coating and substrate (Sahin, 2013) ; analysis of dynamic behavior of an axially functionally graded pipe conveying fluid (An & Su, 2017); the effects of wall shear stress on unsteady MHD conjugate flow (Khan et al, 2014) ; and study of a plane wave diffraction problem by a finite plate having different impedance boundaries (Nawaz, Ayub & Javaid, 2014).

This book is mainly addressed to physicists, advanced undergraduate and graduate students in the Natural Sciences and Mechanical Engineering.

CHAPTER 1

GLOBAL BEHAVIOR OF INTEGRAL TRANSFORMS

Jasson Vindas and Ricardo Estrada

Mathematics Department, Louisiana State University, Baton Rouge, USA

ABSTRACT

We obtain global estimates for various integral transforms of positive differentiable functions that satisfy inequalities of the type $c_1 f(x)/x \leq -f'(x) \leq c_2 f(x)/x$, for x > 0.

Keywords: Differentiable Function, Global Estimate, Global Behavior, Positive Differentiable Function

INTRODUCTION

In a recent article, Berndt [1] obtained the following global estimate for the Fourier sine transform of the function f :

$$\frac{A}{x}f\left(\frac{1}{x}\right) \leq \int_0^\infty f(u)\sin(ux)du \leq \frac{B}{x}f\left(\frac{1}{x}\right), \quad \forall x > 0, \tag{1.1}$$

CITATION (APA): Vindas, J., & Estrada, R. (2006). Global behavior of integral transforms. Journal of Inequalities and Applications, 2006(1), 61980. Pages 1- 16 where A and B are positive constants, provided that f is a differentiable function defined on $(0,\infty)$ that satisfies

$$c_1 \frac{f(x)}{x} \le -f'(x) \le c_2 \frac{f(x)}{x}, \tag{1.2}$$

where c_1 and c_2 are constants with

$$0 < c_1 \le c_2 < 2. \tag{1.3}$$

It should be remarked that asymptotic estimates of the behavior of the sine and of other integral transforms of regularly varying functions [6] in terms of the function $f(1/x)$ had been obtained before [7–9], both as $x \to 0$ and as $x \to \infty$. However, (1.1) is a global estimate that not only considers the endpoint behavior but also holds for all $x > 0$.

Our aim in this article is to generalize (1.1) in two directions. On the one hand, we want to consider other kernels than sine, so we will give conditions on the kernel $k(x)$ such that an estimate of the form

$$\frac{A'}{x} f\left(\frac{1}{x}\right) \le \int_0^\infty f(u)k(ux)du \le \frac{B'}{x} f\left(\frac{1}{x}\right), \quad \forall x > 0, \tag{1.4}$$

holds if f satisfies (1.2).

On the other hand, we will remove the condition $c_2 < 2$ for the sine transform. Actually, this condition was imposed by Berndt to guarantee the integrability of $\sin(ux)f'(x)$ at $x = 0$; if $c_2 \ge 2$, this function might not be integrable near $x = 0$. In such a case, the ordinary sine transform of f will not exist, but one may consider regularizations of f which are tempered distribution of the space $\mathcal{S}'$, and whose Fourier sine transforms satisfy a global estimate as in (1.1), modulo a polynomial. In this way, we remove the problem of nonintegrability at $x = 0$. We are also able to remove the integrability condition (in general, if $c_2 \ge 1$, f may not be integrable at 0) and obtain global estimates modulo a polynomial for the Laplace transform of f.

Our analysis is based on a characterization of the class of function $\mathfrak{V}$, which consists of those differentiable functions that satisfy (1.2). This characterization is given in Section 3. Using this characterization we are able to give several global estimates for integral transforms of elements of $\mathfrak{V}$, both for general oscillatory kernels, particularly for the sine transform, and for the Laplace transform in Sections 4 and 5.

PRELIMINARIES

In this section, we explain the spaces of test functions and distributions employed in this article. We also present some of the properties of these spaces that will be needed in our analysis.

The space of test functions of rapid decay and its dual space $\mathcal{S}$', space of tempered distributions, are well known [4, 5, 10, 11].

We will discuss the concept of regularization [4, 5, 11]. If f is a function, denote by supp f the closure of the set of points for which f does not vanish. Let f be a real-valued function, which we assume to be locally integrable in R \ {0}; we say that a distribution $\tilde{f}$ ∈$\mathcal{S}$' is a regularization of f at 0 if for all φ∈$\mathcal{S}$ with sup φ ⊆ (−∞,0) ∪ (0,∞), we have

$$\langle \tilde{f}(x), \phi(x) \rangle = \int_{-\infty}^{\infty} f(x)\phi(x)dx. \tag{2.1}$$

Of course, we assume that the integral in the right of the last equality makes sense. The function f has a regularization at x = 0 if and only if it has an algebraic growth near the origin in the Cesaro sense [` 3](see also [4, pages 297–332] for a complete discussion of Cesaro behavior of distributions). If a function ` f has a regularization at 0, then it has infinitely many regularizations at 0, and all of them are obtained by adding a linear combination of the Dirac delta function and its derivatives concentrated at 0 [4, 10, 11]. Thus, given $\tilde{f}$ and $\tilde{f}_1$, two regularizations of f at 0, they satisfy

$$\tilde{f}_1(x) = \tilde{f}(x) + \sum_{i=0}^{n} a_i \delta^{(i)}(x), \tag{2.2}$$

for some real constants $a_0,\ldots,a_n$.

Let T be a linear continuous operator on $\mathcal{S}$. We define its transpose as the linear operator ⊥T defined on given by g ↦ ⊥ Tg, where ⊥Tg is the tempered distribution defined by

$$\langle ({}^{\perp}Tg)(x), \phi(x) \rangle = \langle g(x), (T\phi)(x) \rangle. \tag{2.3}$$

The Fourier transform of a tempered distribution is defined as the transpose operator of the Fourier transform on the space of test functions of rapid decay at ∞ [10]. If φ ∈$\mathcal{S}$, then its Fourier transform is again an element of $\mathcal{S}$ [4, 5, 10, 11]. Therefore, if g ∈ $\mathcal{S}$', we define its Fourier transform G ∈$\mathcal{S}$' as

$$\langle G(x), \phi(x) \rangle = \left\langle g(x), \int_{-\infty}^{\infty} \phi(u) e^{ixu} du \right\rangle. \tag{2.4}$$

We will define the sine transform of a tempered distribution in the same way as we defined the Fourier transform. Note that if φ ∈ $\mathcal{S}$, then its sine transform, defined as

$$\int_{0}^{\infty} \phi(u) \sin(xu) du, \tag{2.5}$$

is also an element of $\mathcal{S}$'. We define the sine transform on as the transpose of the sine transform on $\mathcal{S}$.

The Laplace transform of a tempered distribution cannot be defined in every case. However, it can be defined for tempered distributions whose support is bounded on the left [11, pages 222–224]. In fact, if g ∈$\mathcal{S}$ with sup g ⊆ [0,∞), we define L, the Laplace transform of g, as the function

$$L(x) = \langle g(u), \lambda(u) e^{-xu} \rangle, \tag{2.6}$$

where λ is any infinitely smooth function with support bounded on the left, which equals one over a neighborhood of the support of g [11]. This definition is independent of the choice of λ.

CHARACTERIZATION OF THE CLASS $\mathfrak{V}$

In this section, we will define and characterize the class of functions V. The study of integral transforms of elements in this class will be the central subject of this paper.

Definition 3.1. A positive, differentiable function f defined on (0,∞) is said to be an element of $\mathfrak{V}$ if it satisfies

$$c_1 \frac{f(x)}{x} \leq -f'(x) \leq c_2 \frac{f(x)}{x}, \tag{3.1}$$

where c_1 and c_2 are positive numbers.

We will prove that the functions in $\mathfrak{V}$ satisfy a variational property. Let us start by setting

$$\epsilon(t) = \frac{-t f'(t)}{f(t)}. \tag{3.2}$$

It follows that satisfies

$$c_1 \leq \epsilon(t) \leq c_2, \quad \forall t > 0. \tag{3.3}$$

By integrating −(t)/t, we obtain

$$\log f(x) = -\int_1^x \frac{\epsilon(t)}{t}dt + \log f(1), \tag{3.4}$$

and hence

$$f(x) = f(1)\exp\left\{-\int_1^x \frac{\epsilon(t)}{t}dt\right\}, \tag{3.5}$$

which gives us a representation formula for f. Conversely, if (3.3) and (3.5) hold, then f satisfies (3.1). This fact is stated in the following lemma.

Lemma 3.2. A function f defined on $(0,\infty)$ belongs to the class $\mathfrak{V}$ if and only if it satisfies (3.5), where satisfies (3.3).

In fact, the last lemma was obtained by Berndt independently in his dissertation [2, Lemma 1.4]. We now give another characterization of the elements of $\mathfrak{V}$.

Theorem 3.3. A function f defined on $(0,\infty)$ belongs to $\mathfrak{V}$ if and only if it is a positive differentiable function and satisfies

$$\frac{1}{u^{c_1}} \leq \frac{f(ux)}{f(x)} \leq \frac{1}{u^{c_2}}, \quad \forall x \in (0,\infty),\ \forall u \in (0,1], \tag{3.6}$$

$$\frac{1}{u^{c_2}} \leq \frac{f(ux)}{f(x)} \leq \frac{1}{u^{c_1}}, \quad \forall x \in (0,\infty),\ \forall u \in [1,\infty). \tag{3.7}$$

Proof. We assume that $f \in \mathfrak{V}$. By Lemma 3.2,

$$f(x) = f(1)\exp\left\{-\int_1^x \frac{\epsilon(t)}{t}dt\right\}, \tag{3.8}$$

where $c_1 \leq (t) \leq c_2$. Therefore,

$$\frac{f(ux)}{f(x)} = \exp\left\{\int_1^x \frac{\epsilon(t)}{t}dt - \int_1^{xu} \frac{\epsilon(t)}{t}dt\right\}. \tag{3.9}$$

Let us take $u \in (0, 1]$. Then we have

$$\int_1^x \frac{\epsilon(t)}{t}dt - \int_1^{xu} \frac{\epsilon(t)}{t}dt = \int_{xu}^x \frac{\epsilon(t)}{t}dt. \tag{3.10}$$

Moreover,

$$\log\left(\frac{1}{u^{c_1}}\right) = c_1\int_{xu}^x \frac{dt}{t} \leq \int_{xu}^x \frac{\epsilon(t)}{t}dt \leq c_2\int_{xu}^x \frac{dt}{t} = \log\left(\frac{1}{u^{c_2}}\right). \tag{3.11}$$

Therefore, (3.6) holds. By using a similar argument, we can see that (3.7) follows.

Let us now assume the converse. First of all, we will show that f is a decreasing function. Let us take $y \geq x$; by setting $u = x/y$ in (3.6), we obtain

$$\frac{f(x)}{f(y)} = \frac{f(y(x/y))}{f(y)} \geq \left(\frac{x}{y}\right)^{-c_1} \geq 1, \tag{3.12}$$

and so f is a decreasing function. Set now $g(y) = \log f(e^y)$; by (3.6), we have

$$-c_1 u \leq g(y+u) - g(y) \leq -c_2 u, \quad \forall u < 0, \tag{3.13}$$

or

$$-c_2 \leq \frac{g(y+u) - g(x)}{u} \leq -c_1, \quad \forall u < 0. \tag{3.14}$$

Taking $u \to 0^-$, we obtain

$$-c_2 \leq g'(y) \leq -c_1, \tag{3.15}$$

and hence

$$c_1 \leq \frac{-f'(e^y)}{f(e^y)} e^y \leq c_2. \tag{3.16}$$

Therefore,

$$\frac{c_1 f(x)}{x} \leq -f'(x) \leq \frac{c_2 f(x)}{x}, \tag{3.17}$$

and thus $f \in \mathfrak{V}$.

Corollary 3.4. If f belongs to $\mathfrak{V}$, with constants c_1 and c_2, then

$$f(t) = O\left(\frac{1}{t^{c_2}}\right), \quad t \longrightarrow 0^+. \tag{3.18}$$

Proof. According to Theorem 3.3,

$$t^{-c_1} \leq \frac{f(t)}{f(1)} \leq t^{-c_2}, \quad \forall t \in (0,1]. \tag{3.19}$$

Thus,

$$0 < t^{c_2} f(t) \leq f(1), \quad \forall t \in (0,1], \tag{3.20}$$

as required.

Note that the last corollary implies the integrability of $f(u)\sin(ux)$ (with respect to u), in any interval $(0,a)$, $a < \infty$, only for $c_2 < 2$. Moreover, if k is continuous on $(0,\infty)$ and

$$k(t) = O(t^{\alpha}), \quad \text{as } t \longrightarrow 0, \tag{3.21}$$

then for the integrability of f (u)k(ux) at 0 it is sufficient to have $c_2 < \alpha + 1$. We observe also that the corollary implies that any $f \in \mathfrak{V}$ admits regularizations in the space $\mathcal{S}$' since f (t) is bounded by a power of t as $t \to 0^+$.

It is interesting that one may obtain inequalities similar to (3.6) and (3.7) for functions that do not belong to V. Indeed, the following result applies to oscillatory functions like $f(x) = x^{-c}(2 + \sin\ln x)$.

Theorem 3.5. Let f be a positive function defined in $(0,\infty)$. Suppose that for each compact set $J \subset (0,\infty)$ there are constants m = m(J) and M = M(J) with 0 <m

$$m \le \frac{f(ux)}{f(x)} \le M, \quad \forall x \in (0,\infty),\ \forall u \in J. \tag{3.22}$$

Then there exist constants K_q, $1 \le q \le 4$, and c_1, c_2 such that

$$\frac{K_1}{u^{c_1}} \le \frac{f(ux)}{f(x)} \le \frac{K_2}{u^{c_2}}, \quad \forall x \in (0,\infty),\ \forall u \in (0,1], \tag{3.23}$$

$$\frac{K_3}{u^{c_2}} \le \frac{f(ux)}{f(x)} \le \frac{K_4}{u^{c_1}}, \quad \forall x \in (0,\infty),\ \forall u \in [1,\infty). \tag{3.24}$$

Proof. Let

$$M_+(u) = \sup\left\{\frac{f(ux)}{f(x)} : x \in (0,\infty)\right\}. \tag{3.25}$$

Then M_+ is locally bounded in $(0,\infty)$ and satisfies

$$M_+(uv) \le M_+(u)M_+(v). \tag{3.26}$$

If we now write $\ln u = n+\theta$, where $n \in \mathbb{N}$ and where $0 \le \theta < 1$, for $u \ge 1$, we obtain

$$M_+(u) \le \sup\{M_+(e^{\theta}) : 0 \le \theta \le 1\} M_+(e)^{\ln u}, \tag{3.27}$$

whenever $u \ge 1$, and thus the right inequality in (3.23) follows with $K_2 = \sup\{M+(e^{\theta}) : 0 \le \theta \le 1\}$ and $c_2 = -\ln\max\{M+(e),1\}$. This also gives us the left inequality in (3.24) with $K_3 = 1/K_2$. The proof of the other two inequalities is similar (or can be obtained by applying what we already proved to the function 1/ f).

OSCILLATORY KERNELS

Let $f \in \mathfrak{V}$. Suppose that $c_2 < 2$ in Definition 3.1. It was proved by Berndt [1] that its sine transform satisfies

$$\frac{A}{x} f\left(\frac{1}{x}\right) \leq \int_0^\infty f(u)\sin(ux)du \leq \frac{B}{x} f\left(\frac{1}{x}\right), \quad \forall x > 0. \tag{4.1}$$

The previous inequality provides us an estimate of the global behavior for the sine transform of f in terms of f (1/x).

$$F(x) = \int_0^\infty k(xu)f(u)du. \tag{4.2}$$

Our aim is to generalize (4.1) in two directions. First, we want to consider other kernels than sine, so we will give conditions on the kernel such that an estimate similar to (4.1) holds. Second, we will remove the condition $c_2 < 2$ for the sine transform; in such a case, the sine transform of f will exist as a tempered distribution satisfying a global estimate as in (4.1), modulo a polynomial.

For our first goal, we define the k transform of f as the function F given by

$$\left| \int_{\lambda_n}^{\lambda_{n+1}} k(t)dt \right| \geq \left| \int_{\lambda_{n+1}}^{\lambda_{n+2}} k(t)dt \right|; \tag{4.3}$$

We will assume that k satisfies

(1) k is continuous on $[0,\infty)$;

(2) k has only simple zeros, located at $t = \lambda_n$, where $\{\lambda_n\}_{n=0}^{\infty}$ satisfies that $\lambda_0 = 0$, and $\lambda_0 < \lambda_1 < \cdots < \lambda_n < \cdots$, where $\lambda_n \to \infty$ as $n \to \infty$; k changes sign at every λ_n, being positive on (λ_0, λ_1), and

(3) $k(t) = O(t^\alpha)$, $\alpha \geq 0$, $t \to 0$.

We can now state our first theorem.

Theorem 4.1. Let f be an element of the class $\mathfrak{V}$. If k satisfies (1), (2), and (3), and $c_2 < \alpha + 1$, then

$$F(x) = \frac{1}{x} f\left(\frac{1}{x}\right) h(x), \quad \forall x > 0, \tag{4.4}$$

where h is continuous and bounded above and below by positive constants. Hence there exist positive constants A and B such that

$$\frac{A}{x} f\left(\frac{1}{x}\right) \leq F(x) \leq \frac{B}{x} f\left(\frac{1}{x}\right), \quad \forall x > 0. \tag{4.5}$$

Note that Theorem 4.1 is applicable to a wide class of kernels. For example, it applies to the Hankel kernel defined by

$$k(t) = t^{1/2}J_\nu(t), \quad \nu > -\frac{1}{2}, \tag{4.6}$$

under the assumption $c_2 < \nu + 3/2$. Let us consider the proof of Theorem 4.1.

Proof. If we perform a change of variables we obtain

$$F(x) = x^{-1}\int_0^\infty f\left(\frac{u}{x}\right)k(u)du. \tag{4.7}$$

Let

$$d_n(x) = \int_{\lambda_n}^{\lambda_{n+1}} f\left(\frac{u}{x}\right)k(u)du. \tag{4.8}$$

It follows that

$$F(x) = x^{-1}\sum_{n=0}^{\infty} d_n(x). \tag{4.9}$$

Since $\sum_{n=0}^{\infty} d_n(x)$ is an alternating series and $|d_n(x)|$ decreases to zero as $n \to \infty$, we have

$$x^{-1}\sum_{j=0}^{2n+1} d_j(x) \le F(x) \le x^{-1}\sum_{j=0}^{2n} d_j(x), \quad n \ge 0, \tag{4.10}$$

which is equivalent to

$$\int_0^{\lambda_{2n+2}} \frac{f(u/x)}{f(1/x)}k(u)du \le \frac{F(x)}{x^{-1}f(1/x)} \le \int_0^{\lambda_{2n+1}} \frac{f(u/x)}{f(1/x)}k(u)du. \tag{4.11}$$

In particular, for $n = 0$,

$$\int_0^{\lambda_2} \frac{f(u/x)}{f(1/x)}k(u)du \le \frac{F(x)}{x^{-1}f(1/x)} \le \int_0^{\lambda_1} \frac{f(u/x)}{f(1/x)}k(u)du. \tag{4.12}$$

Next, we will find positive constants $A, B < \infty$ such that

$$\int_0^{\lambda_1} \frac{f(u/x)}{f(1/x)}k(u)du \le B, \quad \forall x > 0, \tag{4.13}$$

$$\int_0^{\lambda_2} \frac{f(u/x)}{f(1/x)}du \ge A, \quad \forall x > 0, \tag{4.14}$$

and then (4.5) will follow. By Theorem 3.3,

$$\frac{f(u/x)}{f(1/x)} \le \max\left\{\frac{1}{u^{c_1}}, \frac{1}{u^{c_2}}\right\}, \tag{4.15}$$

and hence

$$\int_0^{\lambda_1} \frac{f(u/x)}{f(1/x)} k(u)du \le \int_0^{\lambda_1} \max\left\{\frac{1}{u^{c_1}}, \frac{1}{u^{c_2}}\right\} k(u)du. \tag{4.16}$$

If we set

$$B = \int_0^{\lambda_1} \max\left\{\frac{1}{u^{c_1}}, \frac{1}{u^{c_2}}\right\} k(u)du, \tag{4.17}$$

then (4.13) follows. Since f is a decreasing function and k is negative on (λ_1, λ_2),

$$\begin{aligned}
&\int_0^{\lambda_1} \frac{f(u/x)}{f(1/x)} k(u)du + \int_{\lambda_1}^{\lambda_2} \frac{f(u/x)}{f(1/x)} k(u)du \\
&\ge \int_0^{\lambda_1} \frac{f(u/x)}{f(1/x)} k(u)du + \int_{\lambda_1}^{\lambda_2} \frac{f(\lambda_1/x)}{f(1/x)} k(u)du \\
&= \int_0^{\lambda_1} \frac{(f(u/x) - f(\lambda_1/x))}{f(1/x)} k(u)du + \frac{f(\lambda_1/x)}{f(1/x)} \int_0^{\lambda_2} k(u)du,
\end{aligned} \tag{4.18}$$

so that

$$\int_0^{\lambda_2} \frac{f(u/x)}{f(1/x)} k(u)du \ge \int_0^{\lambda_1} \frac{(f(u/x) - f(\lambda_1/x))}{f(1/x)} k(u)du. \tag{4.19}$$

Therefore, applying the mean value theorem, we obtain

$$f\left(\frac{u}{x}\right) - f\left(\frac{\lambda_1}{x}\right) = -f'\left(\frac{\eta}{x}\right)\left(\frac{\lambda_1 - u}{x}\right), \tag{4.20}$$

for some point $\eta \in (u, \lambda_1)$. Then, by the left inequality in Definition 3.1,

$$f\left(\frac{u}{x}\right) - f\left(\frac{\lambda_1}{x}\right) \ge c_1 f\left(\frac{\eta}{x}\right)\frac{\lambda_1 - u}{\eta}. \tag{4.21}$$

Since $(1/\eta)f(\eta/x) \ge (1/\lambda_1)f(\lambda_1/x)$, we have

$$f\left(\frac{u}{x}\right) - f\left(\frac{\lambda_1}{x}\right) \ge f\left(\frac{\lambda_1}{x}\right)\frac{c_1(\lambda_1 - u)}{\lambda_1} \ge c_1 f\left(\frac{\lambda_1}{x}\right). \tag{4.22}$$

Combining (4.19) and the last inequality, it follows that

$$\int_0^{\lambda_2} \frac{f(u/x)}{f(1/x)} k(u)du \ge \frac{f(\lambda_1/x)}{f(1/x)} \int_0^{\lambda_1} c_1 k(u)du. \tag{4.23}$$

By Theorem 3.3, this implies that

$$\int_0^{\lambda_2} \frac{f(u/x)}{f(1/x)} k(u)du \ge c_1 \min\left\{\frac{1}{\lambda_1^{c_1}}, \frac{1}{\lambda_1^{c_2}}\right\} \int_0^{\lambda_1} k(u)du. \tag{4.24}$$

Setting A equal to the right-hand side of the last inequality, the relation (4.14) has been proved.

Set now

$$h(x) = \frac{F(x)}{x^{-1} f(1/x)}, \quad x > 0, \tag{4.25}$$

so that

$$h(x) = \lim_{n\to\infty} \sum_{j=0}^{2n} \frac{d_j(x)}{f(1/x)}. \tag{4.26}$$

We will show that each d_j is continuous. Pick $x_0 \in (0,\infty)$ and choose a such that $a > \max\{x_0, 1\}$. By Theorem 3.3

$$\left| f\left(\frac{u}{x}\right) k(u) \right| \le \max\{x^{c_1}, x^{c_2}\} f(u) k(u), \tag{4.27}$$

so that, for any $x \in (0,a]$, it follows that

$$\left| f\left(\frac{u}{x}\right) k(u) \right| \le a^{c_2} f(u) \,|k(u)|. \tag{4.28}$$

We have found an integrable function that dominates $f(u/x)k(u)$ for $x \in (0,a]$, this implies that

$$\lim_{x\to x_0} d_j(x) = d_j(x_0). \tag{4.29}$$

Finally, we show that h is continuous. We claim that the convergence in (4.26) is uniform on each interval $[a,b]$, $0 < a < b < \infty$. By (4.10),

$$\left| h(x) - \sum_{j=0}^{2n} \frac{d_j(x)}{f(1/x)} \right| \le \frac{|d_{2n+1}(x)|}{f(1/x)}. \tag{4.30}$$

We also have

$$\frac{|d_{2n+1}(x)|}{f(1/x)} = \int_{\lambda_{2n+1}}^{\lambda_{2n+2}} \frac{f(u/x)}{f(1/x)} |k(u)|\,du \le \frac{1}{f(1/a)} \int_{\lambda_{2n+1}}^{\lambda_{2n+2}} f\left(\frac{u}{x}\right) |k(u)|\,du$$

$$\le \frac{f(\lambda_{2n+1}/b)}{f(1/a)} \int_{\lambda_{2n+1}}^{\lambda_{2n+2}} |k(u)|\,du \le \frac{f(\lambda_{2n+1}/b)}{f(1/a)} \int_0^{\lambda_1} k(u)\,du. \tag{4.31}$$

Since the last term approaches to 0 as $n \to \infty$, the convergence in (4.26) is uniform on any interval $[a,b]$, $0 < a < b < \infty$. Therefore, h is continuous.

We now consider the second generalization of the estimate (4.1). We want to emphasize that the sine transform in this analysis will be considered as a tempered distribution, so that we will take a regularization of f, instead of f. If we let $c_2 > 2$ with no restriction, the sine transform of f may not exist, as we remarked at the end of Section 3. In order to define a regularization of f, we need to extend f to the whole real line; we do this by setting f (x) =

0 for x < 0; for the sake of simplicity, we will keep denoting this extension by f.

We state our second result.

Theorem 4.2. Let $f \in \mathfrak{V}$. Suppose that $\tilde{f}$ is any regularization of f in $\mathcal{S}'$ and denote the sine transform of $\tilde{f}$ by F. Then for all x > 0 either

$$F(x) = \frac{h(x)}{x} f\left(\frac{1}{x}\right) + P(x), \tag{4.32}$$

or

$$F(x) = -\frac{h(x)}{x} f\left(\frac{1}{x}\right) + P(x), \tag{4.33}$$

where h is continuous and bounded above and below by positive constants and P is a polynomial.

Proof. It is known that any two regularizations of f, say $\tilde{f}$ and $\tilde{f}_1$, satisfy

$$\tilde{f}(x) = \tilde{f}_1(x) + \sum_{i=0}^{m} a_i \delta^{(i)}(x), \tag{4.34}$$

where $a_0, a_1, \ldots, a_m$ are some real constants. Observe that the sine transform of the sum of delta functions and its derivatives on the right-hand side is a polynomial. To see this fact, let φ be a test function of the space $\mathcal{S}$, k ∈ N; then,

$$\left\langle \delta^{(k)}(x), \int_0^\infty \phi(u)\sin(ux)du \right\rangle = 0, \quad \text{if } k \text{ is even};$$

$$\left\langle \delta^{(k)}(x), \int_0^\infty \phi(u)\sin(ux)du \right\rangle = \int_0^\infty (-x)^k \phi(x)dx, \quad \text{if } k = 4j+1;$$

$$\left\langle \delta^{(k)}(x), \int_0^\infty \phi(x)\sin(ux)du \right\rangle = \int_0^\infty x^k \phi(x)dx, \quad \text{if } k = 4j+3. \tag{4.35}$$

Therefore, it suffices to work with any particular regularization of f. So we will find a regularization of f for which the conclusion of the theorem holds. We will suppose that $c_2 \geq 2$; otherwise, the conclusion of this theorem would be a consequence of Theorem 4.1. Let n be the unique natural number such that

$$2n + 1 \leq c_2 < 2n + 3. \tag{4.36}$$

We will divide the proof into two cases. We consider the cases when n is odd and then when n is even.

Assume first that n is odd. Define now $\tilde{f}$ as

$$\langle \tilde{f}(x), \phi(x)\rangle = \int_0^{2\pi} f(x)\left(\phi(x) - \sum_{i=0}^{2n+1} \frac{\phi^{(i)}(0)}{i!} x^i\right) dx + \int_{2\pi}^{\infty} f(x)\phi(x)dx, \tag{4.37}$$

for $\varphi \in \mathcal{S}$. We will prove that $\tilde{f}$ is well defined. Let $\varphi \in \mathcal{S}$, then by Corollary 3.4,

$$f(x)\left(\phi(x) - \sum_{i=0}^{2n+1} \frac{\phi^{(i)}(0)}{i!} x^i\right) = O(x^{2n+2-c_2}), \quad x \longrightarrow 0, \tag{4.38}$$

and so, by (4.36), it is integrable on $(0,2\pi)$. The integrability on $(2\pi,\infty)$ is clear since $\varphi \in \mathcal{S}$. By a standard argument, $\tilde{f} \in \mathcal{S}'$.

We will prove the formula for the sine transform of $\tilde{f}$. Denote by $\tilde{F}$ the sine transform of $\tilde{f}$. Let us now set

$$K(x) = \sin x - \sum_{i=0}^{n} \frac{(-1)^i}{(2i+1)!} x^{2i+1}. \tag{4.39}$$

Since n is odd,

$$K(x) \geq 0, \quad \text{for } x \geq 0. \tag{4.40}$$

Using the definition of $\tilde{F}$, we have for $x > 0$,

$$\begin{aligned}\langle \tilde{F}(x), \phi(x)\rangle &= \left\langle \tilde{f}(x), \int_0^{\infty} \phi(u)\sin(xu)du\right\rangle = \int_0^{2\pi} f(x)\left(\int_0^{\infty} \phi(u)K(xu)du\right)dx \\ &\quad + \int_{2\pi}^{\infty} f(x)\left(\int_0^{\infty} \phi(u)\sin(xu)du\right)dx \\ &= \int_0^{\infty} \frac{\phi(x)}{x}\left(\int_0^{2\pi} f\left(\frac{u}{x}\right)K(u)du + \int_{2\pi}^{\infty} f\left(\frac{u}{x}\right)\sin u\, du\right)dx,\end{aligned} \tag{4.41}$$

for every $\varphi \in \mathcal{S}$. It follows that if $x > 0$,

$$\tilde{F}(x) = \frac{1}{x}\left[\int_0^{2\pi} f\left(\frac{u}{x}\right)K(u)du + \int_{2\pi}^{\infty} f\left(\frac{u}{x}\right)\sin u\, du\right]. \tag{4.42}$$

Hence $\tilde{F}$ can be identified with a classical function, in the sense that $\tilde{F}$ is the distribution generated by the function given by (4.42).

Next we set

$$h(x) = \frac{\widetilde{F}(x)}{x^{-1} f(1/x)}, \quad \text{for } x > 0. \tag{4.43}$$

We will find two constants, A and B, so that

$$A \le h(x) \le B, \quad x > 0. \tag{4.44}$$

Notice that

$$h(x) - \int_0^{2\pi} \frac{f(u/x)}{f(1/x)} K(u)du = \int_{2\pi}^{\infty} \frac{f(u/x)}{f(1/x)} \sin u\, du. \tag{4.45}$$

We also have that

$$\int_{2\pi}^{4\pi} \frac{f(u/x)}{f(1/x)} \sin u\, du \le \int_{2\pi}^{\infty} \frac{f(u/x)}{f(1/x)} \sin u\, du \le \int_{2\pi}^{3\pi} \frac{f(u/x)}{f(1/x)} \sin u\, du. \tag{4.46}$$

We can apply the argument that we used in Theorem 4.1 to find positive constants A' and B' such that

$$\int_{2\pi}^{3\pi} \frac{f(u/x)}{f(1/x)} \sin u\, du \le B', \qquad A' \le \int_{2\pi}^{4\pi} \frac{f(u/x)}{f(1/x)} \sin u\, du, \tag{4.47}$$

for all x ∈ (0,∞). Using the last inequalities, we obtain that

$$A' \le \int_{2\pi}^{\infty} \frac{f(u/x)}{f(u/x)} \sin u\, du \le B'. \tag{4.48}$$

It follows that

$$\int_0^{2\pi} \min\left\{\frac{1}{u^{c_1}}, \frac{1}{u^{c_2}}\right\} K(u)du + A' \le h(x), \qquad h(x) \le \int_0^{2\pi} \max\left\{\frac{1}{u^{c_1}}, \frac{1}{u^{c_2}}\right\} K(u)du + B', \tag{4.49}$$

which shows that h is bounded above and below by positive constants.

We now prove the continuity of h. The continuity of

$$\int_{2\pi}^{\infty} f\left(\frac{u}{x}\right) \sin u\, du \tag{4.50}$$

follows from the proof of Theorem 4.1. Moreover, since

$$\frac{f(u/x)}{f(1/x)} K(u) \le \max\left\{\frac{1}{u^{c_1}}, \frac{1}{u^{c_2}}\right\} K(u), \tag{4.51}$$

it follows by the Lebesgue dominated convergence theorem that

$$h(x) - \int_{2\pi}^{\infty} \frac{f(u/x)}{f(1/x)} \sin u\, du \tag{4.52}$$

is continuous, and so is h(x). This completes the proof for the odd case.

We now assume that n is an even number. Define $\tilde{f}$ as

$$\langle \tilde{f}(x), \phi(x) \rangle = \int_0^{3\pi} f(x) \left(\phi(x) - \sum_{i=0}^{2n+1} \frac{\phi^{(i)}(0)}{i!} x^i \right) dx + \int_{3\pi}^{\infty} f(x)\phi(x)dx, \tag{4.53}$$

for $\varphi \in \mathcal{S}$. It follows that $\tilde{f} \in \mathcal{S}'$. Set

$$J(x) = \sum_{i=0}^{n} \frac{(-1)^i}{(2i+1)!} x^{2i+1} - \sin x, \tag{4.54}$$

which is a positive function, since n is an even number. Let $\tilde{F}$ be the sine transform of $\tilde{f}$. We have that if $x > 0$,

$$\tilde{F}(x) = \frac{1}{x}\left[-\int_0^{3\pi} f\left(\frac{u}{x}\right) J(u)du + \int_{3\pi}^{\infty} f\left(\frac{u}{x}\right) \sin u\, du \right]. \tag{4.55}$$

Set

$$h(x) = -\frac{\tilde{F}(x)}{x^{-1} f(1/x)}, \qquad x > 0. \tag{4.56}$$

It follows that

$$h(x) = \int_0^{3\pi} \frac{f(u/x)}{f(1/x)} J(u)du - \int_{3\pi}^{\infty} \frac{f(u/x)}{f(1/x)} \sin u\, du, \tag{4.57}$$

for $x > 0$. We can find two positive constants, A'' and B'', such that

$$-\int_{3\pi}^{4\pi} \frac{f(u/x)}{f(1/x)} \sin u\, du \le B'', \qquad -\int_{3\pi}^{5\pi} \frac{f(u/x)}{f(1/x)} \sin u\, du \ge A''. \tag{4.58}$$

From these inequalities, it follows that

$$\int_0^{3\pi} \min\left\{\frac{1}{u^{c_1}}, \frac{1}{u^{c_2}}\right\} J(u)du + A'' \le h(x), \qquad h(x) \le \int_0^{3\pi} \max\left\{\frac{1}{u^{c_1}}, \frac{1}{u^{c_2}}\right\} J(u)du + B'', \tag{4.59}$$

which proves the required inequalities. The continuity of h can be established as in the odd case.

LAPLACE TRANSFORM

In this section, we will give a result analogous to Theorem 4.2 for the Laplace transform. The estimate is as follows.

Theorem 5.1. Let $f \in \mathfrak{V}$. Suppose that $\tilde{f}$ is any regularization of f in $\mathcal{S}'$ and denote its Laplace transform by L. Then for all $x > 0$, either

$$L(x) = \frac{h(x)}{x} f\left(\frac{1}{x}\right) + P(x), \tag{5.1}$$

or

$$L(x) = -\frac{h(x)}{x} f \frac{1}{x} + P(x), \tag{5.2}$$

where h is continuous and bounded above and below by positive constants, and P is a polynomial.

Proof. We proceed as in Theorem 4.2. It suffices to consider a particular regularization of f . Let n be the integer part of c_2. We will consider two cases. First, we assume that n is odd, and then we consider the even case.

Assume that n is odd. Define $\tilde{f}$ as

$$\langle \tilde{f}(x), \phi(x) \rangle = \int_0^1 f(x)\left(\phi(x) - \sum_{i=0}^{n} \frac{\phi^i(0)}{i!}\right) dx + \int_1^\infty f(x)\phi(x)dx, \tag{5.3}$$

for $\varphi \in \mathcal{S}$. Then, $\tilde{f}$ is a regularization of f in $\mathcal{S}$'. Since supp $\tilde{f}$ = $[0,\infty)$, its Laplace transform is well defined. Let us denote its Laplace transform by $\tilde{L}$, so that

$$\begin{aligned}\tilde{L}(x) &= \int_0^1 f(u)\left(e^{-ux} - \sum_{i=0}^{n} \frac{(-ux)^i}{i!}\right) du + \int_1^\infty f(u)e^{-ux} du \\ &= \frac{1}{x}\left[\int_0^1 f\left(\frac{u}{x}\right)\left(e^{-u} - \sum_{i=0}^{n} \frac{(-u)^i}{i!}\right) du + \int_1^\infty f\left(\frac{u}{x}\right)e^{-u} du\right].\end{aligned} \tag{5.4}$$

We now consider the following inequality:

$$e^{-x} - \sum_{i=0}^{n} \frac{(-x)^i}{i!} > 0, \quad \text{for } x > 0. \tag{5.5}$$

Set

$$h(x) = \frac{\tilde{L}(x)}{x^{-1} f(1/x)}, \qquad K(x) = e^{-x} - \sum_{i=0}^{n} \frac{(-x)^i}{i!}. \tag{5.6}$$

Then, we have

$$\int_0^1 \frac{K(u)}{u^{c_1}} du + \int_1^\infty \frac{e^{-u}}{u^{c_2}} du \le h(x) \le \int_0^1 \frac{K(u)}{u^{c_2}} du + \int_0^\infty \frac{e^{-u}}{u^{c_1}} du. \tag{5.7}$$

This completes the proof for the odd case.

Assume now that n is even. Set

$$J(x) = \sum_{i=0}^{n} \frac{(-x)^i}{i!} - e^{-x}; \tag{5.8}$$

it follows that

$$J(x) > 0, \quad \text{for } x > 0. \tag{5.9}$$

Take A > 1 such that

$$\int_0^1 \frac{J(u)}{u^{c_1}} du - \int_A^\infty \frac{e^{-u}}{u^{c_1}} du > 0, \qquad \int_0^1 \frac{J(u)}{u^{c_2}} du - \int_A^\infty \frac{e^{-u}}{u^{c_2}} du > 0. \tag{5.10}$$

We define $\tilde{f}$, a regularization of f, as

$$\langle \tilde{f}(x), \phi(x) \rangle = \int_0^A f(x)\left(\phi(x) - \sum_{i=0}^{n} \frac{\phi^{(i)}(0)}{i!}\right)dx + \int_A^\infty f(x)\phi(x)dx. \tag{5.11}$$

It follows that $\tilde{L}$, the Laplace transform of $\tilde{f}$, is given by

$$\tilde{L}(x) = \frac{1}{x}\left(-\int_0^A f\left(\frac{u}{x}\right)J(u)du + \int_A^\infty f\left(\frac{u}{x}\right)e^{-u}du\right). \tag{5.12}$$

Define now h by

$$h(x) = \frac{-\tilde{L}(x)}{x^{-1}f(1/x)}. \tag{5.13}$$

We have that

$$\int_0^1 \frac{J(u)}{u^{c_1}} du + \int_1^A \frac{J(u)}{u^{c_2}} du - \int_A^\infty \frac{e^{-u}}{u^{c_1}} du \le h(x),$$

$$h(x) \le \int_0^1 \frac{J(u)}{u^{c_2}} du + \int_1^A \frac{J(u)}{u^{c_1}} du - \int_A^\infty \frac{e^{-u}}{u^{c_2}} du, \tag{5.14}$$

so h is bounded above and below by positive constants.

REFERENCES

1. R. Berndt, A formula for the Fourier transform of certain odd differentiable functions, Journal of Mathematical Analysis and Applications 285 (2003), no. 2, 349–355.
2. ______, Singular integrals with new singularities, Dissertation, University of Minnesota, Minnesota, 2003.
3. R. Estrada, Regularization of distributions, International Journal of Mathematics and Mathematical Sciences 21 (1998), no. 4, 625–636.
4. R. Estrada and R. P. Kanwal, A Distributional Approach to Asymptotics. Theory and Applications, 2nd ed., Birkhauser Advanced Texts: Basel Textbooks, Birkh ¨ auser Boston, Massachusetts, 2002. ¨
5. R. P. Kanwal, Generalized Functions. Theory and Technique, 2nd ed., Birkhauser Boston, Massachusetts, 1998.
6. E. Seneta, Regularly Varying Functions, Lecture Notes in Mathematics, vol. 508, Springer, Berlin, 1976.
7. K. Soni and R. P. Soni, Slowly varying functions and asymptotic behavior of a class of integral transforms. I, Journal of Mathematical Analysis and Applications 49 (1975), no. 1, 166–179.
8. ______, Slowly varying functions and asymptotic behavior of a class of integral transforms. II, Journal of Mathematical Analysis and Applications 49 (1975), no. 1, 477–495.
9. ______, Slowly varying functions and asymptotic behavior of a class of integral transforms. III, Journal of Mathematical Analysis and Applications 49 (1975), no. 3, 612–628.
10. F. Treves, ` Topological Vector Spaces, Distributions and Kernels, Academic Press, New York, 1967.
11. A. H. Zemanian, Distribution Theory and Transform Analysis. An Introduction to Generalized Functions, with Applications, 2nd ed., Dover, New York, 1987.

CHAPTER 2

THE INTRINSIC STRUCTURE AND PROPERTIES OF LAPLACE-TYPED INTEGRAL TRANSFORMS

Hwajoon Kim

Faculty of General Education, Kyungdong University, Yangju, Gyeonggi 11458, Republic of Korea

ABSTRACT

We would like to establish the intrinsic structure and properties of Laplace-typed integral transforms. The methodology of this article is done by a consideration with respect to the common structure of kernels of Laplace-typed integral transform, and G-transform, the generalized Laplace-typed integral transform, is proposed with the feature of inclusiveness. The proposed -transform can provide an adequate transform in a number of engineering problems.

CITATION (APA): Kim, H. (2017). The intrinsic structure and properties of Laplace-typed integral transforms. Mathematical Problems in Engineering, 2017. (8 pages)

INTRODUCTION

The key motivation for pursuing theories for integral transforms is that it gives a simple tool which is represented by an algebraic problem in the process of solving differential equations [1]. In the most theories for integral transforms, the kernel is doing the important role which transforms one space to the other space in order to solve the solution. The main reason to transform is because it is not easy to solve the equation in the given space, or it is easy to find a characteristic for the special purpose. For example, in computed tomography (CT) or magnetic resonance imaging (MRI), we obtain the projection data by integral transform and produce the image with the inverse transform. This is the strong point of integral transform.

To begin with, let us see the intrinsic structure of Laplace-typed integral transforms. Of course, the structure is dependent on the kernel, and the form of the kernel in Laplace-typed integral transforms is as follows. Laplace transform is defined by

$$\pounds(f) = \int_0^\infty e^{-st} f(t)\,dt, \tag{1}$$

Sumudu one is

$$S(u) = \frac{1}{u}\int_0^\infty e^{-t/u} f(t)\,dt, \tag{2}$$

and Elzaki one is

$$E(u) = u\int_0^\infty e^{-t/u} f(t)\,dt. \tag{3}$$

Since the Laplace transform $\pounds(f)$ can be rewritten as

$$\int_0^\infty e^{-t/u} f(t)\,dt \tag{4}$$

for $s = 1/u$, we can naturally consider that the form of Laplace-typed integral transform G is

$$F(u) = G(f) = u^\alpha \int_0^\infty e^{-t/u} f(t)\,dt \tag{5}$$

for an suitable integer : for example, if

$$G(1) = u^\alpha \int_0^\infty e^{-t/u} dt = u^\alpha(-u)\left[e^{-t/u}\right]_0^\infty = u^{\alpha+1} \tag{6}$$

for u > 0.

Normally, integral transforms have a base of exponential function and it gets along with how to integrate from 0 to ∞ in order to utilize the property to $e\ -st$ or $e\ -t/u$ converges to 0 when t approaches ∞. Sumudu/Elzaki transform is a kind of modified Laplace one introduced by Watugala [2] in 1993/Elzaki et al. [3] in 2012 to solve initial value problems in engineering problems [4]. Belgacem et al. [5, 6] are mentioning that Sumudu transform ($\alpha = -1$) has scale and unit-preserving properties, and it may be used to solve problems without resorting to a new frequency domain. Elzaki et al. [3, 7, 8] insists that Elzaki transform (α=1) should be easily applied to the initial value problems with less computational work and solve the various examples which are not solved by the Laplace or the Sumudu transform. As an application, Agwa et al. [9] deal with Sumudu transform on time scales and its applications, Eltayeb and Kilicman [10] have checked some applications of Sumudu one, and Eltayeb et al. are highlighting the importance of fractional operators of integral transform and their applications in [11]. The shifted data problems, shifting theorems, and the forms of solutions of ODEs with variable coefficients can be found in [4, 12, 13].

On the other hand, Kreyszig [1] says that Laplace transform (α=0) has a strong point in the transforms of derivatives; that is, the differentiation of a function (t) corresponds to multiplication of its transform m(f) by s. In the other view, if we want the inverse case, the transform giving a simple tool for transforms of integrals, then we can choose a suitable form of integral transform such as

$$\frac{1}{u^2}\int_0^{\infty} e^{-t/u} f(t)\, dt. \tag{7}$$

This means that the integer α is applicable to −2. As we checked above, the comprehensive transform in Laplacetyped ones is needed, and thus we would like to propose G- transform, a generalized Laplace-typed integral transform, which is more comprehensive and intrinsic than the existing transforms.

This intrinsic structure in Laplace-typed integral transforms has a meaning which can be directly applied to any situation by choosing an appropriate integer α. The main objective of this paper is to construct the generalized form of Laplace-typed integral transforms and establish the properties of it, and, to the author's knowledge, the proposed G-transform is the first attempt to generalize Laplace-typed integral transforms. Finally, we would like to mention that Laplace transform gave many considerations to this article.

THE PROPERTIES OF LAPLACE-TYPED INTEGRAL TRANSFORMS

The Definition and the Table of Generalized Integral Transform G, Shifting Theorems

As mentioned before, let us rewrite the definition of Laplace-typed integral transforms, and we would like to call it G-transform.

Definition 1. . If $f(t)$ is an integrable function defined for all $t \geq 0$, its generalized integral transform G is the integral of $f(t)$ times $u^{\alpha} \cdot e^{-t/u}$ from $t=0$ to ∞. It is a function of u, say (u), and is denoted by $G(f)$; thus

$$F(u) = G(f) = u^{\alpha} \int_0^{\infty} e^{-t/u} f(t)\, dt. \tag{8}$$

Let us first check the shifting theorems.

Theorem 2. (1) (u-shifting) If (t) has the transform $F(u)$, then $e^{at} f(t)$ has the transform

$$F\left(\frac{u}{1-au}\right). \tag{9}$$

That is,

$$G\left[e^{at} f(t)\right] = F\left(\frac{u}{1-au}\right). \tag{10}$$

(2) (t-shifting) If (t) has the transform $F(u)$, then the shifted function $f(t - a)h(t - a)$ has the transform $e^{-a/u}F(u)$. In formula,

$$G\left[f(t-a)\, h(t-a)\right] = e^{-a/u} F(u) \tag{11}$$

for $h(t - a)$ is Heaviside function (we write h since we need u to denote u-space). Additionally, $[h(t - a)] = u^{\alpha+1} e^{-a/u}$ holds.

Proof. (1) From

$$G\left[e^{at} f(t)\right] = u^{\alpha} \int_0^{\infty} e^{-t/u} e^{at} f(t)\, dt,$$

$$e^{-t/u} e^{at} = e^{(a-1/u)t} = e^{-t/(u/(1-au))}, \tag{12}$$

we obtain the result.

(2)

$$e^{-a/u}F(u) = e^{-a/u} \cdot u^{\alpha} \int_0^{\infty} e^{-\tau/u} f(\tau)\, d\tau$$

$$= u^{\alpha} \int_0^{\infty} e^{-(a+\tau)/u} f(\tau)\, d\tau. \tag{13}$$

Substituting $\tau+a=t$, we obtain

$$e^{-a/u}F(u) = u^{\alpha} \int_a^{\infty} e^{-t/u} f(t-a)\, dt$$

$$= u^{\alpha} \int_0^{\infty} e^{-t/u} f(t-a)\, h(t-a)\, dt$$

$$= G[f(t-a)\, h(t-a)] \tag{14}$$

for h is Heaviside function.

By the similar way, we have $[h(t-a)] = u^{\alpha+1}e^{-a/u}$ for h is Heaviside function.

Using $G(f) = u^{\alpha} \cdot F(1/u)$ for Laplace transform $m(f) = F(s)$, we can obtain the table of generalized integral transform G as shown in Table 1. In the table, we regard Laplacetyped integral transform G as a transform. However, we can choose an appropriate constant according to each situation. For example, the choice of $\alpha=0$ has a merit in the transforms of derivatives, and $\alpha = -2$ has a strong point in the transforms of integrals.

If (t) is defined and is piecewise continuous on $t \geq 0$ and satisfies $|f(t)| \leq Me^{kt}$ for all $t \geq 0$, then $G(f)$ exists for all $u < 1/k$. Since

$$|G(f)| = \left| u^{\alpha} \int_0^{\infty} e^{-t/u} f(t)\, dt \right|$$

$$\leq u^{\alpha} \int_0^{\infty} e^{-t/u} |f(t)|\, dt \leq u^{\alpha} \int_0^{\infty} e^{-t/u} Me^{kt}\, dt$$

$$= \frac{Mu^{\alpha+1}}{ku-1}, \tag{15}$$

the statement is valid.

Table 1: Table of Laplace-typed integral transform

	$f(t)$	$G(f)$
1	1	$u^{\alpha+1}$
1	t	$u^{\alpha+2}$
3	t^n	$n!\cdot u^{n+\alpha+1}$
4	e^{at}	$\frac{u^{\alpha+1}}{1-au}$
5	$\sin at$	$\frac{au^{\alpha+2}}{1+u^2a^2}$
6	$\cos at$	$\frac{u^{\alpha+1}}{1+u^2a^2}$
7	$\sinh at$	$\frac{au^{\alpha+2}}{1-u^2a^2}$
8	$\cosh at$	$\frac{u^{\alpha+1}}{1-u^2a^2}$
9	$e^{at}\cos bt$	$\frac{u^{\alpha}(1/u-a)}{(1/u-a)^2+b^2}$
10	$e^{at}\sin bt$	$\frac{bu^{\alpha}}{(1/u-a)^2+b^2}$

Transforms of Derivatives and Integrals

Theorem 3. Let a function f be n-th differentiable. Then the transforms of the first, second, and n-th derivatives of (t) satisfy

(1) $$G(f')=\frac{1}{u}G(f)-u^{\alpha}f(0) \tag{16}$$

(2) $$G(f'')=\frac{1}{u^2}G(f)-\frac{1}{u}f(0)u^{\alpha}-f'(0)u^{\alpha} \tag{17}$$

(3) $$(f^{(n)})=\frac{1}{u^n}G(f)-\frac{1}{u^{n-1}}f(0)u^{\alpha}-\frac{1}{u^{n-2}}f'(0)u^{\alpha}-\cdots-f^{(n-1)}(0)u^{\alpha} \tag{18}$$

for α is an arbitrary integer.

(4) Let (t) be piecewise continuous for $t\geq 0$ and integrable. Then

$$G\left[\int_0^t f(\tau)\,d\tau\right]=uF(u) \tag{19}$$

holds for $F(u)=G[f(t)]$.

Proof. By the integration by parts,

$$G(f') = u^{\alpha} \int_0^{\infty} e^{-t/u} f'(t)\, dt$$

$$= u^{\alpha} \left[e^{-t/u} f(t)\right]_0^{\infty} + u^{\alpha-1} \int_0^{\infty} e^{-t/u} f(t)\, dt$$

$$= \frac{1}{u} G(f) - u^{\alpha} f(0) \qquad (20)$$

and, similarly,

$$G(f'') = \frac{1}{u} G(f') - f'(0)\, u^{\alpha}$$

$$= \frac{1}{u}\left[\frac{1}{u} G(f) - u^{\alpha} f(0)\right] - f'(0)\, u^{\alpha}$$

$$= \frac{1}{u^2} G(f) - \frac{1}{u} f(0)\, u^{\alpha} - f'(0)\, u^{\alpha} \qquad (21)$$

follows.

Continuing this process by substitution as the above and using induction, (3) follows. Let us minutely establish the validity of the statement of (3) by the mathematical induction. For n=1, it clearly follows. Next, we suppose that

$$G(f^{(n)}) = \frac{1}{u^n} G(f) - \frac{1}{u^{n-1}} f(0)\, u^{\alpha} - \frac{1}{u^{n-2}} f'(0)\, u^{\alpha}$$

$$- \cdots - f^{(n-1)}(0)\, u^{\alpha}, \qquad (22)$$

and we show that $G(f^{(n+1)})$ can be expressed by

$$G(f^{(n+1)}) = \frac{1}{u^{n+1}} G(f) - \frac{1}{u^n} f(0)\, u^{\alpha}$$

$$- \frac{1}{u^{n-1}} f'(0)\, u^{\alpha} - \cdots - f^{(n)}(0)\, u^{\alpha}. \qquad (23)$$

In statement (1), the proof of statement (3) follows by applying (1) to (n).

Finally, (4) follows as below.

Show that $G[\int_0^t f(\tau) d\tau] = u(u)$ holds for t>0. Using Theorem 3 and putting $(t) = \int_0^t f(\tau) d\tau$,

$$G(f(t)) = G(g'(t)) = \frac{1}{u} G(g) - u^{\alpha} g(0)$$

$$= \frac{1}{u} G(g). \qquad (24)$$

Moreover, F(u) clearly satisfies a growth restriction.

Let us check an example in [1] by the G-transform.

Example 4. Solve $y'' - y = t$, $(0) = 1$, and $y'(0) = 1$.

Solution. Taking -transform on both sides, we have

$$\frac{1}{u^2}G(f) - \frac{1}{u}y(0)u^\alpha - y'(0)u^\alpha - G(f) = u^{\alpha+2}. \tag{25}$$

Organizing this equality, we have

$$\left(\frac{1}{u^2} - 1\right)G(f) = \left(u^{\alpha+2} + u^\alpha + u^{\alpha-1}\right). \tag{26}$$

Simplification gives

$$G(f) = \left(u^{\alpha+2} + u^\alpha + u^{\alpha-1}\right) \cdot \frac{u^2}{1-u^2}$$
$$= \frac{u^{\alpha+4}}{1-u^2} + \frac{u^{\alpha+2}}{1-u^2} + \frac{u^{\alpha+1}}{1-u^2}. \tag{27}$$

Since

$$\frac{u^{\alpha+4}}{1-u^2} = -u^{\alpha+2} + \frac{u^{\alpha+2}}{1-u^2}, \tag{28}$$

from Table 1, we have the solution

$$y(t) = -t + 2\sin ht + \cos ht \tag{29}$$

for h is hyperbolic functions. From the substitution, above (t) is exactly a solution of the given equation. Of course, by a simple calculation, the above answer is equal to the solution of [1] which is $e^t \sin ht - t$.

Convolution and Integral Equations for -Transform

It is a well-known fact that £(fg)≠£(f) £(g) and £(f) £(g) = £$(f*g)$for$f*g$ is the convolution of $f$and g.This means that convolution has to do with the multiplication of transforms in Laplace transform. Here, we investigate the change at G-transform. If two functions f and g are integrable, the following theorem is held.

Lemma 5 (Lebesgue's dominated convergence theorem (LDCT) [14]). Let (X, M, μ) be a measure space and suppose that $\{fn\}$ is a sequence of extended real-valued measurable functions defined on X such that

(a) $\lim_{n\to\infty} f(x) = f(x)$ exists μ-a.e

(b) there is an integrable function g so that, for each n, $|f_n| \leq g\mu$-a.e. Then f is integrable and

$$\lim_{n\to\infty} \int_X f_n d\mu = \int_X f\, d\mu. \tag{30}$$

We note that the above lemma gives validity to the following equality:

$$\int \sum_{n=1}^{\infty} g_n d\mu = \sum_{n=1}^{\infty} \int g_n d\mu \tag{31}$$

for (g_n) is a nondecreasing sequence.

Theorem 6.

$$u^\alpha G(f * g) = G(f) G(g). \tag{32}$$

Proof. Let us put

$$G(f) = u^\alpha \int_0^\infty e^{-\tau/u} f(\tau)\, d\tau \tag{33}$$

and put

$$G(g) = u^\alpha \int_0^\infty e^{-v/u} g(v)\, dv. \tag{34}$$

Then

$$G(f) G(g) = u^\alpha \int_0^\infty e^{-\tau/u} f(\tau) \cdot u^\alpha \int_0^\infty e^{-v/u} g(v)\, dv\, d\tau. \tag{35}$$

Let us put $t - V + \iota$, where τ is at first constant. Then $V = t - \tau$ and so we get

$$G(g) = u^\alpha \int_\tau^\infty e^{-(t-\tau)/u} g(t-\tau)\, dt = e^{\tau/u} u^\alpha \int_\tau^\infty e^{-t/u} g(t-\tau)\, dt. \tag{36}$$

Since the function f is integrable, we can change the order of integration by using Lebesgue's dominated convergence theorem. Hence

$$G(f)G(g) = u^{\alpha}$$

$$\cdot u^{\alpha} \int_0^{\infty} \int_{\tau}^{\infty} f(\tau) e^{-t/u} g(t-\tau)\, dt\, d\tau, \tag{37}$$

and when t varies τ to ∞, τ varies 0 to t. Hence

$$G(f)G(g) = u^{\alpha}$$

$$\cdot u^{\alpha} \int_0^{\infty} e^{-t/u} \int_0^{t} f(\tau) g(t-\tau)\, d\tau\, dt$$

$$= u^{\alpha} \cdot u^{\alpha} \int_0^{\infty} e^{-t/u} (f * g)(t)\, dt$$

$$= u^{\alpha} G(f * g). \tag{38}$$

It is a well-known fact that convolution helps us to solve integral equations of certain type, mainly Volterra integral equation. Hence, we would like to check the theorem by means of some examples using G-transform.

Example 7. Solve the Volterra integral equation of the second kind

$$y(t) - \int_0^t y(\tau) \sin(t-\tau)\, d\tau = t. \tag{39}$$

Solution. This is rewritten as a convolution:

$$y(t) - y * \sin t = t. \tag{40}$$

Taking G-transform on both sides and applying Theorem 6, we have

$$Y(u) - u^{-\alpha} Y(u) \frac{u^{\alpha+2}}{1+u^2} = Y(u)\left(1 - \frac{u^2}{1+u^2}\right)$$

$$= u^{\alpha+2} \tag{41}$$

for $Y = G(y)$. The solution is

$$Y(u) = \left(1+u^2\right) u^{\alpha+2} \tag{42}$$

and gives the answer

$$y(t) = t + \frac{1}{6} t^3 \tag{43}$$

by Table 1.

Example 8. Solve the Volterra integral equation of the second kind

$$y(t) - \int_0^t (1+\tau)\, y(t-\tau)\, d\tau = 1 - \sinh t. \tag{44}$$

Solution. In a way similar to Example 7, the given equation is same as $y - (1 + t) * y = 1 - \sinh t$. Taking G-transform, we have

$$Y(u) - u^{-\alpha}\left(u^{\alpha+1} + u^{\alpha+2}\right) Y(u) = u^{\alpha+1} - \frac{u^{\alpha+2}}{1-u^2}; \tag{45}$$

hence,

$$Y(u)\left[1 - u - u^2\right] = \frac{u^{\alpha+1}\left(1 - u - u^2\right)}{1-u^2}. \tag{46}$$

Simplification gives

$$Y(u) = \frac{u^{\alpha+1}}{1-u^2}, \tag{47}$$

and so we obtain the answer

$$y(t) = \cosh t \tag{48}$$

by Table 1.

Example 9. Find the solution of

$$y(t) + \int_0^t (t-\tau)\, y(\tau)\, d\tau = 1. \tag{49}$$

Solution. Since this equation is $y+y*t = 1$, taking G- transform on both sides, we have

$$Y + u^{-\alpha}\left(Y \cdot u^{\alpha+2}\right) = u^{\alpha+1} \tag{50}$$

for $Y = G(y)$. Thus

$$Y = \frac{u^{\alpha+1}}{1+u^2}, \tag{51}$$

and so we obtain the solution $y = \cos t$.

Let us check this by the direct calculation. Expanding the given equation, we have

$$y(t) + t \cdot \int_0^t y(\tau)\, d\tau - \int_0^t \tau y(\tau)\, d\tau = 1. \tag{52}$$

Differentiating both sides twice with respect to t, we have $y''(t) + y(t) = 0$. Thus, from $(0) = 1$ and $y'(0) = 0$ obtained by calculating course, we have the solution $y = \cos t$.

Similarly, we can easily obtain the solution of integral equations by using $\alpha = -2$. For example, let us consider

$$y(t) - \int_0^t y(\tau)\, d\tau = 1. \tag{53}$$

By the G-transform, we have

$$Y - uY = \frac{1}{u} \tag{54}$$

and so, we have the solution $y = e^t$ for $Y = G(y)$. Of course, this result is the same as the result

$$y - y * 1 = 1 \tag{55}$$

by using convolution, and the result of the direct calculation

$$y'(t) - y(t) = 0 \tag{56}$$

is so as well. Similarly, since

$$G\left(te^{at}\right) = \frac{u^{\alpha+2}}{(1-au)^2}, \tag{57}$$

the solution of

$$y(t) + 2e^t \int_0^t e^{-\tau} y(\tau)\, d\tau = te^t \tag{58}$$

is $y = \sinh t$. Here, we note that the G-transform of $y + 2(y * e\ ) = te^t$ is

$$Y + 2u^{-\alpha} Y \cdot \frac{u^{\alpha+1}}{u(1-u)} = \frac{u^{\alpha+2}}{(1-u)^2} \tag{59}$$

for $Y = G(y)$.

DIFFERENTIATION AND INTEGRATION OF TRANSFORMS: ODES WITH VARIABLE COEFFICIENTS

Theorem 10. Let us put $Y = (u)$. Then

(1) $F'(u) = Y(\alpha/u + t/u^2)$,

(2) $F''(u) = Y((\alpha^2 - \alpha)/u^2 + 2t(\alpha - 1)/u^3 + t^2/u^4)$,

(3) $\int_u^\infty F(s)ds = \int_0^\infty f(t) \int_u^\infty s^\alpha e^{-t/s} ds\, dt$,

(4) $G(ty) = -\alpha uY - (d/du)G(y)$, $G(ty') = -\alpha u[(1/u)Y - y(0)u^\alpha] - (d/du)G(y')$,

(5) $G(ty'') = -\alpha u[(1/u^2)Y - (1/u)y(0)u^\alpha - y'(0)u^\alpha] - (d/du)G(y'')$.

Proof. (1) Since

$$F(u) = G(f) = u^\alpha \int_0^\infty e^{-t/u} f(t)\, dt,$$

$$F'(u) = \alpha u^{\alpha-1} \int_0^\infty e^{-t/u} f(t)\, dt + u^\alpha \frac{t}{u^2} \int_0^\infty e^{-t/u} f(t)\, dt = Y\left(\frac{\alpha}{u} + \frac{t}{u^2}\right) \tag{60}$$

for $Y = F(u)$.

(2) Since

$$F''(u) = \frac{d}{du} F'(u) = \frac{d}{du}\left[Y\left(\frac{\alpha}{u} + \frac{t}{u^2}\right)\right], \tag{61}$$

we have

$$F''(u) = Y'\left(\frac{\alpha}{u} + \frac{t}{u^2}\right) + Y\left(\frac{\alpha}{u} + \frac{t}{u^2}\right)' = Y\left(\frac{\alpha}{u} + \frac{t}{u^2}\right)^2 - Y\left(\frac{\alpha}{u^2} + \frac{2t}{u^3}\right). \tag{62}$$

Organizing this equality, we have

$$F''(u) = Y\left(\frac{\alpha^2 - \alpha}{u^2} + \frac{2t(\alpha \quad 1)}{u^3} + \frac{t^2}{u^4}\right). \tag{63}$$

(3)

$$\int_u^\infty F(s)\, ds = \int_u^\infty \left[s^\alpha \int_0^\infty e^{-t/s} f(t)\, dt\right] ds = \int_u^\infty \int_0^\infty s^\alpha e^{-t/s} f(t)\, dt\, ds = \int_0^\infty f(t) \int_u^\infty s^\alpha e^{-t/s} ds\, dt \tag{64}$$

thanks to Lebesgue's dominated convergence theorem.

(4) In the definition of

$$F(u) = G(y) = u^{\alpha} \int_0^{\infty} e^{-t/u} y(t)\, dt, \tag{65}$$

let us put $1/u = s$. Then

$$F(u) = G(y) = \frac{1}{s^{\alpha}} \int_0^{\infty} e^{-st} y(t)\, dt,$$

$$F'(u) = -\alpha s^{-\alpha-1} \int_0^{\infty} e^{-st} y(t)\, dt$$

$$-\frac{1}{s^{\alpha}} \int_0^{\infty} e^{-st} ty(t)\, dt = \frac{-\alpha}{s} G(y) - G(ty). \tag{66}$$

Since $1/s$ is the same as u, we have

$$G(ty) = -\alpha uY - F'(u) = -\alpha uY - \frac{d}{du} G(y) \tag{67}$$

for $Y = G(y)$.

In the above theorem, we note that $(ty^{(n)})$ can be represented by

$$G\left(ty^{(n)}\right) = -\alpha u G\left(y^{(n)}\right) - \frac{d}{du} G\left(y^{(n)}\right). \tag{68}$$

(5) From (4), the statement is held for an arbitrary integer n.

G-TRANSFORMS OF HEAVISIDE FUNCTION AND DIRAC'S DELTA FUNCTION: SHIFTED DATA PROBLEMS

Heaviside Function

$$G[h(t-a)] = u^{\alpha} \int_0^{\infty} e^{-t/u} h(t-a)\, dt$$

$$= u^{\alpha} \int_a^{\infty} e^{-t/u} dt = -u^{\alpha+1} \left[e^{-t/u}\right]_a^{\infty}$$

$$= u^{\alpha+1} e^{-a/u}, \tag{69}$$

where h is Heaviside function.

Dirac's Delta Function

We consider the function $f(t-a) = 1/k$ if $a \le t \le a+k$ and 0 otherwise. Taking$G$-transform, we have

$$G\left[f_k(t-a)\right] = u^{\alpha}\int_0^{\infty} e^{-t/u} f_k(t-a)\,dt$$

$$= -\frac{u^{\alpha+1}}{k}\left[e^{-t/u}\right]_a^{a+k}$$

$$= -\frac{u^{\alpha+1}}{k}\left(e^{-(a+k)/u} - e^{-a/u}\right)$$

$$= -\frac{u^{\alpha+1}}{k}e^{-a/u}\cdot\left(e^{-k/u} - 1\right). \tag{70}$$

If we denote the limit of fk as $(t-a)$, then

$$\delta(t-a) = \lim_{k\to 0} f_k(t-a) = u^{\alpha}e^{-a/u}. \tag{71}$$

Shifted Data Problems

For given differential equations $y'' + ay' + by = (t)$, $y(t_0)=c_0$, and $y'(t_0)=c_1$, where $t_0 \neq 0$ and a and b are constant, we can set $t=t_1 + t_0$. Then $t=t_0$ gives $t_1 = 0$ and so we have

$$y_1'' + ay_1' + by_1 = r(t_1 + t_0),$$

$$y_1(0) = c_0,$$

$$y_1'(0) = c_1 \tag{72}$$

for input $r(t)$. Taking the transform, we can obtain the output (t).

THE SOLUTION OF SEMI-INFINITE STRING BY -TRANSFORM

Let us check the solution of semi-infinite string by G-transform in terms of a typical example as given in [1].

Example 11 (semi-infinite string). Find the displacement (x, t) of an elastic string subject to the following conditions:

- The string is initially at rest on x-axis from $x=0$ to ∞

- For $t>0$, the left end of the string is moved in a given fashion, namely, according to a single sine wave $(0, t) = f(t) = \sin t$, if $0 \leq t \leq 2\pi$ and zero otherwise.
- Furthermore, $\lim (x, t) = 0$ as $x \to \infty$ for $t \geq 0$.

Of course there is no infinite string, but our model describes a long string or rope (of negligible weight) with its right end fixed far out on x-axis [1].

Solution. It is a well-known fact that the equation of semi-infinite string can be expressed by

$$\frac{\partial^2 w}{\partial t^2} = c^2 \frac{\partial^2 w}{\partial x^2} \tag{*}$$

subject to $w(0, t) = f(t)$, $\lim w(x, t) = 0$ as $x \to \infty$, $w(x, 0) = 0$, and $w_t(x, 0) = 0$. Taking G-transform with respect to t and by Theorem 3, we have

$$G\left[\frac{\partial^2 w}{\partial t^2}\right] = \frac{1}{u^2}W - \frac{1}{u}w(x,0)\,u^\alpha - w_t(x,0)\,u^\alpha$$

$$= \frac{1}{u^2}W \tag{73}$$

for $W(x, u) = G[w(x, t)]$. On the other hand,

$$G\left[\frac{\partial^2 w}{\partial x^2}\right] = u^\alpha \int_0^\infty e^{-t/u} \frac{\partial^2 w}{\partial x^2} dt$$

$$= \frac{\partial^2}{\partial x^2} u^\alpha \int_0^\infty e^{-t/u} w(x,t)\, dt$$

$$= \frac{\partial^2}{\partial x^2} G[w(x,t)] = \frac{\partial^2 W}{\partial x^2}. \tag{74}$$

Thus, the equation $(*)$ becomes

$$c^2 \frac{\partial^2 W}{\partial x^2} - \frac{1}{u^2}W = 0. \tag{75}$$

Since this equation may be regarded as an ODE for (x, u) considered as a function of x, its general solution can be represented by

$$W(x,u) = A(u)\,e^{x/cu} + B(u)\,e^{-x/cu}. \tag{76}$$

From the initial conditions, we have $(0, u) = G[w(0, t)] = G[f(t)] = F(u)$. In [15, 16], we have dealt with the validity on exchangeability of integral and limit in the solving process of PDEs by using dominated convergence theorem. Hence, we have

$$\lim_{x\to\infty} W(x,u) = \lim_{x\to\infty} u^{\alpha} \int_0^{\infty} e^{-t/u} w(x,t)\, dt$$
$$= u^{\alpha} \int_0^{\infty} e^{-t/u} \lim_{x\to\infty} w(x,t)\, dt = 0, \tag{77}$$

implying that $A(u) = 0$ and so $W(0, u) = B(u) = F(u)$. Thus,

$$W(x,u) = F(u)\, e^{-x/cu}. \tag{78}$$

By the t-shifting theorem (Theorem 2), we obtain the inverse transform

$$w(x,t) = f\left(t - \frac{x}{c}\right) h\left(t - \frac{x}{c}\right) = \sin\left(t - \frac{x}{c}\right) \tag{79}$$

for $x/c < t < x/c + 2\pi$ and zero otherwise, where h is Heaviside function.

CONCLUSION

This paper has constructed the generalized form of Laplace typed integral transforms and has established the properties of the generalized Laplace-typed integral transform, G- transform. The transform is comprehensive form, and it has been well adapted in a number of situations of engineering problems by choosing adequate values α in kernel, and we newly presented the value $\alpha = -2$ which is suitable for transforms of integrals. And the future work is to find the other values of α which are suitable for each situation. The strong point of this article is in the high applicability to engineering problems.

ACKNOWLEDGMENTS

This research was supported by Kyungdong University Research Grant.

REFERENCES

1. E. Kreyszig, Advanced Engineering Mathematics, Wiley, Singapore, 2013.
2. G. K. Watugala, "Sumudu transform—a new integral transform to solve differential equations and control engineering problems," International Journal of Mathematical Education in Science and Technology, vol. 24, no. 1, pp. 35–43, 1993.
3. T. M. Elzaki, S. M. Elzaki, and E. M. A. Hilal, "Elzaki and Sumudu transforms for solving some differential equations," Global Journal of Pure & Applied Mathematics, vol. 8, no. 2, pp. 167–173, 2012.
4. Hj. Kim, "The time shifting theorem and the convolution for Elzaki transform," Global Journal of Pure and Applied Mathematics, vol. 87, pp. 261–271, 2013.
5. F. B. Belgacem and S. Sivasundaram, "New developments in computational techniques and transform theory applications to nonlinear fractional and stochastic differential equations and systems," Nonlinear Studies. The International Journal, vol. 22, no. 4, pp. 561–563, 2015.
6. F. B. M. Belgacem and A. Karaballi, "Sumudu transform fundamental properties investigations and applications," Journal of Applied Mathematics and Stochastic Analysis, vol. 2006, Article ID 91083, 23 pages, 2006.
7. T. M. Elzaki and J. Biazar, "Homotopy perturbation method and Elzaki transform for solving system of nonlinear partial differential equations," World Applied Sciences Journal, vol. 24, no. 7, pp. 944–948, 2013.
8. T. M. Elzaki and H. Kim, "The solution of radial diffusivity and shock wave equations by Elzaki variational iteration method," International Journal of Mathematical Analysis, vol. 9, no. 21-24, pp. 1065–1071, 2015.
9. H. A. Agwa, F. M. Ali, and A. Kilicman, "A new integral transform on time scales and its applications," Advances in Difference Equations, vol. 60, pp. 1–14, 2012.
10. H. Eltayeb and A. Kilicman, "On some applications of a new integral transform," International Journal of Mathematical Analysis, vol. 4, no. 1-4, pp. 123–132, 2010.

11. H. Eltayeb, A. Kilicman, and M. B. Jleli, "Fractional Integral Transform and Application," Abstract and Applied Analysis, vol. 2015, Article ID 150863, 1 page, 2015.
12. H. Kim, "The form of solution of ODEs with variable coefficients by means of the integral and Laplace transform," Global Journal of Pure and Applied Mathematics, vol. 12, pp. 2901–2904, 2016.
13. H. Kim, "The shifted data problems by using transform of derivatives," Applied Mathematical Sciences, vol. 8, no. 149-152, pp. 7529–7534, 2014.
14. D. L. Cohn, Measure Theory, Birkhäauser, Boston, Mass, USA, 1980.
15. H. C. Chae and H. Kim, "The validity checking on the exchange of integral and limit in the solving process of PDEs," International Journal of Mathematical Analysis, vol. 8, no. 22, pp. 1089–1092, 2014.
16. J. Jang and H. Kim, "An application of monotone convergence theorem in pdes and fourier analysis," Far East Journal of Mathematical Sciences, vol. 98, no. 5, pp. 665–669, 2015.

CHAPTER 3

ON THE Q-LAPLACE TRANSFORM AND RELATED SPECIAL FUNCTIONS

Shanoja R. Naik[1], **and Hans J. Haubold**[2]

[1]Department of Mathematics and Statistics, University of Regina, Regina, SK S4S 0A2, Canada

[2]Office for Outer Space Affairs, United Nations, Vienna International Centre, Vienna 1400, Austria

ABSTRACT

Motivated by statistical mechanics contexts, we study the properties of the q-Laplace transform, which is an extension of the well-known Laplace transform. In many circumstances, the kernel function to evaluate certain integral forms has been studied. In this article, we establish relationships between q-exponential and other well-known functional forms, such as Mittag–Leffler functions, hypergeometric and H-function, by means of

CITATION (APA): Naik, S. R., & Haubold, H. J. (2016). On the q-Laplace Transform and Related Special Functions. Axioms, 5(3), 24. Pages 1- 16

the kernel function of the integral. Traditionally, we have been applying the Laplace transform method to solve differential equations and boundary value problems. Here, we propose an alternative, the q-Laplace transform method, to solve differential equations, such as the fractional space-time diffusion equation, the generalized kinetic equation and the time fractional heat equation.

Keywords: convolution property, G-transform, Gauss hypergeometric function, generalized kinetic equation, Laplace transform, Mittag–Leffler function, versatile integral

INTRODUCTION

The classical Laplace, Fourier and Mellin transforms have been widely used in mathematical physics and applied mathematics. The theory of the Laplace transform is well-known [1], and its generalization was considered by many authors [2,3,4,5,6]. Various existence conditions and detailed study about the range and invertibility were studied by Rooney [7]. The Laplace transform and Mellin transform are widely used together to solve the fractional kinetic equations and thermonuclear equations [8,9]. Different types of integral transforms, like the Hankel transform, Erdély–Kober type fractional integration operators, the Gauss hypergeometric function as a kernel, the Bessel-type integral transform, etc. [10], are introduced in the literature to solve the boundary value problems for models of ordinary and partial differential equations. In some situations, the solutions of the differential equation cannot be tractable using the classical integral transforms, but may be characterized by many integral transforms with various special functions as kernels. Many of the integral transforms can be interpreted in terms of the G-transform and H-transform [11,12,13,14,15,16].

In physical situations when an appropriate density is selected, the best practice is to maximize the entropy. Mathai and Rathie [17] considered various generalizations of the Shannon entropy measure and describe various properties, including additivity, the characterization theorem, etc. Mathai and Haubold [18] introduced a new generalized entropy measure, which is a generalization of the Shannon entropy measure. For a multinomial population $P=(p_1,\ldots,p_k), p_i\geq 0, i=1,\ldots,k, p_1+p_2+\cdots+p_k=1$, the Mathai's entropy measure (discrete case) is given by the relation:

$$M_{k,\alpha}(P) = \frac{\sum_{i=1}^{k} p_i^{2-\alpha} - 1}{\alpha - 1}, \ \alpha \neq 1, \ -\infty < \alpha < 2.$$

When $\alpha \to 1$, the above measure goes to the Shannon entropy measure, and this is a variant of Havrda–Charvat entropy and Tsallis entropy. One can derive Tsallis statistics and superstatistics [19,20,21,22] by using Mathai's entropy. By optimizing Mathai's entropy measure, a new pathway model, which consists of many of the standard distributions in the statistical literature as special cases (see [23]), is derived. The main idea behind the derivation of this model is the switching properties of the special functions, like ${}_1F_1$ and ${}_0F_1$, which means the binomial to exponential function. Thus, the pathway between the exponential function e^{-cx} and the binomial function $[1-c(1-\alpha)x]^{1/1-\alpha}$ can be created with the parameter α named as the pathway parameter. For the real scalar case, the pathway density can be written in the form:

$$f_1(x)=c|x|^{\gamma}[1-a(1-\alpha)|x|^{\delta}]^{\eta/1-\alpha},\ a>0, 1-a(1-\alpha)|x|^{\delta}\geq 0,\ \eta>0,\ \alpha<1$$

where c is the normalizing constants. One can assume the Type 2 model by replacing $(1-\alpha)$ by $-(\alpha-1)$. These distributions include Type 1 beta, Type 2 beta, gamma, Weibull, Gaussian, Cauchy, exponential, Rayleigh, Student *t*, Fermi–Dirac, chi-square, logistic, etc. The corresponding asymmetric generalization was introduced and studied in the paper [24]. By representing the entropy function in terms of a density function $f(\cdot)$ for the continuous case and giving the suitable constraints therein, the generalized entropy is maximized. There are restrictions, such as the $[(\gamma-1)(1-\alpha)]$-th moment, and the $[(\gamma-1)(1-\alpha)+\delta]$-th moments are constants for fixed $\gamma>0$ and $\delta>0$. Maximizing Mathai's entropy by using the calculus of variations, we get the basic function of the model, and when the range of $x$is restricted over the positive real line and by evaluating the normalizing constant, we get the pathway model introduced by Mathai [23]. As $q\to 1$,$f_1(x)$ tend to $f_2(x)$, which is the generalized gamma distribution, where $f_2(x)$ is given by:

$$f_2(x) = \frac{\delta(a\beta)^{\frac{\alpha}{\delta}}}{2\Gamma\left(\frac{\alpha}{\delta}\right)} \mid x \mid^{\alpha-1} \exp(-a\beta \mid x \mid^{\delta}); \quad -\infty < x < \infty; a, \alpha, \beta, \delta > 0. \qquad (1)$$

For different values of parameters in the pathway model, we get different distributions like Weibull, gamma, beta Type 1, beta Type 2, etc. By taking $\delta=\alpha$, $\beta=1$,$a=\lambda^{\alpha}$ in $f_1(x)$, the pathway model reduces to the *q*-Weibull distribution, which facilitates a transition to the Weibull distribution [25]. The connection of pathway models and Tsallis statistics with the *q*-extended versions of various functions is also considered. To this extent, we generalize the Laplace transform using the switching property of ${}_0F_0$ to ${}_1F_0$. Here, the q-exponential function is the kernel, and we call the extension as the q-Laplace transform; as q approaches to unity, we get the Laplace transform of the original function.

The article is organized as follows. In Section 2, we introduce the q-Laplace transform and the obtained various properties of the transform. Section 3 deals with the q-Laplace transform of some basic functions, which includes special functions, like the hypergeometric function, the Mittag–Leffler function and the H-function. In Section 4, this transform is connected to other known integral transforms, like the Mellin transform, the G-transform and the Henkel transforms. In Section 5, we obtain the solution of the fractional space-time diffusion equation, the generalized kinetic equation and the time fractional heat equation through the q-Laplace transform in terms of the Mittag–Leffler function.

THE *Q*-LAPLACE TRANSFORM AND BASIC PROPERTIES

The Laplace transform *L* of a function f(·) is given by:

$$L[f(x)](s) \equiv \int_0^{\infty} f(x)\, e^{-sx}\, dx$$

where f(x) is defined over the positive real line and s∈C,R(s)>0, R(·) denotes the real part of (·). This Laplace transform plays a major role in pure and applied analysis, especially in solving differential equations. Now, we define the extended Laplace transform concept, namely the q-Laplace transform of a function, which will play a similar role in mathematical analysis, as well as mathematical physics. Instead of the exponential function, here, we consider the e^{-sx}_q the q-exponential defined as:

$$e_q^{-x} \equiv c \begin{cases} [1-(1-q)x]^{\frac{1}{1-q}} \text{ for } 0 < x < \frac{1}{1-q}, q < 1 \\ [1+(q-1)x]^{-\frac{1}{q-1}} \text{ for } x \geq 0, q > 1 \end{cases} \tag{2}$$

with $e^x_1 \equiv e^x$ and c is the normalizing constant. More precisely, for given function f(·)and for s∈C with support over (0,∞), we define its q-Laplace transform as:

$$L_q[f(x)](s) = \int_0^{\infty} [e_q^{-sx}] f(x) dx \text{ for }, \Re(s) > 0 \tag{3}$$

where e_q^{-x} is defined as in Equation (2). This Laplace transform can be written in the form,

$$L_q[f(x)](s) = \begin{cases} \int_0^{\frac{1}{(1-q)s}} [1-(1-q)sx]^{\frac{1}{1-q}} f(x)dx & \text{for } \Re(1-(1-q)sx) > 0, \Re(s) > 0 \\ \int_0^{\infty} [1+(q-1)xs]^{-\frac{1}{q-1}} f(x)dx & \text{for } \Re(s) > 0. \end{cases}$$

The q-Laplace transform of a function f(·) is valid at every point at which f(·) is continuous provided that the function is defined in (0,∞), is piecewise continuous and of bounded variation in every finite subinterval in (0,∞), and the integral is finite. Some basic properties of the q-Laplace transform are given below.

- Scaling: For a real constant k, $L_q[kf(x)](s)=kL_q[f(x)](s)$.
- Linearity : $L_q[mf(x)+ng(x)](s)=mL_q[f(x)](s)+nL_q[g(x)](s)$, where m, n∈R.
- Transform of derivatives: For R(s)>0, $L_q[\frac{d}{dx}f(x)](s)=sL_q(f)(sq)$ for all q∈R/{1}.

Proof.

Let $g(x) = \frac{d}{dx}f(x)$. Then:

$$L_q[g(x)](s) = \int_0^{\infty} [1+(q-1)xs]^{-\frac{1}{q-1}} \frac{d}{dx}[f(x)]dx.$$

By applying integration by parts, we get:

$$\int_0^{\infty} [1+(q-1)xs]^{-\frac{1}{q-1}} \frac{d}{dx}[f(x)]dx = f(x)[1+(q-1)xs]^{-\frac{1}{q-1}}\Big|_0^{\infty} - \int_0^{\infty} f(x)(-s)[1+(q-1)xs]^{-\frac{1}{q-1}-1}dx$$

which implies:

$$L_q[g(x)] = -f(0) + sL_{\frac{2q-1}{q}}(f)(sq).$$

As a consequence, we get:

$$D^n\{L_q[f(x)](s)\} = s^n L_q[f(x)](sA_n(q)) - \sum_{j=1}^{n} s^{n-j}D^j f(0) \tag{4}$$

where $A_n(q) = \prod_{j=1}^{n}(j+(j-1)q)$, for $q > 1$, $D = \frac{d}{dx}f(x)$.

- Derivatives of transforms: The nth derivative of the q-Laplace transform is given by $D^n[L_q(f)(s)]=A_{n-1}(q)\{L_q[(-x)^n{}_nA^{-1}(q)\, f(x_nA^{-1})](s)\}$ where ${}_nA^{-1}$ is the reciprocal of the nth term of $A_n(q)$.

Proof.

For q>1:

$$D^2(e_q^{-sx}) = qx^2[e_q^{-sx}]^{1-2q}$$
$$D^3(e_q^{-sx}) = q(1-2q)(-x^3)[e_q^{-sx}]^{2-3q}$$
$$\vdots$$
$$D^n(e_q^{-sx}) = \prod_{j=1}^{n-1} A_j \int_0^\infty (-x)^n e_q^{-s\,{}_nA^{-1}(q)x} f({}_nA^{-1}(q)x)dx$$
$$= A_{n-1}(q)\{L_q\left[(-x)^n\ {}_nA^{-1}(q)f(x\,{}_nA^{-1})\right]\}(s).$$

- Transforms of integrals: For R(s)>0, $L_q\left[\int_0^x f(t)dt\right](s)=1/sL_q(f)(s)$.

Proof.

For q>1, we have:

$$L_q\left[\int_0^x f(t)dt\right](s) = \int_0^\infty [e_q^{-sx}]\{\int_0^x f(t)dt\}dx$$
$$= -\frac{1}{s}\int_0^\infty \{\int_0^x f(t)dt\}\frac{d}{dx}[e_q^{-s(2-q)x}]dx$$
$$= -\frac{1}{s}\left\{\int_0^x f(t)dt[e_q^{-s(2-q)x}]\Big|_0^\infty - \int_0^\infty f(x)e_q^{-sx}dx\right\}$$
$$= \frac{1}{s}L_q(f)(s) \text{ for } \Re(s) > 0.$$

- Convolution property: Let $f_1(x)$ and $f_2(x)$ be two positive real scalar functions of *x*, and let $g_1(t)$ and $g_2(t)$ be their *q*-Laplace transform. Then,

$L_q[f_1(x)*f_2(x)](s)=g_1(x)g_2(x)$
where $f_1(x) * f_2(x) = \int_0^x f_1(t)f_2(x-t)dt$.

Proof.

$$L_q[f_1(x) * f_2(x)](s) = \int_0^\infty e_q^{-sx}\left\{\int_0^x f_1(t)f_2(x-t)dt\right\}dx$$
$$= \int_0^\infty \int_0^x [1+(q-1)xs]^{-\frac{1}{q-1}} f_1(t)f_2(x-t)dtdx$$
$$= \int_{t=0}^\infty f_1(t)\left\{\int_{x=t}^\infty [1+(q-1)xs]^{-\frac{1}{q-1}} f_2(x-t)dx\right\}dt.$$

Now, let us consider the integral $I=\int_{x=t}^\infty[1+(q-1)xs]^{-1/q-1}f_2(x-t)dt$. Substitute x−t=u, and manipulate the integral; we get:

$$I = [1+(q-1)ts]^{-\frac{1}{q-1}} \quad \int_{x=0}^{\infty} \quad [1+(q-1)us]^{-\frac{1}{q-1}}[1+(q-1)ts]^{\frac{1}{q-1}}[1+(q-1)us]^{\frac{1}{q-1}}$$
$$\times \quad [1+(q-1)(t+u)s]^{-\frac{1}{q-1}} f_2(u)\mathrm{d}u.$$

Let $[1+(q-1)ts]^{\frac{1}{q-1}}[1+(q-1)us]^{\frac{1}{q-1}}[1+(q-1)(t+u)s]^{-\frac{1}{q-1}} f_2(u) = f_2^*(u)$, then $f_2(x-t) = [1+(q-1)ts]^{-\frac{1}{q-1}}[1+(q-1)(x-t)s]^{-\frac{1}{q-1}}[1+(q-1)xs]^{\frac{1}{q-1}} f_2^*(x-t)$. Then:

$$L_q[f_1(x)*f_2(x)](s) = \int_{t=0}^{\infty}[1+(q-1)ts]^{-\frac{1}{q-1}} f_1(t)$$
$$\left\{\int_{x=t}^{\infty}[1+(q-1)(x-t)s]^{-\frac{1}{q-1}} f_2^*(x-t)\mathrm{d}x\right\}\mathrm{d}t.$$

On substituting x−t=u, the integral can be separated, and hence, we have: $L_q[f_1(x)*f_2(x)](s) = L_q[f_1(x)]L_q[f_2^*(x)]$.

THE *Q*-LAPLACE TRANSFORM OF SOME BASIC FUNCTIONS

Let us introduce a new notation, $\Gamma_{(q)}(\alpha)$, such that:

$$\Gamma_{(q)}(\alpha) = \int_0^{\frac{1}{1-q}} x^{\alpha-1}[1-(1-q)x]^{\frac{1}{1-q}}\mathrm{d}x \text{ for } \Re(\alpha) > 0, q < 1.$$

If we replace (1−q) by −(q−1), then the function assumes the form:

$$\Gamma_{(q)}(\alpha) = \int_0^{\infty} x^{\alpha-1}[1+(q-1)x]^{-\frac{1}{q-1}}\mathrm{d}x \text{ for } \Re(\alpha) > 0, q > 1$$

and for q=1 in the sense q→1, the *q*-gamma function is the usual classical gamma function defined as

$$\Gamma(\alpha) = \int_0^{\infty} x^{\alpha-1}\mathrm{e}^{-x}\mathrm{d}x.$$

Now, the *q*-gamma function can be explicitly written as:

$$\Gamma_{(q)}(\alpha) = \begin{cases} \frac{1}{(1-q)^{\alpha}} \frac{\Gamma(\alpha)\,\Gamma(\frac{1}{1-q}+1)}{\Gamma(\frac{1}{1-q}+\alpha+1)} & \text{for } q<1 \\ \Gamma(\alpha) & \text{for } q=1 \\ \frac{1}{(q-1)^{\alpha}} \frac{\Gamma(\alpha)\,\Gamma(\frac{1}{q-1}-\alpha)}{\Gamma(\frac{1}{q-1})} & \text{for } q>1, \frac{1}{q-1}-\alpha>0 \end{cases} \tag{5}$$

for R(α)>0. Here, q=1 in the sense q→1 the *q*-gamma function $\Gamma_{(q)}(\cdot)\to\Gamma(\cdot)$, which can be easily proven using the asymptotic expansion of the gamma function:

$\Gamma(z+a) \approx \sqrt{2\pi z} z^{z+a-\frac{1}{2}} e^{-z}.$

Mathai [26] introduced a general class of integrals, known as the versatile integrals, which are connected to the reaction rate in kinetic theory. The integral is in the form:

$$I = \int_0^\infty x^{\gamma-1}[1+z_1^\delta(\alpha-1)x^\delta]^{-\frac{1}{\alpha-1}}[1+z_2^\delta(\beta-1)x^{-\rho}]^{-\frac{1}{\beta-1}} \tag{6}$$

for $\alpha,\beta>1, z_1, z_2 \geq 0, \delta,\rho>0, R(\gamma+1)>0,$ $\Re(\frac{1}{\alpha-1}-\frac{\gamma+1}{\delta})>0, \Re(\frac{1}{\beta-1}-\frac{1}{\rho})>0,$ and the solution is obtained in terms of the *H*-function as follows:

$$I = c\ H_{2,2}^{2,2}\left[z_1 z_2 (\alpha-1)^{\frac{1}{\delta}}(\beta-1)^{\frac{1}{\rho}}\middle|^{(1-\frac{1}{\alpha-1}+\frac{\gamma}{\delta},\frac{1}{\delta}),(1-\frac{1}{\beta-1},\frac{1}{\rho})}_{(\frac{\gamma}{\delta},\frac{1}{\delta}),(0,\frac{1}{\rho})}\right]$$

where $c=\delta\rho z_1^\gamma(\alpha-1)^{\gamma/\delta}$ and $H^{m,n}_{p,r}$ is a *H*-function. Here, we provide the definition of *H*-function as follows:

$$H_{p,q}^{m,n}\left[z\middle|^{(a_1,\alpha_1),(a_2,\alpha_2),\cdots,(a_k,\alpha_k)}_{(b_1,\beta_1),(b_2,\beta_2),\cdots,(b_q,\beta_q)}\right] = \frac{1}{2\pi i}\int_L h(s) z^{-s} ds$$

where:

$$h(s) = \frac{\left\{\prod_{j=1}^{m}\Gamma(b_j+\beta_j s)\right\}\left\{\prod_{j=1}^{n}\Gamma(1-a_j-\alpha_j s)\right\}}{\left\{\prod_{j=m+1}^{q}\Gamma(1-b_j-\beta_j s)\right\}\left\{\prod_{j=n+1}^{p}\Gamma(a_j+\alpha_j s)\right\}}$$

and L is a suitable path. An empty product is interpreted as unity, and it is assumed that the poles of $\Gamma(b_j+\beta_j s)$, $j=1,2,\ldots,m$ are separated from the poles of $\Gamma(1-a_j-\alpha_j s)$, $j=1,2,\ldots,n$. Here, $a_1,a_2,\ldots,a_p$; $b_1,b_2,\ldots,b_q$ are complex numbers and $\alpha_1,\alpha_2,\ldots,\alpha_p$, $\beta_1,\beta_2,\ldots,\beta_q$ are positive real numbers. The poles of $\Gamma(b_j+\beta_j s)$, $j=1,2,\ldots,m$ are at the points $s=-b_j+\nu\beta_j$ where $j=1,2,\ldots,m$, $\nu=0,1,\ldots$, and the poles of $\Gamma(1-a_j-\alpha_j s)$, $j=1,2,\ldots,n$ are at $s=1-a_k+\lambda / \alpha_k$ where $k=1,2,\ldots,n, \lambda=0,1,\ldots$. For more details about the theory and applications, refer to [27]. This integral includes the *q*-Laplace transform of gamma function and *q*-gamma function as special cases. Now, as $q\to1$ in any of the functions $[1+z_1^\delta(\alpha-1)x^\delta]^{-\frac{1}{\alpha-1}}$ or $[1+z_2^\delta(\beta-1)x^{-\rho}]^{-\frac{1}{\beta-1}}$, we get the *q*-Laplace transform of some basic functions. The following table gives the *q*-Laplace transform of some basic functions with $q>1$, which are special cases of the above integral. The results are obtained in terms of hypergeometric function. The Gaussian hyper geometric function is defined as:

$$ {}_mF_n\left[\begin{matrix} a_1,a_2,\cdots,a_m \\ b_1,b_2,\cdots,b_n \end{matrix}\middle| x\right] = \sum_{k=0}^{\infty} \frac{(a_1)_k(a_2)_k\cdots(a_m)_k}{(b_1)_k(b_2)_k\cdots(b_n)_k}\frac{x^k}{k!} $$

where (a)m denotes the Pochhammer symbol expressed in the form:

$$ (a)_m = a(a+1)...(a+m-1),\ a\neq 0,\ m=1,2... \tag{7} $$

Lemma 1.

For $\alpha, s \in C$, $R(s)>0$ *and for* $q\neq 1$, *the q- Laplace transform of* $x^{\alpha-1}$ *is given by* $L_q[x^{\alpha-1}](s)=\Gamma_q(\alpha)/s^{\alpha}$ *for* $q\neq 1$.

Proof.

For q>1,

$$ \begin{aligned} L_q[x^{\alpha-1}](s) &= \int_0^{\infty} x^{\alpha-1}e_q^{-sx}dx \\ &= \int_0^{\infty} x^{\alpha-1}[1+(q-1)sx]^{-\frac{1}{q-1}}dx. \end{aligned} $$

Now, substitute (q−1)sx=t, and dx=1 / s(q−1)du. Then:

$$ L_q[x^{\alpha-1}](s) = \frac{\Gamma_q(\alpha)}{s^{\alpha}},\quad \alpha, s\in C, \Re(s)>0. $$

Lemma 2.

For $s\in C$, $R(s)>0$, *there holds the formula:*

$$ L_q[e^{-ax}](s) = \frac{1}{(2-q)s}\,{}_1F_1\left[1;\frac{2q-3}{q-1};\frac{a}{s(q-1)}\right] $$

for $a>0$, $3/2<q<2$.

Proof.

For q>1,

$$ \begin{aligned} L_q[e^{-ax}](s) &= \int_0^{\infty}[1+(q-1)sx]^{-\frac{1}{q-1}}e^{-ax}dx \\ &= \sum_{k=0}^{\infty}\frac{(-a)^k}{k!}\int_0^{\infty}x^k[1+(q-1)sx]^{-\frac{1}{q-1}}dx \\ &= \frac{1}{s}\sum_{k=0}^{\infty}\frac{(-\frac{a}{s})^k}{k!}\frac{\Gamma_{(q)}(k+1)}{s^{k+1}}, \Re(\frac{1}{q-1}-k-1)>0 \\ &= \frac{1}{s(2-q)}\,{}_1F_1\left[1;\frac{2q-3}{q-1};\frac{a}{s(q-1)}\right] \text{ for } \frac{3}{2}<q<2, \Re(s)>0, a>0. \end{aligned} $$

Lemma 3.

For a∈R,R(s)>0, *the q-Laplace transform of the function* e_q^{-ax} *is given by* $L_q\left[e_q^{-ax}\right](s) = \frac{1}{(s+a)(2-q)}{}_2F_1\left[1,\frac{1}{2};\frac{2q-3}{q-1};-\frac{4as}{(a+s)^2}\right]$ for $\frac{3}{2}<q<2,\left|\frac{4as}{(a+s)^2}\right|<1$.

Proof.

$$\begin{aligned} L_q[e_q^{-ax}](s) &= \int_0^\infty [1+(q-1)sx]^{-\frac{1}{q-1}}[1+(q-1)ax]^{-\frac{1}{q-1}}\,dx \\ &= \int_0^\infty \{[1+(q-1)sx][1+(q-1)ax]\}^{-\frac{1}{q-1}}\,dx \\ &= \int_0^\infty \left[1+(q-1)sx+(q-1)ax+(q-1)^2asx^2\right]^{-\frac{1}{q-1}}\,dx \\ &= \frac{1}{(s+a)(2-q)}{}_2F_1\left[1,\frac{1}{2};\frac{2q-3}{q-1};-\frac{4as}{(a+s)^2}\right] \\ &\quad \text{provided } \frac{3}{2}<q<2,\left|\frac{4as}{(a+s)^2}\right|<1. \end{aligned}$$

Lemma 4.

For R(s)>0 *and for* 3/2<q<2, a∈R, *there holds the formula,* $L_q[\cos(ax)]$ $\frac{1}{s(2-q)}\,{}_1F_2\left(1;\frac{2q-3}{2(q-1)},\frac{3q-4}{2(q-1)};-\frac{a^2}{[2s(q-1)]^2}\right)$.

Proof.

For q>1 , R(s)>0, a∈R, the q-Laplace transform of the trigonometric function cos(ax) is given by:

$$\begin{aligned} L_q[\cos(ax)](s) &= \int_0^\infty [1+(q-1)sx]^{-\frac{1}{q-1}}\cos(ax)\,dx \\ &= \sum_{k=0}^{\infty}\frac{(-1)^k a^{2k}}{(2k)!}\int_0^\infty x^{2k}[1+(q-1)sx]^{-\frac{1}{q-1}}\,dx \end{aligned}$$

$$= \frac{1}{s(2-q)}\sum_{k=0}^{\infty}\frac{(a^2)^k}{[4(q-1)^2s^2]^k}\frac{(1)_k}{k!\left(\frac{2q-3}{2(q-1)}\right)_k\left(\frac{3q-4}{2(q-1)}\right)_k}\quad \text{for } q>\frac{3}{2}.$$

By applying the properties of the beta function and integral evaluations, we get:

$$\begin{aligned} L_q[\cos(ax)](s) &= \frac{1}{s(2-q)}\,{}_1F_2\left(1;\frac{2q-3}{2(q-1)},\frac{3q-4}{2(q-1)};-\frac{a^2}{[2s(q-1)]^2}\right), \\ &\quad \text{for } \frac{3}{2}<q<2, q>1, \Re(s)>0, a\in\Re. \end{aligned}$$

One can easily check that as q→1, the above function gives a direct connection to the Laplace transforms of the original function simply by applying Sterling's approximation for the gamma function involved in the hypergeometric function involved in the equation.

Lemma 5.

The q-Laplace transform of the Gauss hypergeometric function is given by:

$$L_q[{}_mF_n](s) = \frac{1}{s(2-q)}\ {}_{m+1}F_{n+1}\left[\begin{matrix}a_1,a_2,\cdots,a_m,1\\ b_1,b_2,\cdots,b_n,\frac{2q-3}{q-1}\end{matrix}\middle|\frac{1}{(q-1)s}\right]$$

for R(s)>0,3/2<q<2.

Proof.

For q>1, the q-Laplace transform of the Gauss hyper geometric function is given by:

$$\begin{aligned} L_q[{}_mF_n](s) &= \sum_{k=0}^{\infty}\frac{(a_1)_k(a_2)_k\cdots(a_m)_k}{(b_1)_k(b_2)_k\cdots(b_n)_k}\frac{1}{k!}\int_0^{\infty}x^k[1+(q-1)sx]^{-\frac{1}{q-1}} \\ &= \frac{1}{(2-q)s}\sum_{k=0}^{\infty}\frac{(a_1)_k(a_2)_k\cdots(a_m)_k,(1)_k}{(b_1)_k(b_2)_k\cdots(b_n)_k,(\frac{2q-3}{q-1})_k}\frac{(\frac{1}{(q-1)s})^k}{k!} \\ &= \frac{1}{s(2-q)}\ {}_{m+1}F_{n+1}\left[\begin{matrix}a_1,a_2,\cdots,a_m,1\\ b_1,b_2,\cdots,b_n,\frac{2q-3}{q-1}\end{matrix}\middle|\frac{1}{(q-1)s}\right]\ \text{for } \Re(s)>0, \frac{3}{2}<q<2. \end{aligned}$$

Corollary: When m=n=0, we get the exponential function, and the q-Laplace transform is the confluent hypergeometric function F11.

The q-Laplace Transform of the Mittag–Leffler Function

The single parameter Mittag–Leffler function is defined as follows:

$$E_\alpha(z) = \sum_{k=0}^{\infty}\frac{z^k}{\Gamma(1+\alpha k)}, \text{for } \alpha\in C, \Re(\alpha)>0.$$

Lemma 6.

For q>1,R(s)>0, *the q-Laplace transform of* $E_\alpha(x^\alpha)$ *is given by:*

$$L_q[E_\alpha(x^\alpha)](s) = \frac{1}{s(q-1)\Gamma(\frac{1}{q-1})}\ H_{1\,2}^{2\,1}\left[\frac{1}{s^\alpha(q-1)^\alpha}\middle|\begin{matrix}(0,1)\\ (0,1),(\frac{2-q}{q-1},\alpha)\end{matrix}\right]$$

with suitable restrictions for the existence of Mittag–Leffler function.

Proof.

For q>1,R(s)>0, the q-Laplace transform of the Mittag–Leffler function is given by:

$$L_q[E_\alpha(x^\alpha)](s) = \sum_{k=0}^{\infty} \frac{1}{\Gamma(1+\alpha k)} \int_0^\infty x^{\alpha k}[1+(q-1)sx]^{-\frac{1}{q-1}} dx$$

$$= \sum_{k=0}^{\infty} \frac{\Gamma\left(\frac{2-q}{q-1} - \alpha k\right)}{\Gamma(\frac{1}{q-1})} \left(\frac{1}{s(q-1)}\right)^{\alpha k+1}$$

$$= \frac{1}{s(q-1)\Gamma(\frac{1}{q-1})} H_{1\,2}^{2\,1}\left[\frac{1}{s^\alpha(q-1)^\alpha}\middle|_{(0,1),(\frac{2-q}{q-1},\alpha)}^{(0,1)}\right].$$

The generalized Mittag–Leffler function introduced by Prabhakar is defined as follows:

$$E_{\beta,\gamma}^{\delta}(z) = \sum_{k=0}^{\infty} \frac{(\delta)_n z^k}{\Gamma(\beta k+\gamma)}, \text{for } \beta,\gamma,\delta \in C, \Re(\gamma) > 0, \Re(\delta) > 0.$$

Lemma 7.

Let β,γ,δ∈C, R(β)>0,R(γ)>0,R(δ)>0,R(1/q−1−γ)>0, *and for* 1<q<2, *there holds the formula:*

$$L_q[E_{\beta,\gamma}^{\delta}(ax^\beta)](s) = \frac{1}{s^\gamma(q-1)^\gamma\Gamma(\delta)\Gamma(\frac{1}{q-1})} H_{1\,2}^{2\,1}\left[{}^{(1-\delta,1)}_{(0,1),(\frac{1}{q-1}-\gamma,\beta)}\middle|\frac{1}{s^\beta(q-1)^\beta}\right]$$

for 1<q<2,R(β)>0,R(γ)>0,R(δ)>0,R(1/q−1−γ)>0.

The proof is similar to Lemma (7).

The details of the existence conditions, various properties and applications of H-functions are available in [27].

The q-Laplace Transform of the Fox H Function

Lemma 8.

For q<1, *consider the following restrictions. Let* $a^* = \sum_{i=1}^{n} a_i - \sum_{i=n+1}^{p} \alpha_i + \sum_{j=1}^{m} \beta_j - \sum_{j=m+1}^{r} \beta_j$, $\Delta = \sum_{j=1}^{r} \beta_j - \sum_{i=1}^{p} \alpha_i$ *and* $\mu = \sum_{j=1}^{r} b_j - \sum_{i=1}^{p} a_i + \frac{p-r}{2}$ *from the basic definition of the H-function.*

$$\textit{If either } a^* > 0, a^* = 0, \Re(\mu) < -1, \min_{1\le j\le m} \frac{\Re(b_j)}{\beta_j} > -1$$

$$\textit{when } a^* > 0, a^* = 0, \Delta \ge 0, \min_{1\le j\le m}\left[\frac{\Re(b_j)}{\beta_j}, \frac{\Re(\mu)+\frac{1}{2}}{\Delta}\right] > -1$$

when $a^*=0,\Delta<0$, *then for* $1<q<2$, *the q-Laplace transform of the H-function exists, and the formula:*

$$L_q[H_{p,r}^{m,n}](s) = \frac{1}{s(q-1)\Gamma(\frac{1}{q-1})} H_{p+1,r+1}^{m+1,n+1}\left[\frac{1}{s(q-1)}\Big|_{(b_1,\beta_1),(b_2,\beta_2),\cdots,(b_r,\beta_r),(\frac{2-q}{q-1},1)}^{(a_1,\alpha_1),(a_2,\alpha_2),\cdots,(a_k,\alpha_k),(0,1)}\right]$$

holds for $s\in C, R(s)>0$.

Proof.

For $q>1$,

$$L_q[H_{p,r}^{m,n}](s) = \frac{1}{2\pi i}\int_L h(t)\int_0^\infty x^{-t}[1+(q-1)sx]^{-\frac{1}{q-1}}dx\wedge dt$$

$$= \frac{1}{s(q-1)\Gamma(\frac{1}{q-1})}\frac{1}{2\pi i}\int_L h(t)\,\Gamma(1-t)\,\Gamma\left(\frac{2-q}{q-1}+t\right)dt$$

$$= \frac{1}{s(q-1)\Gamma(\frac{1}{q-1})} H_{p+1,r+1}^{m+1,n+1}\left[\frac{1}{s(q-1)}\Big|_{(b_1,\beta_1),(b_2,\beta_2),\cdots,(b_r,\beta_r),(\frac{2-q}{q-1},1)}^{(a_1,\alpha_1),(a_2,\alpha_2),\cdots,(a_k,\alpha_k),(0,1)}\right]$$

with suitable existing conditions.

CONNECTION TO OTHER INTEGRAL TRANSFORMS

In this section, we consider connections of the q-Laplace transform of a function f(·) to other integral transforms. The following theorem gives a relation between the Mellin transform of the q-Laplace transform of a function, where the Mellin transform of the function f(x) for x>0 is defined by $(Mf)(t)=\int_0^\infty x^{t-1}f(x)dx$, $t\in C$.

Theorem 1.

For $t\in C$, $R(t)<1/q-1, q>1$, *the Mellin transform* $L_q[x^{\gamma-1}f(x)](s)$ *is given by:*

$$\mathfrak{M}(L_q(x^{\gamma-1}f);t) = \frac{\Gamma(t)\Gamma(\frac{1}{q-1}-t)}{(q-1)^t\Gamma(\frac{1}{q-1})}\mathfrak{M}(f;\gamma-t).$$

Proof.

For $q>1$:

$$\mathfrak{M}L_q[x^{\gamma-1}f(x)](s) = \int_0^\infty s^{t-1}\int_0^\infty x^{\gamma-1}e_q^{-sx}f(x)dx$$

$$= \int_0^\infty x^{\gamma-1}f(x)\frac{1}{(q-1)^t x^t}\frac{\Gamma(t)\Gamma(\frac{1}{q-1}-t)}{\Gamma(\frac{1}{q-1})};\ \Re(\frac{1}{q-1}-t)>0$$

$$= \frac{\Gamma(t)\Gamma(\frac{1}{q-1}-t)}{\Gamma(\frac{1}{q-1})}\mathfrak{M}(f;\gamma-t);\ \Re(t)<\frac{1}{q-1}$$

hence the result.

Remark 1.

For γ=1 *and* t∈C, *it directly implies that the Mellin transform of the q-Laplace transform is given by:*

$$\mathfrak{M}(L_q(f);t) = \frac{\Gamma(t)\Gamma(\frac{1}{q-1}-t)}{(q-1)^t\Gamma(\frac{1}{q-1})}\mathfrak{M}(f;1-t) \; for \; q>1, \Re(t) < \frac{1}{q-1}.$$

The *G*-transform of the function f(x) is given in the form:

$$(Gf)(t) = \int_0^\infty G_{p,r}^{m,n}\left[xt\Big|_{(b_i)_{1,r}}^{(a_i)_{1,p}}\right] f(x)\mathrm{d}x$$

where the Meijers *G*-function is considered as the kernel, with suitable existence conditions. The following theorem helps to evaluate the *G*-transform of L_q(f(x)).

Theorem 2.

The G-transform of L_q(f(x)) *is given by the following relation:*

$$\mathbb{G}_{p,r}^{m,n}\{L_q[f(x)]\}(t) \;=\; \frac{1}{(q-1)\Gamma(\frac{1}{q-1})}\mathbb{G}_{p+1,r+1}^{m+1,n+1}[f(x)](t)$$

with suitable existing conditions.

Proof.

$$\mathbb{G}_{p,r}^{m,n}\{L_q[f(x)]\}(t) \;=\; \int_0^\infty G_{p,r}^{m,n}\left[st\Bigg|_{(b_j)_{1,r}}^{(a_i)_{1,p}}\right] f(s)\mathrm{d}s$$

$$= \frac{1}{2\pi i}\int_L\int_0^\infty h(\omega)(st)^{-\omega}L_q[f(x)](s)\mathrm{d}s\mathrm{d}\omega$$

$$= \frac{1}{2\pi i}\int_L\int_0^\infty\int_0^\infty h(\omega)(st)^{-\omega}[1+(q-1)sx]^{-\frac{1}{q-1}}f(x)\mathrm{d}x\mathrm{d}s\mathrm{d}\omega$$

$$= \frac{1}{2\pi i}\int_L\int_0^\infty h(\omega)x^{\omega-1}\frac{\Gamma(1-\omega)\Gamma(\frac{1}{q-1}-1+\omega)}{[x(q-1)]^{1-\omega}\Gamma(\frac{1}{q-1})}f(x)\mathrm{d}x\mathrm{d}\omega$$

$$= \frac{1}{(q-1)\Gamma(\frac{1}{q-1})}\int_0^\infty G_{p+1,r+1}^{m+1,n+1}\left[st\Bigg|_{(b_j)_{1,r},1}^{(a_i)_{1,p},\frac{1}{q-1}}\right] f(s)\mathrm{d}s$$

$$= \frac{1}{(q-1)\Gamma(\frac{1}{q-1})}\mathbb{G}_{p+1,r+1}^{m+1,n+1}[f(x)](t).$$

Remark 2.

The q-Laplace transform can be converted in terms of the G-transform in the sense that the q-exponential can be converted as $\frac{1}{\Gamma(\frac{1}{q-1})} G_{0,1}^{1,0}\left[-(q-1)sx\Big|^{1-\frac{1}{q-1}}\right]$ *for* $|(q-1)sx| \leq 1$. *That is:*

$$L_q[f(x)](s) = \int_0^\infty G_{0,1}^{1,0}\left[-(q-1)sx\Big|_0^{1-\frac{1}{q-1}}\right] f(x)\mathrm{d}x \text{ for } s > 0, |(q-1)sx| \leq 1. \tag{8}$$

Now, the integral transforms is of the form:

$$(Hf)(t) = \int_0^\infty H_{p,r}^{m,n}\left[xt\Big|_{(b_i,\beta_i)_{1,r}}^{(a_i,\alpha_i)_{1,p}}\right] f(x)\mathrm{d}x$$

which is known as the *H*-transform with suitable existence conditions.

The Hankel transform of a function f(x) for x>0 is defined by:

$$(H_n f)(t) = \int_0^\infty (xt)^{\frac{1}{2}} J_n(xt) f(x)\mathrm{d}x$$

where Jn(z) is the Bessel function of the first kind of order η∈C, such that R(η)>−1, which is given by:

$$J_n(z) = \sum_{k=0}^{\infty} \frac{(-1)^k}{\Gamma(\eta+k+1)k!}\left(\frac{z}{2}\right)^{2k+\eta}.$$

Theorem 3.

The Hankel transform of the q-Laplace transform $(H_n L_q(f))(t)$ *can be expressed in terms of the H-transform.*

Proof.

The integral transform with the Hankel kernel, which is operated on the *q*-Laplace transform, is given by:

$$\begin{aligned} H_n L_q(f)(t) &= \int_0^\infty (st)^{\frac{1}{2}} J_n(st) \int_0^\infty e_q^{-sx} f(x)\mathrm{d}x\mathrm{d}s \\ &= \int_0^\infty t^{\frac{1}{2}} \sum_{k=0}^{\infty} \frac{(-1)^k}{\Gamma(\eta+k+1)k!}\left(\frac{t}{2}\right)^{2k+\eta} \frac{\Gamma(2k+\eta+\frac{3}{2})\Gamma(\frac{1}{q-1}-2k-\eta-\frac{3}{2})}{[(q-1)x]^{2k+\eta+\frac{3}{2}}\Gamma(\frac{1}{q-1})} f(x)\mathrm{d}x \\ &= \frac{t^{\eta+\frac{1}{2}}}{2^\eta (q-1)^{\eta+\frac{3}{2}}\Gamma(\frac{1}{q-1})} \int_0^\infty H_{1\,3}^{2\,1}\left[\left(\frac{t}{2(q-1)x}\right)^2 \Big|_{(-\eta-\frac{1}{2},2)}^{(0,1),(\frac{1}{q-1}-\eta-\frac{3}{2}),(\eta,1)}\right] x^{-\eta-\frac{3}{2}} f(x)\mathrm{d}x \end{aligned}$$

which is the *H*-transform of $x^{-\eta-3/2}f(x)$.

Remark 3.

The q-Laplace transform of f(·) *for* q<1 *can be considered as a general case of the Riemann–Liouville integral operator, since for* q=0 *and for* x=u / t, *we get the general form of the Riemann–Liouville operator.*

Remark 4.

We can extend the q-Laplace transform to its generalized version by considering the function f(·) *with support over* (0,∞) *with:*

$$L_q[f(x)](s) = \int_0^\infty (xs)^{-\alpha}[e_q^{-sx}]f(x)dx \text{ for } \Re(s) > 0, \Re(\alpha) > 0 \tag{9}$$

where e^{-x}_q *is defined as in 2. Now, as* q→1, *we get the generalized Laplace transform of the function f, with support over the positive real line defined as:*

$$(Lf)(t) = \int_0^\infty (xt)^{-\alpha} e^{-(tx)^k} f(x)dx$$

that has interesting application in various fields.

DIFFERENTIAL EQUATIONS BY MEANS OF THE *Q*-LAPLACE TRANSFORM

In this section, we apply the properties of the *q*-Laplace transform to solve the fractional space-time diffusion equation, the kinetic equation and the time-fractional heat equation.

Fractional Space-Time Diffusion: Laplace Transform and H-Function

We consider the following diffusion model with fractional-order spatial and temporal derivatives:

$${}_0D_t^{\beta} N(x,t) = \eta \, {}_xD_\theta^{\alpha} N(x,t), \tag{10}$$

with the initial conditions ${}_0D_t^{\beta-1}N(x,0)=\sigma(x), 0\leq\beta\leq1, \lim_{x\to\pm\infty} N(x,t)=0$, where η is a diffusion constant; $\eta, t>0, x \in R; \alpha, \theta, \beta$ are real parameters with the constraints:

$$0 < \alpha \leq 2, |\theta| \leq min(\alpha, 2-\alpha),$$

and δ(x) is the Dirac-delta function. Then, for the fundamental solution of (1) with initial conditions, there holds the formula:

$$N(x,t) = \frac{t^{\beta-1}}{\alpha|x|} H_{3,3}^{2,1}\left[\frac{|x|}{(\eta t^{\beta})^{1/\alpha}} \middle| \begin{matrix} (1,1/\alpha),(\beta,\beta/\alpha),(1,\rho) \\ (1,1/\alpha),(1,1),(1,\rho) \end{matrix}\right], \alpha > 0 \tag{11}$$

where $\rho=\alpha-\theta/2\alpha$. The following special cases of (1) are of special interest for fractional diffusion models:

(i) For α=β, the corresponding solution of (1), denoted by N^{θ}_{α}, can be expressed in terms of the H-function as given below and can be defined for x>0:

Non-diffusion: 0<α=β<2;θ≤min{α,2−α},

$$N_{\alpha}^{\theta}(x) = \frac{t^{\alpha-1}}{\alpha|x|} H_{3,3}^{2,1}\left[\frac{|x|}{t\eta^{1/\alpha}}\left|\begin{matrix}(1,1/\alpha),(\alpha,1),(1,\rho)\\(1,1/\alpha),(1,1),(1,\rho)\end{matrix}\right.\right], \rho = \frac{\alpha-\theta}{2\alpha}. \tag{12}$$

(ii) When β=1,0<α≤2;θ≤min{α,2−α}, then (1) reduces to the space-fractional diffusion equation, which is the fundamental solution of the following space-time fractional diffusion model:

$$\frac{\partial N(x,t)}{\partial t} = \eta\, {}_xD_{\theta}^{\alpha} N(x,t), \eta > 0, x \in R, \tag{13}$$

with the initial conditions N(x,t=0)=σ(x),$\lim_{x\to\pm\infty}$N(x,t)=0, where η is a diffusion constant and σ(x) is the Dirac-delta function. Hence, for the solution of (1), there holds the formula:

$$L_{\alpha}^{\theta}(x) = \frac{1}{\alpha(\eta t)^{1/\alpha}} H_{2,2}^{1,1}\left[\frac{(\eta t)^{1/\alpha}}{|x|}\left|\begin{matrix}(1,1),(\rho,\rho)\\(\frac{1}{\alpha},\frac{1}{\alpha}),(\rho,\rho)\end{matrix}\right.\right], 0 < \alpha < 1, |\theta| \leq \alpha, \tag{14}$$

where ρ=α−θ/2α. The density represented by the above expression is known as α-stable Lévy density. Another form of this density is given by:

$$L_{\alpha}^{\theta}(x) = \frac{1}{\alpha(\eta t)^{1/\alpha}} H_{2,2}^{1,1}\left[\frac{|x|}{(\eta t)^{1/\alpha}}\left|\begin{matrix}(1-\frac{1}{\alpha},\frac{1}{\alpha}),(1-\rho,\rho)\\(0,1),(1-\rho,\rho)\end{matrix}\right.\right], 1 < \alpha < 2, |\theta| \leq 2-\alpha. \tag{15}$$

(iii) Next, if we take α=2,0<β<2;θ=0, then we obtain the time-fractional diffusion, which is governed by the following time-fractional diffusion model:

$$\frac{\partial^{\beta} N(x,t)}{\partial t^{\beta}} = \eta \frac{\partial^2}{\partial x^2} N(x,t), \eta > 0, x \in R, \ 0 < \beta \leq 2, \tag{16}$$

with the initial conditions ${}_0D_t^{\beta-1}$N(x,0)=σ(x), ${}_0D_t^{\beta-2}$N(x,0)=0, for x∈r, limx→±∞N(x,t)=0, where η is a diffusion constant and σ(x) is the Dirac-delta function, whose fundamental solution is given by the equation:

$$N(x,t) = \frac{t^{\beta-1}}{2|x|} H_{1,1}^{1,0}\left[\frac{|x|}{(\eta t^{\beta})^{1/2}}\left|\begin{matrix}(\beta,\beta/2)\\(1,1)\end{matrix}\right.\right]. \tag{17}$$

(iv) If we set α=2,β=1 and θ→0, then for the fundamental solution of the standard diffusion equation:

$$\frac{\partial}{\partial t} N(x,t) = \eta \frac{\partial^2}{\partial x^2} N(x,t), \tag{18}$$

with initial condition:

$$N(x,t=0)=\sigma(x),\ \lim_{x\to\pm\infty} N(x,t)=0, \tag{19}$$

there holds the formula:

$$N(x,t)=\frac{1}{2|x|}H_{1,1}^{1,0}\left[\frac{|x|}{\eta^{1/2}t^{1/2}}\left|\begin{matrix}(1,1/2)\\(1,1)\end{matrix}\right.\right]=(4\pi\eta t)^{-1/2}\exp[-\frac{|x|^2}{4\eta t}], \tag{20}$$

which is the classical Gaussian density.

Solution of the Generalized Kinetic Equation

Consider the generalized kinetic equation derived by Haubold and Mathai [8],

$$N(t)-N_0=-c_0\ {}_0D_t^{-\alpha}N(t) \text{ for } \alpha>0, \tag{21}$$

where ${}_0D_t^{-\alpha}N(t)$ is the Riemann–Liouville integral operator, in the form:

$${}_0D_t^{-\alpha}N(t)=\frac{1}{\Gamma(\alpha)}\int_0^t (t-u)^{\alpha-1}f(u)\mathrm{d}u$$

with the assumption that ${}_aD_t^0g(t)=g(t)$.

Lemma 9.

The solution of the kinetic Equation (21) is given by:

$$N(t)=\frac{N(0)}{2-q}E_\alpha\left(\frac{c_0\Gamma_{(q)}(\alpha)t^\alpha}{\Gamma(\alpha)(2-q)}\right)$$

where Eα(·) *represents the two parameter Mittag–Leffler function.*
Proof.

The q-Laplace transform of the Riemann–Liouville integral operator is given by $L_q[f(s)]=\frac{\Gamma_{(q)}(\alpha)\tilde{f}(u)}{s^\alpha(2-q)\Gamma(\alpha)}$ using the convolution property of the q-Laplace transform, and $\tilde{f}(u)$ is the q-Laplace transform of f(u). Now, by applying the q-Laplace transform on both sides of (21), we get:

$$\tilde{N}(t)-\frac{N(0)}{s(2-q)}=-c_0\frac{\Gamma_{(q)}(\alpha)}{s^\alpha(2-q)\Gamma(\alpha)}\tilde{N}(t)$$

where $\tilde{N}(t)=L_q[N(t)]$, the q-Laplace transform of N(t). Simplifying the equation we get

$$\tilde{N}(t) = \frac{N(0)}{s(2-q)}\left\{1 + \frac{c_0\Gamma_{(q)}(\alpha)}{s^\alpha(2-q)\Gamma(\alpha)}\right\}^{-1}$$

This can be expanded as an infinite sum, and on finding the inverse q-Laplace transform, we get:

$$\begin{aligned} N(t) &= \frac{N(0)}{(2-q)}\sum_{k=0}^{\infty}\left[\frac{c_0\Gamma_{(q)}(\alpha)}{\Gamma(\alpha)(2-q)}\right]^k \frac{t^{\alpha k}}{\Gamma(\alpha k+1)} \\ &= \frac{N(0)}{2-q}E_\alpha\left(\frac{c_0\Gamma_{(q)}(\alpha)t^\alpha}{\Gamma(\alpha)(2-q)}\right) \end{aligned}$$

for $\left|\frac{c_0\Gamma_{(q)}(\alpha)}{\Gamma(\alpha)(2-q)}\right| < 1$ where $E_\alpha(\cdot)$ represents the two-parameter Mittag–Leffler function.

Solution of the Time-Fractional Heat Equation

The standard heat equation is:

$$\frac{\partial u(x,t)}{\partial t} = \frac{\partial^2 u(x,t)}{\partial x^2}$$

where u(x,t) represents the temperature, which is a function of time t and space x. Let us write the equation in terms of the derivative operator D, such as:

$$D_t(u) = D_x^2(u) \tag{22}$$

where u=u(x,t). Then, for t≥0, the boundary conditions are that u(t,0)=u(t,L)=0 where L represents the length of a heating rod and an initial condition:

$$u(0,x) = -\frac{4a}{L^2}x^2 + \frac{4a}{L}x$$

where a=u(0,L/2). The general solution for Equation (22) assumed to be in the form u(t,x)=w(t)v(x) yields:

$$D(w(t))v(x) = w(t)D^2(v(x)) \Rightarrow \frac{D_t(w(t))}{w(t)} = \frac{D_x^2(v(x))}{w(x)} = K(say)$$

obtained from the general Equation (22). Let θ be the temperature decaying rate, and let $K=-\theta^2$ for $\theta \in R$; then, the ordinary differential equations $D(w(t))=-\theta^2 w(t)$ and $D^2(v(x))=-\theta^2 v(x)$ provide the general solution of Equation (22) of the form:

$$u(t,x) = K_1\cos(\theta x)e^{-\theta^2 t} + K_2\sin(\theta x)e^{-\theta^2 t}.$$

Now, let us consider the time fractional heat equation of the form:

$$D_t^{\alpha}(u) = D_x^2(u) \; 0 \leq \alpha < 2. \tag{23}$$

By considering similar steps as in the general solution and using the Laplace transform method to solve the differential equation $D^{\alpha}_t(w(t)) = -\theta^2 w(t)$, this yields the Mittag–Leffler function (similar steps as in Section 5.) as in the form:

$$w(t) = \sum_{k=0}^{\infty} \frac{(-\theta^2 t^{\alpha})^k}{\Gamma(\alpha k + 1)}. \tag{24}$$

Now, motivated from the same, we apply the *q*-Laplace transform for Equation (23) to obtain the solution for $D^{\alpha}_t(w(t)) = -\theta^2 w(t)$. The solution turns out to be:

$$w(t) = \frac{1}{(2-q)} \sum_{k=0}^{\infty} \left[\frac{-\theta^2 \Gamma_{(q)}(\alpha)}{\Gamma(\alpha)(2-q)} \right]^k \frac{t^{\alpha k}}{\Gamma(\alpha k + 1)} = \frac{1}{2-q} E_{\alpha} \left(\frac{-\theta^2 \Gamma_{(q)}(\alpha) t^{\alpha}}{\Gamma(\alpha)(2-q)} \right) \tag{25}$$

and hence, the general solution can be derived accordingly. Throughout the derivation, we consider the Laplace transformation for $q>1$. Similar derivation exists, when $q<1$.

CONCLUSIONS

In this article, we have proposed the *q*-Laplace transform as a suitable extension of the well-known Laplace transform. Despite the fact that it is difficult to evaluate some of the *H*-function numerically due to the constraints, the proposed method is an improvement over the regular practice of evaluating the Laplace transform within boundary values. The numerical illustration is not incorporated in this article; however, the methodology proposed here would be to generalize the result obtained in the regular sense of the Laplace transform. Another enhancement in this theory is that we applied the method of *q*-Laplace transforms in the generalized functional forms, such as Mittag–Leffler, hyper geometric, etc., so that applicability for particular functions, such as exponential, gamma, etc., can be easily deductible. The natural extension of the existing methodology explained in this article would further be considered for its generalized form, and it is an avenue for further research that could flow from this work.

ACKNOWLEDGMENTS

The authors would like to thank Arakaparampil Mathai Mathai, Department of Mathematics and Statistics, McGill University, Canada, Jose Kanichukattu, Department of Mathematics and Statistics, Central University of Rajasthan, India and unknown referees for their valuable suggestions and modifications. Furthermore, the first author would like to thank the Department of Mathematics and Statistics, University of Regina, for all facilities provided during the period of research.

AUTHOR CONTRIBUTIONS

Shanoja R. Naik and Hans J. Haubold contributed materials; Shanoja R. Naik wrote the paper.

REFERENCES

1. Sneddon, I.N. *Fourier Transforms*; McGraw-Hill: New York, NY, USA, 1951.
2. Zemanian, A.H. *Generalized Integral Transforms*; Interscience Publ.: New York, NY, USA, 1968.
3. Ernst, T. The History of q-Calculus and a New Method. Available online: http://www.math.uu.se/thomas/Lics.pdf (accessed on 12 July 2016).
4. Rao, G.L.N. The generalized Laplace transform of generalized functions. *Ranchi Univ. Math. J.* **1974**, *5*, 76–88.
5. Saxena, R.K. Some theorems on generalized Laplace transform, I. *Proc. Natl. Inst. Sci. India Part A* **1960**, *26*, 400–413.
6. Saxena, R.K. Some theorems on generalized Laplace transform, II. *Riv. Mat. Univ. Parma* **1961**, *2*, 287–299.
7. Rooney, P.G. On integral transformation with G-function kernels. *Proc. R. Soc. Edinb. Sect. A* **1982**, *93*, 265–297.
8. Haubold, H.J.; Mathai, A.M. The fractional kinetic equation and thermonuclear functions. *Astrophys. Space Sci.* **2000**, *327*, 53–63.
9. Mathai, A.M.; Haubold, H.J. The fractional kinetic equation and thermonuclear functions. *Astrophys. Space Sci.* **2007**, *273*, 53–63.
10. Samko, S.G.; Kilbas, A.A.; Marichev, O.I. *Fractional Integrals and Derivatives: Theory and Applications*; Gordon and Breach Science Publ.: Yverdon, Switzerland, 1993.
11. Abdi, W.H. On certain q-difference equations and q-Laplace transform. *Proc. Natl. Inst. Sci. India A* **1962**, *28*, 1–15.
12. Borges, E.P. A possible deformed algebra and calculus inspired in nonextensive thermo-statistics. *Phys. A Stat. Mech. Appl.* **2004**, *340*, 95–101.
13. Borges, E.P. A q-generalization of circular and hyperbolic functions. *Phys. A Math. Gen.* **1998**, *31*, 5281–5288.
14. Ferhan, M.A.; Eloe, W.P. Fractional q-Calculus on a time scale. *J. Nonlinear Math. Phys.* **2007**, *14*, 341–352.
15. Kilbas, A.A.; Saigo, M.; Shlapakov, S.A. Integral transforms with Fox's H-function in space of summable functions. *Integral Transform. Spec. Funct.* **1993**, *1*, 87–103.

16. Kilbas, A.A.; Saigo, M.; Shlapakov, S.A. Integral transforms with Fox's H-function in $_{Lv,r}$ -spaces II. *Fukuoka Univ. Sci. Rep.* **1994**, *24*, 13–38.
17. Mathai, A.M.; Rathie, P.N. *Basic Concepts in Information Theory and Statistics: Axiomatic Foundations and Applications*; Wiley Publications: New York, NY, USA, 1975.
18. Mathai, A.M.; Haubold, H.J. Pathway model, superstatistics, Tsallis statistics and a generalized measure of entropy. *Phys. A* **2007**, *375*, 110–122.
19. Beck, C. Stretched exponential from superstatistics. *Phys. A Stat. Mech. Appl.***2006**, *365*, 96–101.
20. Beck, C.; Cohen, E.G.D. Superstatistics. *Phys. A Stat. Mech.* **2003**, *322*, 267–275.
21. Tsallis, C. Possible generalizations of Boltzmann-Gibbs statistics. *J. Stat. Phys.***1988**, *52*, 479–487.
22. Wilk, G.; Wlodarczyk, Z. Non-exponential decays and nonextensivity. *Phys. Lett. A* **2001**, *290*, 55–58.
23. Mathai, A.M. A pathway to matrix-variate gamma and normal densities. *Linear Algebra Appl.* **2005**, *396*, 317–328.
24. Jose, K.K.; Naik, S.R. A class of asymmetric pathway distributions and an entropy interpretation. *Phys. A* **2008**, *387*, 6943–6951.
25. Jose, K.K.; Naik, S.R. On the *q*-Weibull distribution and its applications. *Commun. Stat. Theory Methods* **2009**, *38*, 912–926.
26. Mathai, A.M. *A Versatile Integral*; Centre for Mathamatical and Statistical Sciences: Kottayam, India, 2007.
27. Mathai, A.M. *A Handbook of Generalized Special Functions for Statistical and Physical Sciences*; Oxford University Press: Oxford, UK, 1993.

CHAPTER 4

FOURIER–MELLIN TRANSFORMS FOR CIRCULAR DOMAINS

Darren Crowdy

Department of Mathematics, Imperial College London, 180 Queen's Gate, London SW7 2AZ, UK

ABSTRACT

Generalized Fourier–Mellin transforms for analytic functions defined in simply connected circular domains are derived. Circular domains are taken to be those with boundaries that are a finite union of circular arcs, including straight line edges. The results are an extension to circular domains of the generalized Fourier transforms for convex polygons (having only straight

CITATION (APA): Crowdy, D. (2015). Fourier–Mellin transforms for circular domains. Computational Methods and Function Theory, 15(4), 655-687. (33 pages)

line edges) derived by Fokas and Kapaev (IMA J Appl Math 68:355–408, 2003). First, a new, elementary derivation of the latter result for polygons is given based on Cauchy's integral formula and a spectral representation of the Cauchy kernel. This rederivation extends in a natural way to the case of circular domains once an adapted spectral representation of the Cauchy kernel is established. Domains with boundaries that are a combination of circular arc and straight line edges can be treated similarly. The newly derived transforms are generalizations of the classical Fourier and Mellin transforms to general circular domains. It is shown by example how they can be used to solve boundary value problems for Laplace's equation in such domains. The notions of *spectral matrix* and *fundamental contour*, which arise naturally in the formulation, are also introduced.

Keywords: Transform method, Circular domains, Fourier transform, Mellin transform

INTRODUCTION

The last two decades have seen the emergence of a set of mathematical ideas and techniques which have become known collectively as the *unified transform method*. It has been pioneered by Fokas and collaborators [16, 18] and it is often called the *Fokas method*. For the most recent survey of this method the reader is referred to [17]. Part of this collective involves the reappraisal of classical boundary value problems for linear partial differential equations using the lessons learned from the study of nonlinear integrable systems. While the scope of the unified transform method is broad, the present article has been inspired by novel insights that this impactful method has brought to the area of complex analysis and function theory.

Specifically, in studying mixed boundary value problems for Laplace's equation in a convex polygon, Fokas and Kapaev [14] produced what appears to be a completely novel constructive method for their solution. The heart of the method, which developed ideas set out earlier by Fokas and Gel'fand [12], is the spectral analysis of one of the equations of a certain Lax pair (by means of Riemann–Hilbert techniques) and it leads eventually to the analysis of a so-called *global relation*. Interestingly, the approach also gives rise to a new integral representation for analytic functions in a convex polygon, one that generalizes the classical Fourier transform (naturally associated to a strip geometry) to more complicated domain types. Crowdy and Davis [8] have shown, in the context of a fluid dynamics application and by direct comparison with traditional Fourier methods how, even for the simple strip (or channel) geometry, the new approach of [14] already affords analytical advantages in practice.

The result of Fokas and Kapaev [14]—who derived it from the spectral analysis of a parameter-dependent ordinary differential equation and the use of Riemann–Hilbert methods—has been rederived from various other perspectives, including an approach via the fundamental differential form [13, 15], and the consideration of fundamental solutions and Green's integral representation [24]. The key result of [14], however, is concisely stated as follows. If $f(z)$ is analytic in a bounded, simply connected, convex, N-sided polygon with sides $\{S_j | j=1,\ldots,N\}$ then one can write

$$f(z) = \frac{1}{2\pi} \sum_{j=1}^{N} \int_{\mathcal{L}_j} \tilde{\rho}_j(k) e^{ikz} \, dk, \tag{1}$$

where L_j are a special set of rays in the spectral k-plane (depending on the geometry of the polygon, and defined here in Sect. 2) and where the so-called *spectral functions* are defined as

$$\tilde{\rho}_j(k) = \int_{S_j} f(z') e^{-ikz'} \, dz'. \tag{2}$$

These spectral functions satisfy the *global relation*

$$\sum_{j=1}^{N} \tilde{\rho}_j(k) = 0, \quad k \in \mathbb{C}. \tag{3}$$

In the context of boundary value problems for Laplace's equation for a harmonic function $q(x, y)$—taken to be the real part of some analytic function $f(z)$, say—these global relations can be used to determine the unknown boundary data for $f(z)$ and, hence, the solution for $q(x, y)$ inside the polygonal domain [14, 16, 18].

Ultimately, however, the statements (1)–(3) constitute a basic result in function theory, one that does not appear to have been reported elsewhere in the literature: it is a representation result for analytic functions in simply connected polygons. As such, this author was led to seek out an elementary derivation of (1)–(3) that does not require the language of Lax pairs, differential forms, distribution or spectral theory, or the mathematical technology of Riemann–Hilbert methods. Such a derivation, which we believe to be new, is presented in Sect. 2. It relies on nothing more than elementary properties of complex numbers, of the exponential function, and use of Cauchy's integral formula—all concepts available to a typical undergraduate after a first course in complex analysis.

Significantly the new derivation, which is geometric in nature, points the way to an important generalization: the formulation of the analogous transform methods for analytic functions defined in general *circular domains* (and, hence, to the study of boundary value problems for Laplace's equation in such domains). Where a polygon has N edges that are all straight lines (with zero curvature), we here define a circular domain to be one having N edges that each has constant, generally non-zero, curvature (including the possibility of straight line edges). This includes simple circular discs, of course, but also so-called "polycircular arc" domains which are the natural generalizations of N-sided polygons to regions with N circular arc edges. We derive the natural analogues of (1)–(3) for such domains. In Sects. 5–7 it is shown how these new transform pairs can be used to solve harmonic boundary value problems in a suite of examples which also allow us to outline some computational methods based on our approach.

Spence and Fokas [25] have presented, in the context of boundary value problems for Poisson's equation and several other partial differential equations, generalizations of the unified transform method to "boundary value problems in polar coordinates", including wedge regions. Our work is related to theirs, but our derivation here is more geometrical—not requiring any choice of coordinate system, or separation of variables—and the final statement of the results is different and applies to circular domains that are much more general than the wedge domains considered in [25]. Similarly, Fokas and Pinotsis [15] have developed the ideas of Fokas and Zyskin [13] and given a derivation of the transform pair of Fokas and Kapaev [14] by considering the Cauchy kernel, but the details of their derivation are different to that given here: it is our more geometrical viewpoint that provides the pathway to generalization to circular domains. Our treatment is, however, limited to consideration of analytic functions or, from the point of view of applications to boundary value problems, to Laplace's equation.

This paper treats only simply connected domains. However, the important generalization of the transform method to the case of multiply connected circular domains has been made in a companion paper [9].

A TRANSFORM METHOD FOR CONVEX POLYGONS

This section presents a new and elementary derivation of the transform method for convex polygons first introduced by Fokas and Kapaev [14] who used other techniques. In Sect. 3, it will be generalized to circular domains.

We start with a basic geometrical observation. If z′ lies on some finite length slit on the real axis and z is in the upper half-plane—see Figure 1—then, on geometrical grounds,

$$0 < \arg[z - z'] < \pi. \tag{4}$$

It follows trivially that

$$\int_0^{\infty} e^{ik(z-z')}\, dk = \left[\frac{e^{ik(z-z')}}{i(z-z')}\right]_0^{\infty} = \frac{1}{i(z'-z)} \tag{5}$$

or,

$$\frac{1}{z'-z} = i\int_0^{\infty} e^{ik(z-z')}\, dk, \quad 0 < \arg[z - z'] < \pi, \tag{6}$$

where it is easy to check that the contribution from the upper limit of integration vanishes for the particular choices of z′ and z to which we have restricted consideration.

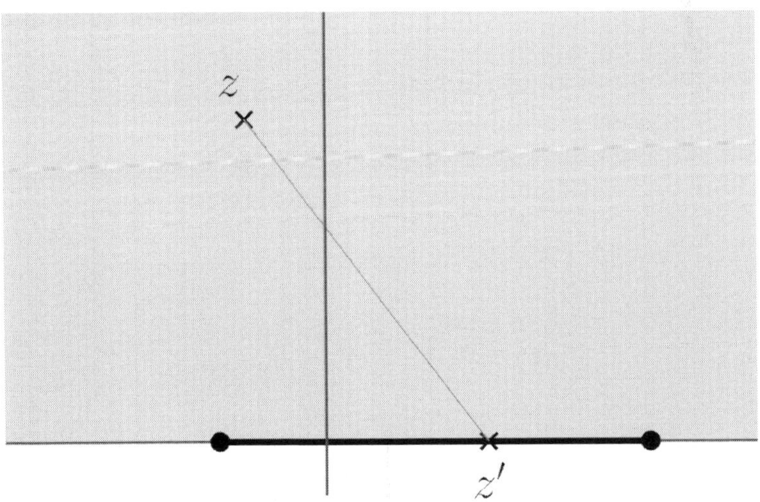

Figure 1: Geometrical positioning of z and z′ for the validity of (6).

On the other hand, suppose z' lies on some other finite length slit now making angle χ with the positive real axis and suppose that z is in the half-plane shown in Figure 2 (the half-plane "to the left" of the slit as one follows its tangent with uniform inclination angle χ). Now the affine transformation

$$z' \mapsto e^{-i\chi}(z'-\alpha), \quad z \mapsto e^{-i\chi}(z-\alpha), \tag{7}$$

for example, where the (unimportant) constant α is shown in Figure 2, returns the slit to the real axis, and z to the upper half-plane, so

$$0 < \arg[e^{-i\chi}(z-\alpha) - e^{-i\chi}(z'-\alpha)] < \pi. \tag{8}$$

(In Figure 2 we have drawn the slit to intersect the imaginary axis, but this is not necessary and our arguments go through for any slit making angle χ to the real axis.) Hence, on use of (6) now with the substitutions (7), we can write

$$\frac{1}{e^{-i\chi}(z'-\alpha) - e^{-i\chi}(z-\alpha)} = i\int_0^\infty e^{ik(e^{-i\chi}(z-\alpha)-e^{-i\chi}(z'-\alpha))}\,dk, \tag{9}$$

or, on cancellation of α and rearrangement,

$$\frac{1}{z'-z} = i\int_0^\infty e^{ie^{-i\chi}k(z-z')}e^{-i\chi}\,dk. \tag{10}$$

This integral identity is valid uniformly for all z and z' having the geometrical positioning depicted in Figure 2.

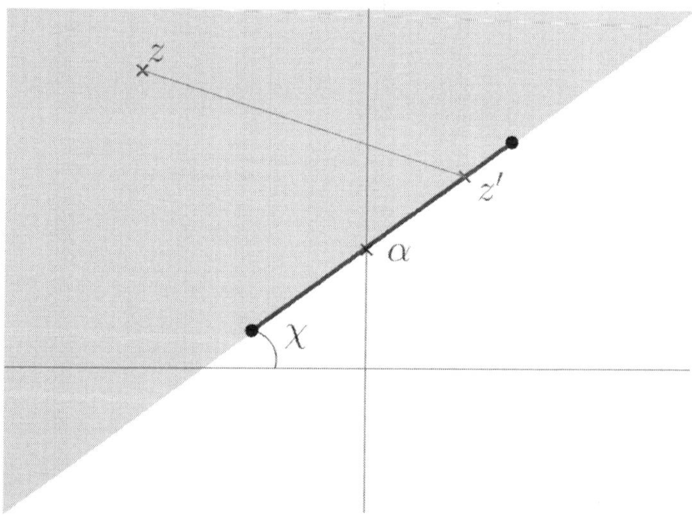

Figure 2: Geometrical positioning of z and z' for the validity of (10).

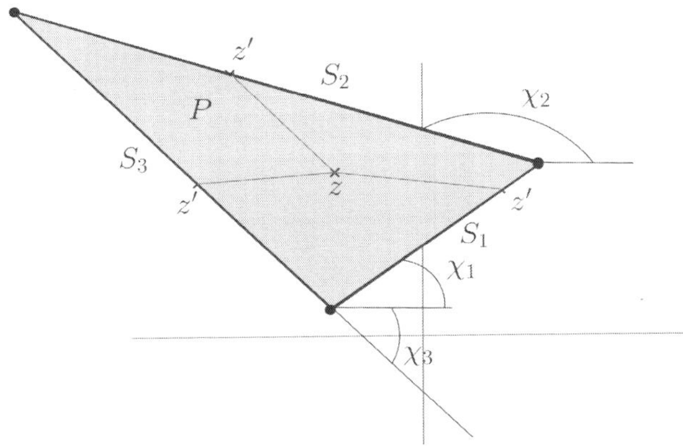

Figure 3: A convex polygon P as an intersection of N=3 half-planes with N angles $\{\chi_j|j=1,2,3\}$. Formula (10) can be used in the Cauchy integral formula with $\chi=\chi_j$ when z′ is on side S_j (for j=1,2,3j=1,2,3).

Now consider a bounded convex polygon P with N sides $\{S_j|j=1,\ldots,N\}$. Figure 3 shows an example with N=3. Geometrically the shaded polygon P can be viewed as the *intersection* of N=3 half-plane regions of the kind just considered. For a function $f(z)$ analytic in P, Cauchy's integral formula provides that for z∈P,

$$f(z) = \frac{1}{2\pi \mathrm{i}} \oint_{\partial P} \frac{f(z')\,\mathrm{d}z'}{z' - z}. \tag{11}$$

We can separate the boundary integral into a sum over the N sides:

$$f(z) = \frac{1}{2\pi \mathrm{i}} \sum_{j=1}^{N} \int_{S_j} f(z') \left\{ \frac{1}{(z' - z)} \right\} \mathrm{d}z'. \tag{12}$$

Now if side S_j has inclination angle χ_j then (10) can be used, with $\chi \mapsto \chi_j$, to re-express the Cauchy kernel uniformly for all z∈P and for z′ positioned on the respective sides:

$$f(z) = \frac{1}{2\pi \mathrm{i}} \sum_{j=1}^{N} \int_{S_j} f(z') \left\{ \mathrm{i} \int_0^{\infty} \mathrm{e}^{\mathrm{i}\mathrm{e}^{-\mathrm{i}\chi_j} k(z-z')} \mathrm{e}^{-\mathrm{i}\chi_j}\, \mathrm{d}k \right\} \mathrm{d}z'. \tag{13}$$

On reversing the order of integration we can write

$$f(z) = \frac{1}{2\pi} \sum_{j=1}^{N} \int_{\mathcal{L}} \rho_{jj}(k) \mathrm{e}^{-\mathrm{i}\chi_j} \mathrm{e}^{\mathrm{i}\mathrm{e}^{-\mathrm{i}\chi_j} kz}\, \mathrm{d}k, \tag{14}$$

where, for integers m, n between 1 and N, we define the *spectral matrix*

$$\rho_{mn}(k) \equiv \int_{S_n} f(z')\mathrm{e}^{-\mathrm{i}\mathrm{e}^{-\mathrm{i}\chi_m}kz'}\,\mathrm{d}z', \tag{15}$$

with L=[0,∞) defined to be the *fundamental contour* for straight line edges. The transform pair for polygons can thus be stated as

$$\begin{cases} f(z) = \frac{1}{2\pi}\sum_{j=1}^{N}\int_{\mathcal{L}}\rho_{jj}(k)\mathrm{e}^{-\mathrm{i}\chi_j}\mathrm{e}^{\mathrm{i}\mathrm{e}^{-\mathrm{i}\chi_j}kz}\,\mathrm{d}k, \\ \rho_{jj}(k) = \int_{S_j} f(z')\mathrm{e}^{-\mathrm{i}\mathrm{e}^{-\mathrm{i}\chi_j}kz'}\,\mathrm{d}z', \quad j = 1,\ldots,N. \end{cases} \tag{16}$$

This is our final result.

The Global Relations

The elements of the spectral matrix, or "spectral functions", have important analytical structure. Observe that, for any k∈C, and for any m=1,…,N,

$$\sum_{n=1}^{N}\rho_{mn}(k) = \sum_{n=1}^{N}\int_{S_n} f(z')\mathrm{e}^{-\mathrm{i}\mathrm{e}^{-\mathrm{i}\chi_m}kz'}\,\mathrm{d}z' = \int_{\partial P} f(z')\mathrm{e}^{-\mathrm{i}\mathrm{e}^{-\mathrm{i}\chi_m}kz'}\,\mathrm{d}z' = 0, \tag{17}$$

where we have used both the fact that $f(z')e^{-ie^{-i\chi_m}kz'}$ (for m=1,…,N) is analytic inside *P*, and Cauchy's theorem. There are *N* such global relations relating different elements of the spectral matrix, but each is an equivalent statement of the analyticity of *f*(*z*) in the domain *P*.

While the global relations relate all elements of the spectral matrix, it is worth emphasizing that only the *diagonal* elements of the spectral matrix appcar in the integral representation (14).

Transform pairs for unbounded polygons, such as semi-strips, can also be written down with minor modifications to the derivation above. The only difference is that the global relations are now valid in restricted sectors of the *k*-plane where the spectral functions are well defined [14].

Connection with Fokas and Kapaev [14]

Our statement (16) of the transform pair differs from (1)–(3) given by Fokas and Kapaev [14] in two ways. First, the notion of a spectral matrix was not introduced there. Second, the rays $\{L_j|j=1,\ldots,N\}$ in the spectral plane of [14] are absent. But it is easy to generate them.

By defining, for each j=1,…,N, the change of spectral variable given by

$$\lambda_j = \mathrm{e}^{-\mathrm{i}\chi_j}k, \quad j = 1,\ldots,N, \tag{18}$$

in both the spectral functions *and* the integral representation for *f*(*z*), then, rather than a sum of *N* contributions over the single fundamental contour in the spectral plane, the transform pair (16) can be written as a sum of contributions from N (generally distinct) rays in a spectral plane:

$$f(z) = \frac{1}{2\pi}\sum_{j=1}^{N}\int_{\mathcal{L}_j}\tilde{\rho}_j(\lambda)e^{i\lambda z}\,d\lambda, \quad \tilde{\rho}_j(\lambda) \equiv \int_{S_j} f(z')e^{-i\lambda z'}\,dz', \tag{19}$$

where $L_j=[0,\infty e^{-i\chi_j})$ is the ray defined by $\arg[\lambda]=-\chi_j$ (these coincide with the definitions given in [14]). Figure 4 shows the rays for the polygon shown in Figure 3. With formulas (16) thus modified they are equivalent to (1) and (2). The changes of variables (18) have the effect of reducing the N-by-N spectral matrix to just N distinct spectral functions. It is therefore tempting to argue that this obviates the need to introduce the spectral matrix. But we will see that this feature of reduction of the spectral matrix to just N spectral functions is *not* shared by more general domain types (notably, the circular domains considered later). Furthermore, it has been observed by previous authors studying the unified transform method (see [10, 16], for example) that for simple polygonal domains, it is possible, using the global relations and their so-called "invariant properties", to express all spectral functions in terms of the given boundary data, using only algebraic manipulations [16]. This procedure of generating new spectral functions by transformations of the spectral argument often generates the other (off-diagonal) elements of the spectral matrix (15) that has arisen naturally in our derivation above. We would argue that the notion of a *spectral matrix* and the *fundamental contour* (for particular edge types) are very natural ones for these transform formulations. Both concepts will appear again in the generalization to circular domains given in Sect. 3.

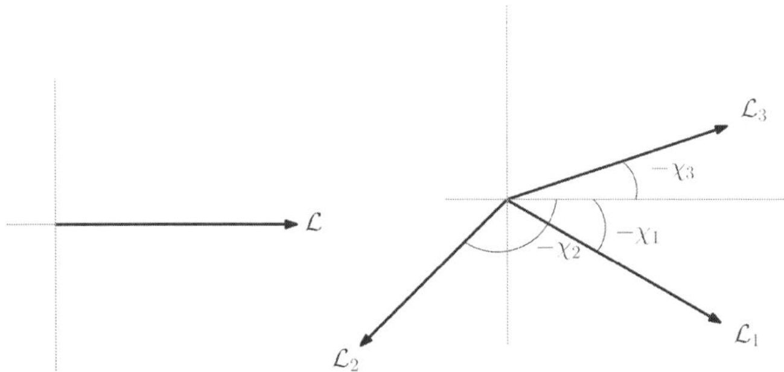

Figure 4: The fundamental contour $L=[0,\infty)$ for straight line edges (shown *left*). The spectral rays $\{L_j|j=1,2,3\}$ from Fokas and Kapaev [14] for the N=3 polygon *P* shown in Figure 3 (*right*). The latter rays are images of LL under the N transformations (18).

Unbounded Polygonal Regions

For *unbounded* regions exterior to some bounded polygon (and, hence, containing the point at infinity), a little thought reveals that it is geometrically impossible to "cover" the entire region as a finite intersection of half-plane regions. The next best thing is to subdivide the domain into an atlas of polygonal subregions that *are* the intersection of half-planes having a representation of the kind introduced above. This realization has already been made, albeit from alternative perspectives, by Charalambopoulos, Dassios and Fokas [4] for the exterior of an equilateral triangle, for example, where an atlas of 6 convex subregions outside the triangle is used.

A TRANSFORM METHOD FOR CIRCULAR DOMAINS

To find generalized transform schemes for circular domains we start by considering D as the simple unit disc; this basic geometrical unit will now replace the "half-plane" regions used for polygons. It should be clear that, to extend our approach to D, we must identify the particular spectral representation of the Cauchy kernel that is uniformly valid for z' on the domain boundary ∂D and for z inside D.

For values $|z|<1$ in D, consider the integral

$$I = \int_{L_1} \frac{1}{1-e^{2\pi i k}} z^k \, dk + \int_{L_2} z^k \, dk + \int_{L_3} \frac{e^{2\pi i k}}{1-e^{2\pi i k}} z^k \, dk. \tag{20}$$

This integral is taken around what we will call the *fundamental contour* for circular arc edges shown in Figure 5; it is the generalization of the fundamental contour for straight line edges shown on the left in Figure 4. To explain it, choose $0<r<1$. The contour L_1 is the union of the negative imaginary axis $(-i\infty,-ir]$ and the arc of the quarter circle $|k|=r$ in the third quadrant traversed in a clockwise sense; the contour L_2 is the real interval $[-r,\infty)$; the contour L_3 is the arc of the quarter circle $|k|=r$ in the second quadrant traversed in a clockwise sense together with the portion of the positive imaginary axis $[ir,i\infty)$. All integrals in (20) are non-singular and it is easy to verify directly that all integrands are exponentially decaying as $|k|\to\infty$ uniformly for all $|z|<1$. (Readers already familiar with the method of Fokas and Kapaev [14] might recognize this fundamental contour as similar to that associated with an infinite semi-strip, and there is a good reason for this that is explained in Appendix A.)

For |z|<1 it can be shown that

$$I = \int_{L_1} \frac{1}{1 - e^{2\pi ik}} z^k \, dk + \int_{L_2} z^k \, dk + \int_{L_3} \frac{e^{2\pi ik}}{1 - e^{2\pi ik}} z^k \, dk = \frac{1}{1 - z}. \tag{21}$$

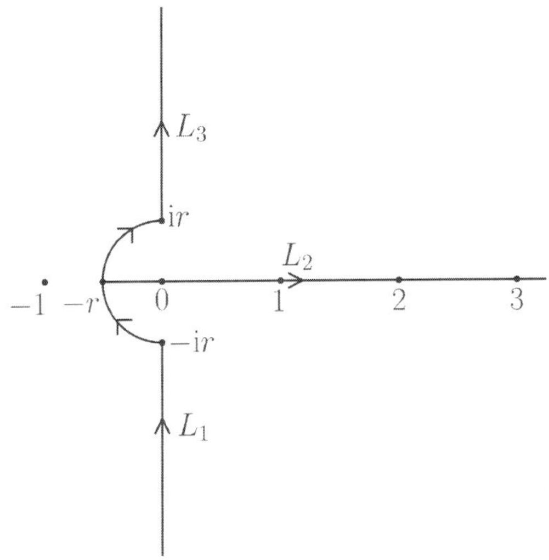

Figure 5: The fundamental contour for circular arc edges with 0<r<1.

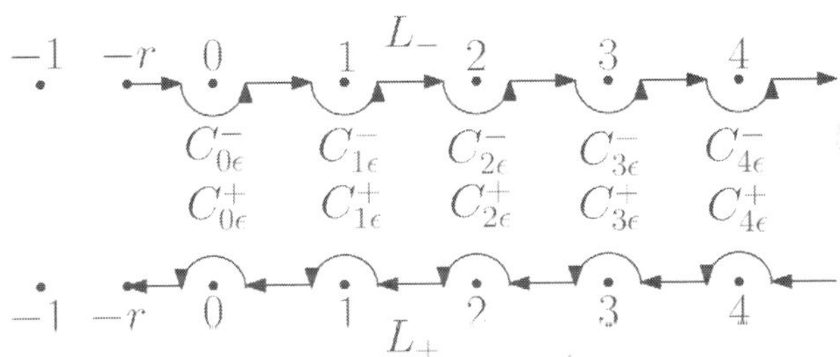

Figure 6: Contours L− and L+ in the spectral *k*-plane used to establish (21).

To derive (21), note that the second integral in *I* can be written as

$$\begin{aligned}\int_{L_2} z^k \, dk &= \mathcal{P} \int_{L_2} \left[\frac{1 - e^{2\pi ik}}{1 - e^{2\pi ik}} \right] z^k \, dk \\ &= \mathcal{P} \int_{L_2} \left[\frac{1}{1 - e^{2\pi ik}} \right] z^k \, dk - \mathcal{P} \int_{L_2} \left[\frac{e^{2\pi ik}}{1 - e^{2\pi ik}} \right] z^k \, dk,\end{aligned} \tag{22}$$

where P denotes the principal value integral. But, with the contours L_- and L_+ shown in Figure 6 for some 0<ε<r,

$$\int_{L_-}\left[\frac{z^k}{1-e^{2\pi ik}}\right]dk = +\mathcal{P}\int_{L_2}\left[\frac{z^k}{1-e^{2\pi ik}}\right]dk + \sum_{n=0}^{\infty}\lim_{\epsilon\to 0}\int_{C^-_{n\epsilon}}\left[\frac{z^k}{1-e^{2\pi ik}}\right]dk$$

$$\int_{L_+}\left[\frac{e^{2\pi ik}z^k}{1-e^{2\pi ik}}\right]dk = -\mathcal{P}\int_{L_2}\left[\frac{e^{2\pi ik}z^k}{1-e^{2\pi ik}}\right]dk + \sum_{n=0}^{\infty}\lim_{\epsilon\to 0}\int_{C^+_{n\epsilon}}\left[\frac{e^{2\pi ik}z^k}{1-e^{2\pi ik}}\right]dk. \tag{23}$$

The contours L_- and L_+ are made up of the union of the radius-ϵ semi-circles $\{C^{\pm}_{n\epsilon}|n\geq 0\}$ centred at $k=n$ (for $n\geq 0$) traversed in an anti-clockwise sense together with the portions of the real *k*-axis between them. Hence, after an explicit computation of the integrals around $\{C^{\pm}_{n\epsilon}|n\geq 0\}$, and on substituting for the principal value integrals in (22) using (23), we can write

$$I = \int_{\mathcal{L}_-}\left[\frac{1}{1-e^{2\pi ik}}\right]z^k\,dk + \sum_{n=0}^{\infty}z^n + \int_{\mathcal{L}_+}\left[\frac{e^{2\pi ik}}{1-e^{2\pi ik}}\right]z^k\,dk, \tag{24}$$

where $L_-\equiv L_1\cup L_-$ and $L_+\equiv L_+\cup L_3$ are shown in Figure 7. But the integral around L_- vanishes because its integrand is analytic in the fourth quadrant; similarly, the integral around L_+ vanishes because its integrand is analytic in the first quadrant. Hence, for $|z|<1$, we have

$$I = \sum_{n=0}^{\infty}z^n = \frac{1}{1-z}, \tag{25}$$

which establishes our result.

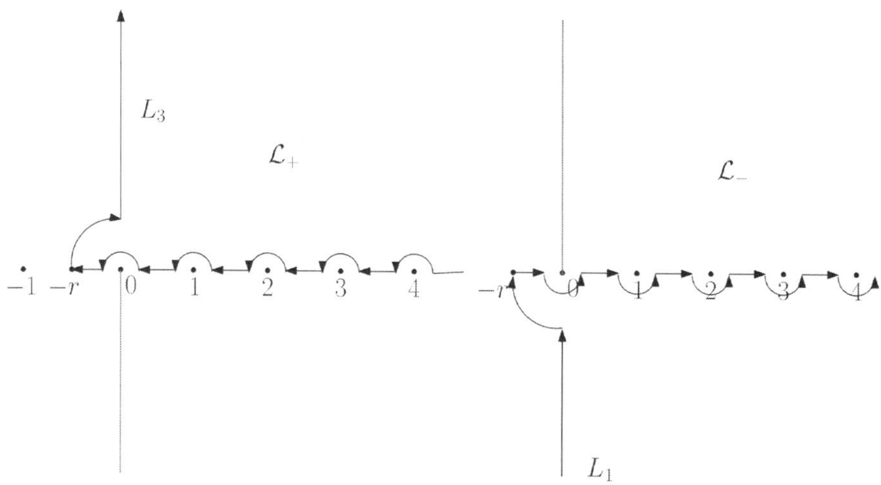

Figure 7: The contours L_+ and L_-.

Several remarks are in order:

- The residue calculus steps just given reveal that the integral (20) around the fundamental contour in Figure 5 can be deformed to the modified integral

$$I = \oint_C \frac{z^k}{1 - e^{2\pi i k}} \, dk, \tag{26}$$

with the contour CC encircling the real axis as shown in Figure 8.

- The reader might recognize the contour C as that appearing in the classical *Watson transformation* (see [25]) and, indeed, our construction of the spectral form (21) of the Cauchy kernel could alternatively have been made by invoking this transformation and giving a modified inversion formula based on (26). But we were led to our preferred form (21) in a quite different way. To see how, next we give the transform pair for the unit disc then, in Appendix A, show how to derive it independently from a logarithmic conformal mapping (a simple Schwarz–Christoffel mapping taking the disc to a semi-strip) coupled with the result for polygons of Sect. 2. Appendix A can be viewed as giving an algorithmic *construction* of the Watson contour C in Figure 8 from the results of Sect. 2.
- While the modified form (26) might, at first sight, appear to be a more concise statement of (20) we continue to use the latter. This is because (20) can be derived, after a logarithmic change of variable, from application of the transform method for a polygon (details are in Appendix A), hence the associated integration contours in Figure 5 are, by the very nature of that construction, the ones along which exponential decay of the integrands is guaranteed. In (26), integration is along the deformed contour C that remains uniformly close to the infinite array of simple pole singularities along the positive real axis and, if high precision is desired, this can compromise accuracy of any numerical quadrature of the inverse transforms. Many such approximation theory questions will be properly addressed in future work.

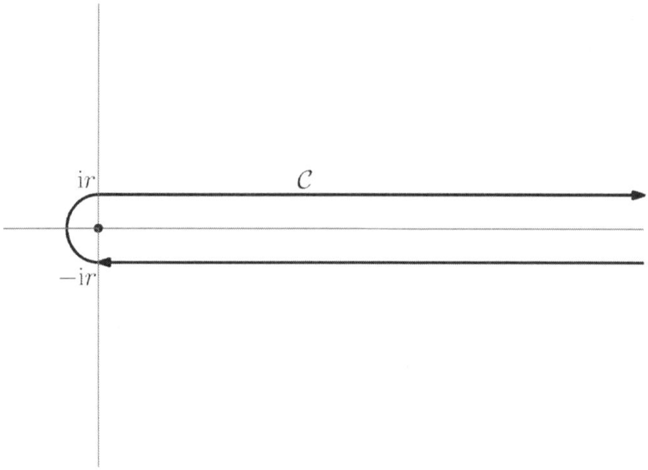

Figure 8: The contour C.

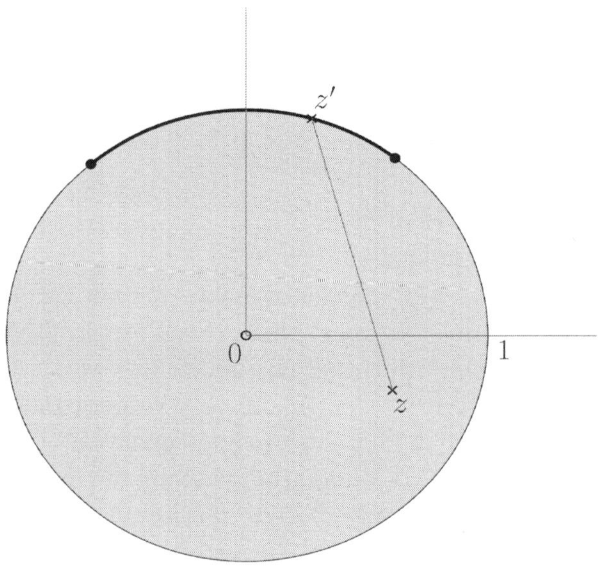

Figure 9: Geometry of z′ and z for derivation of (31).

Transform Pair for Interior of the Unit Disc

Let z′ be a point on the unit circle and let $|z|<1$ be a point inside the unit disc as depicted in Figure 9 (this is the analogue of Figure 1, but now for the unit disc instead of the upper half-plane). Then $|z/z'|<1$ uniformly and it follows on letting $z \mapsto z/z'$ in (21) that

$$\frac{z'}{z'-z} = \int_{L_1} \frac{1}{1-e^{2\pi ik}} \frac{z^k}{z'^k}\, dk + \int_{L_2} \frac{z^k}{z'^k}\, dk + \int_{L_3} \frac{e^{2\pi ik}}{1-e^{2\pi ik}} \frac{z^k}{z'^k}\, dk. \tag{27}$$

The Cauchy kernel is therefore given by the spectral representation

$$\frac{1}{z'-z} = \int_{L_1} \frac{1}{1-e^{2\pi ik}} \frac{z^k}{z'^{k+1}}\, dk + \int_{L_2} \frac{z^k}{z'^{k+1}}\, dk + \int_{L_3} \frac{e^{2\pi ik}}{1-e^{2\pi ik}} \frac{z^k}{z'^{k+1}}\, dk. \tag{28}$$

The Cauchy integral formula for a function $f(z)$ analytic in the unit disc is

$$f(z) = \frac{1}{2\pi i} \oint_{|z'|=1} \frac{f(z')}{z'-z}\, dz'. \tag{29}$$

On substitution of (the uniformly valid) representation (28) for the Cauchy kernel we find

$$f(z) = \frac{1}{2\pi i} \oint_{|z'|=1} f(z') \left[\int_{L_1} \frac{1}{1-e^{2\pi ik}} \frac{z^k}{z'^{k+1}}\, dk + \int_{L_2} \frac{z^k}{z'^{k+1}}\, dk + \int_{L_3} \frac{e^{2\pi ik}}{1-e^{2\pi ik}} \frac{z^k}{z'^{k+1}}\, dk \right] dz'. \tag{30}$$

On swapping the order of integration we arrive at the transform pair

$$\begin{cases} f(z) = \frac{1}{2\pi i} \left\{ \int_{L_1} \frac{\rho(k)}{1-e^{2\pi ik}} z^k\, dk + \int_{L_2} \rho(k) z^k\, dk + \int_{L_3} \frac{\rho(k) e^{2\pi ik}}{1-e^{2\pi ik}} z^k\, dk \right\}, \\ \rho(k) = \oint_{|z'|=1} \frac{f(z')}{z'^{k+1}}\, dz'. \end{cases} \tag{31}$$

The global relation is

$$\rho(k)=0,\ k \in -\mathbb{N} \tag{32}$$

since, for this discrete set of k-values, the integrand of the integral defining $\rho(k)$ is analytic in the unit disc.

It is a simple and instructive exercise to verify the transform pair (31) for simple test analytic functions such as $f(z)=z^p$ for $p \in \mathbb{Z}^+$

Equation (31) provides an alternative way to represent analytic functions in the disc. It is clear from our development that it is the natural generalization of the Fourier transform to the unit disc (or, more properly, the Fourier–Mellin transform—see the discussion in Appendix A and Appendix B, and the later comments at the end of Sect. 6). In complex analysis, it is more common to associate analytic functions in discs with Taylor series but, as we have seen, one can also write down the Fourier–Mellin transform pairs (31) too.

Transform Pair for Exterior of the Unit Disc

The transform pair for the *exterior* of the unit disc follows similarly. For z′ on the unit circle with |z|>1 then |z′/z|<1 uniformly and setting |z′/z|<1 in (21) gives the identity

$$\frac{z}{z-z'} = \int_{L_1} \frac{1}{1-e^{2\pi ik}} \left(\frac{z'}{z}\right)^k dk + \int_{L_2} \left(\frac{z'}{z}\right)^k dk + \int_{L_3} \frac{e^{2\pi ik}}{1-e^{2\pi ik}} \left(\frac{z'}{z}\right)^k dk. \tag{33}$$

Hence we have the spectral representation of the Cauchy kernel given by

$$\frac{1}{z-z'} = \int_{L_1} \frac{1}{1-e^{2\pi ik}} \frac{z'^k}{z^{k+1}} dk + \int_{L_2} \frac{z'^k}{z^{k+1}} dk + \int_{L_3} \frac{e^{2\pi ik}}{1-e^{2\pi ik}} \frac{z'^k}{z^{k+1}} dk. \tag{34}$$

For |z|>1 the Cauchy integral formula

$$f(z) = -\frac{1}{2\pi i} \oint_{|z'|=1} \frac{f(z')}{z'-z} dz' \tag{35}$$

holds for a function analytic outside the unit disc that decays like $1/z$ as z→∞. Use of (the uniformly valid) expression (34) in the Cauchy integral formula then produces the following transform pair for such a function:

$$\begin{cases} f(z) = \frac{1}{2\pi i}\left\{ \int_{L_1} \frac{\rho(k)}{1-e^{2\pi ik}} \frac{dk}{z^{k+1}} + \int_{L_2} \rho(k) \frac{dk}{z^{k+1}} + \int_{L_3} \frac{\rho(k)e^{2\pi ik}}{1-e^{2\pi ik}} \frac{dk}{z^{k+1}} \right\}, \\ \rho(k) = \oint_{|z'|=1} f(z') z'^k dz'. \end{cases} \tag{36}$$

The global relation is

$$\rho(k)=0, \ k \in -\mathbb{N} \tag{37}$$

since, for this discrete set of *k*-values, the integrand of the integral defining ρ(k) is analytic outside the disc and is $O(1/z^2)$ as z→∞

For future use, we prefer to write (36) with the addition of two minus signs:

$$\begin{cases} f(z) = -\frac{1}{2\pi i}\left\{ \int_{L_1} \frac{\rho(k)}{1-e^{2\pi ik}} \frac{dk}{z^{k+1}} + \int_{L_2} \rho(k) \frac{dk}{z^{k+1}} + \int_{L_3} \frac{\rho(k)e^{2\pi ik}}{1-e^{2\pi ik}} \frac{dk}{z^{k+1}} \right\}, \\ \rho(k) = -\oint_{|z'|=1} f(z') z'^k dz', \end{cases} \tag{38}$$

which clearly does not affect the transform pair. The reason for this is to preserve a convention that the spectral function is the integral around the boundary traversed in such a way that the domain of interest is *on the left* as the boundary is traversed.

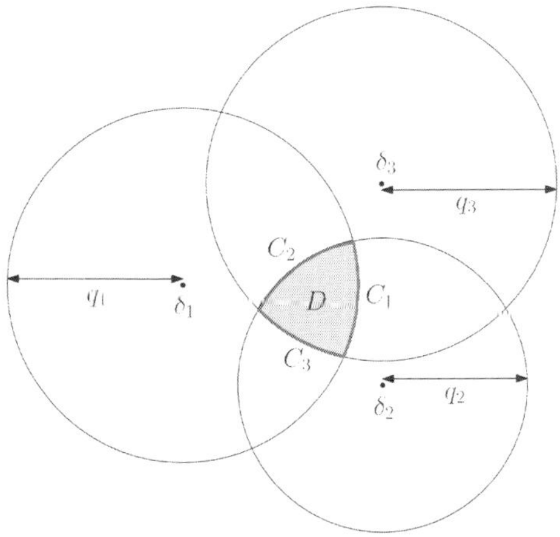

Figure 10: A polycircular domain D with N=3 sides denoted by C_1, C_2 and C_3.

Transform Pair for Polycircular Domains

Just as a convex polygon was understood in Sect. 2 to be the intersection of N half-plane regions, a polycircular domain is now understood to be the intersection of N circular discs.

Consider now the convex simply connected region D bounded by the arcs $\{C_j | j=1,\ldots,N\}$ of N circles $\{|z-\delta_j|=q_j | j=1,\ldots,N\}$ as shown in Figure 10 for N=3. The Cauchy integral formula for a function $f(z)$ analytic in this region is

$$f(z) = \frac{1}{2\pi \mathrm{i}} \oint_{\partial D} \frac{f(z')}{z'-z}\, \mathrm{d}z' = \frac{1}{2\pi \mathrm{i}} \sum_{j=1}^{N} \int_{C_j} \frac{f(z')}{z'-z}\, \mathrm{d}z', \tag{39}$$

where ∂D denotes the boundary of D and where, in the second equality, the integral around ∂D has been separated into the N separate integrals around the individual circular arcs $\{C_j | j=1,\ldots,N\}$.

Focussing on the portion of the integral along C_j for which it holds, uniformly for z in D, that $|z-\delta_j|<|z'-\delta_j|$ the Cauchy kernel for z′ on C_j has the spectral representation

$$
\begin{aligned}
\frac{1}{z'-z} &= \frac{1}{(z'-\delta_j)-(z-\delta_j)} \\
&= \int_{L_1} \frac{1}{1-e^{2\pi ik}} \frac{(z-\delta_j)^k}{(z'-\delta_j)^{k+1}}\, dk + \int_{L_2} \frac{(z-\delta_j)^k}{(z'-\delta_j)^{k+1}}\, dk \\
&\quad + \int_{L_3} \frac{e^{2\pi ik}}{1-e^{2\pi ik}} \frac{(z-\delta_j)^k}{(z'-\delta_j)^{k+1}}\, dk,
\end{aligned} \tag{40}
$$

where we have set $z \mapsto (z-\delta_j)/(z'-\delta_j)$ in (21). It is important to write this in the rescaled form

$$
\begin{aligned}
\frac{1}{z'-z} &= \int_{L_1} \frac{1}{1-e^{2\pi ik}} \frac{1}{q_j} \left(\frac{z-\delta_j}{q_j}\right)^k \left[\frac{z'-\delta_j}{q_j}\right]^{-k-1} dk \\
&\quad + \int_{L_2} \frac{1}{q_j} \left(\frac{z-\delta_j}{q_j}\right)^k \left[\frac{z'-\delta_j}{q_j}\right]^{-k-1} dk \\
&\quad + \int_{L_3} \frac{e^{2\pi ik}}{1-e^{2\pi ik}} \frac{1}{q_j} \left(\frac{z-\delta_j}{q_j}\right)^k \left[\frac{z'-\delta_j}{q_j}\right]^{-k-1} dk.
\end{aligned} \tag{41}
$$

Now for $z \in D$ (see Figure 10) comprising the intersection of the circular discs we can substitute (41) into the Cauchy integral formula (39) when z' sits on each of the separate boundary arcs $\{C_j | j=1,\ldots,N\}$ to find

$$
\begin{aligned}
f(z) &= \frac{1}{2\pi i} \sum_{j=1}^{N} \Bigg\{ \int_{L_1} \frac{\rho_{jj}(k)}{1-e^{2\pi ik}} \left[\frac{z-\delta_j}{q_j}\right]^k dk + \int_{L_2} \rho_{jj}(k) \left[\frac{z-\delta_j}{q_j}\right]^k dk \\
&\quad + \int_{L_3} \frac{\rho_{jj}(k) e^{2\pi ik}}{1-e^{2\pi ik}} \left[\frac{z-\delta_j}{q_j}\right]^k dk \Bigg\},
\end{aligned} \tag{42}
$$

where we have swapped the order of integration and introduced the N-by-N *spectral matrix*

$$
\rho_{mn}(k) \equiv \frac{1}{q_m} \int_{C_n} \left[\frac{z'-\delta_m}{q_m}\right]^{-k-1} f(z')\, dz'. \tag{43}
$$

Global relations for this system are

$$
\sum_{n=1}^{N} \rho_{mn}(k) = 0, \quad k \in -\mathbb{N} \tag{44}
$$

for any $m=1,2,\ldots,N$. There are N such global relations but each is an equivalent statement of the analyticity of $f(z)$ in the domain D.

It is important to remark that, unlike in the case of polygons, it is not clear that it is now possible to invoke any changes of variables in the spectral k-plane, akin to (18), that will reduce the N-by-N spectral matrix to just N spectral functions.

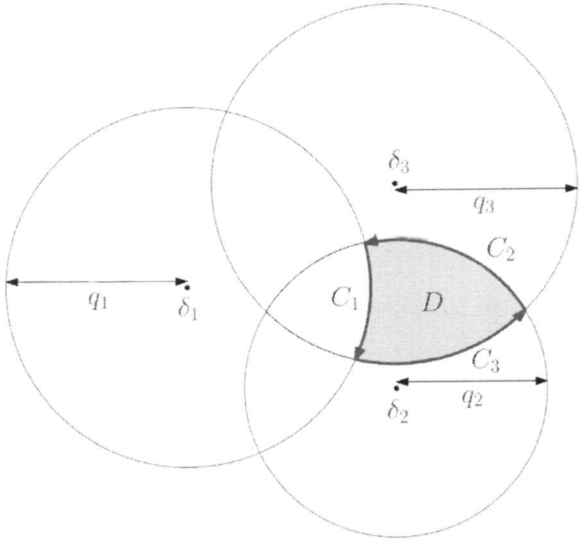

Figure 11: A polycircular domain D with N=3 (directed) sides denoted by C_1, C_2 and C_3. Note that C_1 is an arc of $|z-\delta_1|=q_1$ now traversed in a clockwise sense.

Variants of the above should be obvious. Suppose, on the other hand, that the polycircular domain D is that shown in Figure 11: D is the intersection of the *exterior* of the circle $|z-\delta_1|=q_1$ and the *interior* of the circles $|z-\delta_2|=q_2$ and $|z-\delta_3|=q_3$. Let $\{C_j|j=1,2,3\}$ again denote the boundary arcs making up the boundary ∂D as shown in Figure 11 and such that the domain D remains to the left as ∂D is traversed. By arguments similar to those given above we deduce that the appropriate integral representation is

$$\begin{aligned}f(z) = &-\frac{1}{2\pi i}\left\{\int_{L_1}\frac{\rho_{11}(k)}{1-e^{2\pi ik}}\left(\frac{q_1}{z-\delta_1}\right)^{k+1}dk + \int_{L_2}\rho_{11}(k)\left(\frac{q_1}{z-\delta_1}\right)^{k+1}dk\right.\\ &\left.+\int_{L_3}\frac{\rho_{11}(k)e^{2\pi ik}}{1-e^{2\pi ik}}\left(\frac{q_1}{z-\delta_1}\right)^{k+1}dk\right\}\\ &+\frac{1}{2\pi i}\sum_{j=2}^{3}\left\{\int_{L_1}\frac{\rho_{jj}(k)}{1-e^{2\pi ik}}\left[\frac{z-\delta_j}{q_j}\right]^{k}dk + \int_{L_2}\rho_{jj}(k)\left[\frac{z-\delta_j}{q_j}\right]^{k}dk\right.\\ &\left.+\int_{L_3}\frac{\rho_{jj}(k)e^{2\pi ik}}{1-e^{2\pi ik}}\left[\frac{z-\delta_j}{q_j}\right]^{k}dk\right\},\end{aligned} \tag{45}$$

where the matrix of spectral functions is defined for n=1,2,3 by

$$\rho_{1n}(k) \equiv \frac{1}{q_1} \int_{C_n} \left(\frac{z' - \delta_1}{q_1} \right)^k f(z')\, dz',$$
$$\rho_{mn}(k) \equiv \frac{1}{q_m} \int_{C_n} \left(\frac{z' - \delta_m}{q_m} \right)^{-k-1} f(z')\, dz', \quad m = 2, 3. \tag{46}$$

Global relations are

$$\sum_{n=1}^{3} \rho_{mn}(k) = 0, \quad k \in -\mathbb{N}, \ m = 1, 2, 3. \tag{47}$$

Domains with Both Straight and Circular Edges

It should be clear that our approach generalizes to more general domains with boundaries comprising a *mixture* of circular arc and straight line edges. Correspondingly, the derivation of the relevant transform pairs involves a mixture of the results of Sects. 2 and 3. This is best illustrated by means of an explicit example (given in Sect. 6) from which the general construction can be discerned.

APPLICATIONS TO BOUNDARY VALUE PROBLEMS

In the remainder of the paper we survey just a few of the many possible applications of this new transform perspective for circular domains, in particular to the solution of boundary value problems for Laplace's equation. We proceed under the assumption that solving for the functions in the spectral matrix will lead, on substitution of this spectral data into the integral representations for the relevant analytic function, to the solution of the stated boundary value problems. That this is true is not obvious, but it has been established for boundary value problems for Laplace equations in convex polygons by Fulton, Fokas and Xenophontos [21] and for general elliptic equations in arbitrary convex domains by Ashton [1].

Here we focus on boundary value problems where the effectiveness of the transform approach can be tested against other mathematical schemes. The following examples also give us the opportunity to introduce some numerical schemes we have devised for the analysis of the global relation.

THE PROBLEM OF DUAL FOURIER SERIES

Even for the simple unit disc the new transform approach affords a reappraisal of some boundary value problems previously tackled by other means. A problem considered by Shepherd [20] is to find the set of real coefficients $\{A_n\}$ satisfying the conditions

$$\cos(m\theta) = \sum_{n\geq 0} A_n \cos n\theta, \quad 0 < \theta < \pi/2 \text{ (or } C_1),$$
$$-\sin(m\theta) = \sum_{n\geq 1} A_n \sin n\theta, \quad \pi/2 < \theta < \pi \text{ (or } C_2). \tag{48}$$

where m is some positive integer. This has become known as a problem of *dual Fourier series*[20, 23, 26]; indeed, Shepherd solved this problem, separately for even and odd m, by a sequence of ingenious manipulations of integral representations of the Fourier series coefficients; he also included an application to a problem in potential theory. But the same results follow in a more algorithmic way by an analysis of the global relations for this system as they arise in the new transform method described above.

First note that the problem can be reformulated as a mixed boundary value problem for a function $f(z)$ analytic in the unit z-disc. Let

$$f(z) = \sum_{n\geq 0} A_n z^n \tag{49}$$

be the Taylor expansion of such a function valid convergent for $|z|<1$. Suppose it satisfies the mixed boundary conditions on the two "faces" of the unit circle $|z|=1$ where $z=e^{i\theta}$:

$$\mathrm{Re}[f(z)] = \cos m\theta, \quad -\pi/2 < \theta < \pi/2,$$
$$\mathrm{Im}[f(z)] = -\sin m\theta, \quad \pi/2 < \theta < 3\pi/2. \tag{50}$$

If such a function $f(z)$ can be found then the coefficients in (49) will satisfy (50) as z tends to the boundary.

Since $f(z)$ is analytic in the unit disc we have the following transform pair representation for it:

$$\begin{cases} f(z) = \frac{1}{2\pi i}\left\{\int_{L_1} \frac{\rho(k)}{1-e^{2\pi ik}} z^k\, dk + \int_{L_2} \rho(k) z^k\, dk + \int_{L_3} \frac{\rho(k)e^{2\pi ik}}{1-e^{2\pi ik}} z^k\, dk\right\}, \\ \rho(k) = \oint_{|z'|=1} \frac{f(z')}{z'^{k+1}}\, dz'. \end{cases} \tag{51}$$

The global relation is

$$\rho(k)=0, k\in -\mathbb{N}. \tag{52}$$

Equation (52) can be used to find the spectral function ρ(k), hence $f(z)$, and, in turn, the coefficients $\{A_n\}$.

On C_1 we write

$$f(z) = \cos m\theta + \mathrm{i}\left[a_0 + \sum_{n\geq 1} a_n \mathrm{e}^{2n\theta \mathrm{i}} + \overline{a_n}\mathrm{e}^{-2n\theta \mathrm{i}}\right], \tag{53}$$

where $a_0 \in R$ and $\{an \in C|n \geq 1\}$ are to be determined. Here we have decomposed the unknown imaginary part of $f(z)$ as a period-π Fourier series over the interval [−π/2,π/2]. Similarly, for z on C_2 we write

$$f(z) = \left[b_0 + \sum_{n\geq 1} b_n \mathrm{e}^{2n\theta \mathrm{i}} + \overline{b_n}\mathrm{e}^{-2n\theta \mathrm{i}}\right] - \mathrm{i}\sin m\theta, \tag{54}$$

where $b_0 \in R$ and $\{b_n \in C|n \geq 1\}$ are to be determined and, here, we have decomposed the unknown real part as a period-π Fourier series. On substitution of (53) and (54) into (52) we arrive at the following linear system for the unknown coefficients:

$$\begin{aligned} &a_0\, \mathcal{A}(0,k) + \sum_{n\geq 1} a_n \mathcal{A}(2n,k) + \overline{a_n}\mathcal{A}(-2n,k) \\ &\quad + b_0\, \mathcal{B}(0,k) + \sum_{n\geq 1} b_n \mathcal{B}(2n,k) + \overline{b_n}\mathcal{B}(-2n,k) = r(k), \quad k \in -\mathbb{N}, \\ &a_0\, \overline{\mathcal{A}(0,k)} + \sum_{n\geq 1} \overline{a_n}\, \overline{\mathcal{A}(2n,k)} + a_n\, \overline{\mathcal{A}(-2n,k)} \\ &\quad + b_0\, \overline{\mathcal{B}(0,k)} + \sum_{n\geq 1} \overline{b_n}\, \overline{\mathcal{B}(2n,k)} + b_n\, \overline{\mathcal{B}(-2n,k)} = \overline{r(k)}, \quad k \in -\mathbb{N}, \end{aligned} \tag{55}$$

where

$$\mathcal{A}(n,k) = \begin{cases} \frac{1}{n-k}\left[\mathrm{e}^{\mathrm{i}(n-k)\pi/2} - \mathrm{e}^{-\mathrm{i}(n-k)\pi/2}\right], & n \neq k, \\ \mathrm{i}\pi, & n = k, \end{cases} \tag{56}$$

$$\mathcal{B}(n,k) = \begin{cases} \frac{1}{\mathrm{i}(n-k)}\left[\mathrm{e}^{3\mathrm{i}(n-k)\pi/2} - \mathrm{e}^{\mathrm{i}(n-k)\pi/2}\right], & n \neq k, \\ \pi, & n = k, \end{cases} \tag{57}$$

and

$$r(k) = -\int_{-\pi/2}^{\pi/2} \cos m\theta\, \mathrm{e}^{-\mathrm{i}k\theta}\, \mathrm{d}\theta + \mathrm{i}\int_{\pi/2}^{3\pi/2} \sin m\theta\, \mathrm{e}^{-\mathrm{i}k\theta}\, \mathrm{d}\theta. \tag{58}$$

The system (55) can be solved numerically by truncating the infinite sums at some integer N—so that there is a total of 4N+2 real unknown coefficients—and then picking $N_k > 4N+2$ values of k in (55) to produce an overdetermined system of N_k equations from which the unknown coefficients can be found by a least-squares algorithm. This procedure was carried out and the results checked against those derived by the alternative scheme given by Shepherd [20] (it can also be solved using a method based on conformal slit maps as used recently by the author [7] in a mixed boundary value problem of the same general kind).

While we have used Fourier series representations for the unknown boundary data above, another possibility (akin to Chebyshev-type expansions) is described in the example of Sect. 7and can be applied in the above example too.

A strong argument in favour of the transform approach to this problem just demonstrated is that it is more algorithmic than the rather more ad hoc approach used by Shepherd [20]. Our approach also frames this problem as just another boundary value problem that is amenable to analysis using the same basic mathematical ideas underpinning the unified transform method.

MIXED BOUNDARY VALUE PROBLEM IN A SEMI-DISC

A second boundary value problem that can be used to test the method is to find a function $f(z)$ analytic in D defined to be the upper half unit semi-disc $|z|=1, \mathrm{Im}[z]>0$ as shown in Figure 12. This is an interesting example in that D has both a circular arc boundary and a straight line boundary and it affords an opportunity to see how to combine the ideas given earlier in this paper.

Suppose the boundary conditions on an analytic function $f(z)$ in such a domain are given to be

$$\begin{aligned} &\mathrm{Re}[f(z)] = r(z, \bar{z}), \quad &&\text{on } C, \\ &\mathrm{Im}[f(z)] = 0, &&\text{on } L, \end{aligned} \tag{59}$$

where $r(z, \bar{z})$ is some given real-valued function.

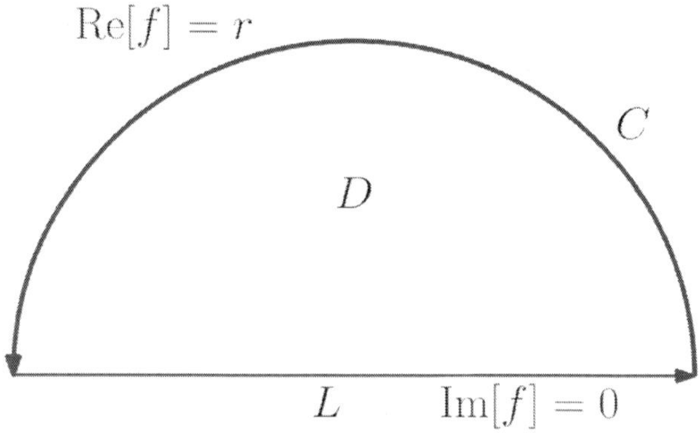

Figure 12: Boundary value problem in D, the upper half semi-disc, with boundary comprising the circular portion C and straight line portion L.

The semi-disc can be understood as the intersection of the upper half-plane and the interior of the unit disc. Hence, by combining the ideas presented earlier for polygons regions (in Sect. 2) and for circular regions (in Sect. 3)—that is, by substituting the appropriate uniformly valid spectral representation of the Cauchy kernel having split the Cauchy integral formula into the two boundary portions—it is straightforward to derive the integral transform representation

$$f(z) = \frac{1}{2\pi \mathrm{i}}\left\{\int_{L_1} \frac{\rho_{11}(k)}{1-\mathrm{e}^{2\pi \mathrm{i}k}} z^k \,\mathrm{d}k + \int_{L_2} \rho_{11}(k) z^k \,\mathrm{d}k \right.$$
$$\left. + \int_{L_3} \frac{\rho_{11}(k)\mathrm{e}^{2\pi \mathrm{i}k}}{1-\mathrm{e}^{2\pi \mathrm{i}k}} z^k \,\mathrm{d}k\right\} + \frac{1}{2\pi}\int_0^\infty \rho_{22}(k)\mathrm{e}^{\mathrm{i}kz}\,\mathrm{d}k, \tag{60}$$

where the spectral matrix has components

$$\rho_{11}(k) = \int_C f(z')z'^{(-k-1)}\,\mathrm{d}z', \quad \rho_{22}(k) = \int_L f(z')\mathrm{e}^{-\mathrm{i}kz'}\,\mathrm{d}z',$$
$$\rho_{21}(k) = \int_C f(z')\mathrm{e}^{-\mathrm{i}kz'}\,\mathrm{d}z', \quad \rho_{12}(k) = \int_L f(z')z'^{(-k-1)}\,\mathrm{d}z'. \tag{61}$$

Two global relations can be expressed as

$$\rho_{11}(k) + \rho_{12}(k) = 0, \quad k \in -\mathbb{N} \tag{62}$$

and

$$\rho_{21}(k) + \rho_{22}(k) = 0, \quad k \in \mathbb{C}. \tag{63}$$

These two conditions are not independent, but both turn out to be helpful in solving the problem.

First we will make use of (62) to find $f(z)$ on C. Some preliminary manipulations reveal that

$$\int_C \overline{f(z')} z'^{(-k-1)}\,dz' = -\overline{\rho_{11}}(-k), \quad \int_L \overline{f(z')} z'^{(-k-1)}\,dz' = \overline{\rho_{12}}(k). \tag{64}$$

The boundary condition on C is

$$f(z) + \overline{f(z)} = 2r(z, \bar{z}), \tag{65}$$

whence, on multiplying it by z^{-k-1} and integrating the resulting relation along C, we find

$$\rho_{11}(k) - \overline{\rho_{11}}(-k) = R_1(k), \quad \forall\, k \in \mathbb{C}, \tag{66}$$

where

$$R_1(k) \equiv \int_C \frac{2r}{z^{k+1}}\,dz. \tag{67}$$

The boundary condition on L is

$$f(z) = \overline{f(z)}. \tag{68}$$

On multiplying this by z^{-k-1} and integrating the resulting relation along L we find

$$\rho_{12}(k) = \overline{\rho_{12}}(k), \quad \forall\, k \in \mathbb{C}. \tag{69}$$

The complex conjugate of (62) is

$$\overline{\rho_{11}}(k) + \overline{\rho_{12}}(k) = 0, \quad k \in -\mathbb{N}. \tag{70}$$

Now (69) implies

$$\overline{\rho_{11}}(k) + \rho_{12}(k) = 0, \quad k \in -\mathbb{N} \tag{71}$$

while (66), with $k \mapsto -k$, gives

$$\overline{\rho_{11}}(k) = \rho_{11}(-k) - R_1(-k) \quad \forall\, k \in \mathbb{C} \tag{72}$$

which can be used in (71) to produce

$$\rho_{11}(-k) - R_1(-k) + \rho_{12}(k) = 0, \quad k \in -\mathbb{N}. \tag{73}$$

Finally (62) can be used to eliminate $\rho_{12}(k)$ to give a relation depending only on $\rho_{11}(k)$:

$$\rho_{11}(-k) = R_1(-k) + \rho_{11}(k), \quad k \in -\mathbb{N}. \tag{74}$$

Equation (74) will be analysed to find *f*(*z*) on *C* and, hence, $\rho_{11}(k)$.

We will represent *f*(*z*) on *C* using a Fourier series on the interval $\theta \in [0,\pi]$ where $z=e^{i\theta}$. For $z \in C$ we therefore write

$$f(z) = r(z, \bar{z}) + \mathrm{i}\left[a_0 + \sum_{n\geq 1} a_n \mathrm{e}^{2in\theta} + \overline{a_n}\mathrm{e}^{-2in\theta}\right], \tag{75}$$

where the set of coefficients $\{a_n | n=0,1,2,\ldots\}$ is to be found. On substitution of (75) into the definition of $\rho_{11}(k)$, and use of the definition of $R_1(k)$, we find

$$\rho_{11}(k) = \frac{R_1(k)}{2} - \int_0^{\pi} \mathrm{e}^{-ik\theta}\left[a_0 + \sum_{n\geq 1} a_n \mathrm{e}^{2in\theta} + \overline{a_n}\mathrm{e}^{-2in\theta}\right] \mathrm{d}\theta. \tag{76}$$

On substitution into (74) we arrive at the system

$$a_0\left[\int_0^{\pi}[-\mathrm{e}^{ik\theta} + \mathrm{e}^{-ik\theta}]\mathrm{d}\theta\right] + \sum_{n\geq 1} a_n\left[\int_0^{\pi}[-\mathrm{e}^{\mathrm{i}(k+2n)\theta} + \mathrm{e}^{\mathrm{i}(2n-k)\theta}]\mathrm{d}\theta\right]$$
$$+ \sum_{n\geq 1} \overline{a_n}\left[\int_0^{\pi}[-\mathrm{e}^{\mathrm{i}(k-2n)\theta} + \mathrm{e}^{\mathrm{i}(-2n-k)\theta}]\mathrm{d}\theta\right] = \frac{1}{2}[R_1(k) + R_1(-k)], \quad k \in -\mathbb{N}. \tag{77}$$

All the matrix elements in this linear system can be determined in closed form. As in Sect. 5, the system can be solved numerically by truncating the infinite sums and forming an overdetermined system by evaluating (77) at a sufficiently large number of choices of $k \in -N$.

Once $\rho_{11}(k)$ has been found the global relation (62) gives ρ12(k), but the latter is not needed in the integral representation (60). But with *f*(*z*) determined on *C*, the spectral function $\rho_{22}(k)$ needed in (60) can be found from the *second* global relation (63). This is because the boundary data on *C* are now known and can be used to compute $\rho_{21}(k)$ with (63) then yielding $\rho_{22}(k)$.

This boundary value problem can be solved using alternative methods based on the Schwarz reflection principle. Suppose we pick the data for the above boundary value problem to be

$$r(z, \bar{z}) = \frac{1}{2}[z + \bar{z}] = \cos\theta, \quad z = \mathrm{e}^{\mathrm{i}\theta}. \tag{78}$$

Then it is clear that the solution is

$$f(z)=z. \tag{79}$$

This, and other simple choices of $f(z)$, provides a check on the transform solution. It should be clear that the transform method will work; however, when the boundary condition on L is more complicated and arguments based on Schwarz reflection fail.

It is the appearance of both e^{ikz} (Fourier-type) and z^k (Mellin-type) in the integral representation (60) that prompts us to refer to the general transform methods of this paper as "Fourier–Mellin transforms for circular domains". We also show, in Appendix B, how to apply the very same method of Appendix A to derive the classical (complex) Mellin transform pair for an infinite wedge region. Together, Appendix A and Appendix B should convince the reader that the transform pairs we have derived above really *are* the natural extensions of the classical Fourier/Mellin transforms to general circular domains.

MIXED BOUNDARY VALUE PROBLEM IN A LENS-SHAPED DOMAIN

To conclude, we solve a boundary value problem for $f(z)$ in the lens-shaped region D shown in Figure 13 having two circular arc boundaries denoted by C_1 and C_2. The boundary conditions are taken to be

$$\begin{aligned} \mathrm{Re}[f(z)] &= r(z,\bar{z}), \quad \text{on } C_1, \\ \mathrm{Im}[f(z)] &= 0, \quad \text{on } C_2, \end{aligned} \tag{80}$$

where $r(z, \bar{z})$ is some given real-valued function.

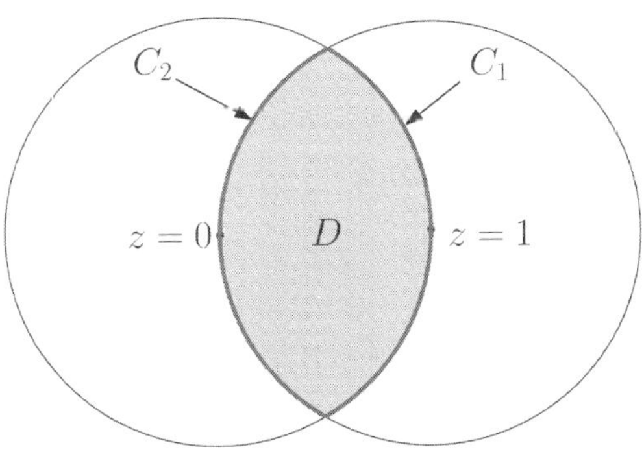

Figure 13: The convex lens-shaped domain D formed by the intersection of the circular discs $|z|<1$ and $|z-1|<1$. The boundary arc of D with $|z|=1$ is denoted by C_1, the arc with $|z-1|=1$ is denoted by C_2.

The general transform pair for an N-sided circular arc polygon was given in (42). In the present example N=2 with

$$q_1 = 1, \quad \delta_1 = 0, \quad q_2 = 1, \quad \delta_2 = 1. \tag{81}$$

We can therefore write

$$f(z) = \frac{1}{2\pi i}\left\{\int_{L_1} \frac{\rho_{11}(k)}{1-e^{2\pi ik}} z^k\, dk + \int_{L_2} \rho_{11}(k) z^k\, dk + \int_{L_3} \frac{\rho_{11}(k)e^{2\pi ik}}{1-e^{2\pi ik}} z^k\, dk \right.$$
$$+ \int_{L_1} \frac{\rho_{22}(k)[z-1]^k}{1-e^{2\pi ik}}\, dk + \int_{L_2} \rho_{22}(k)[z-1]^k\, dk$$
$$\left. + \int_{L_3} \frac{\rho_{22}(k)e^{2\pi ik}[z-1]^k}{1-e^{2\pi ik}}\, dk \right\}, \tag{82}$$

where

$$\rho_{11}(k) = \oint_{C_1} [z']^{-k-1} f(z')\, dz', \quad \rho_{22}(k) = \oint_{C_2} [z'-1]^{-k-1} f(z')\, dz'. \tag{83}$$

The other two elements of the spectral matrix are

$$\rho_{12}(k) = \oint_{C_2} [z']^{-k-1} f(z')\, dz', \quad \rho_{21}(k) = \oint_{C_1} [z'-1]^{-k-1} f(z')\, dz'. \tag{84}$$

The global relation we will analyse is

$$\rho_{11}(k) + \rho_{12}(k) = 0, \quad k \in -\mathbb{N}. \tag{85}$$

We expect singularities of $f(z)$ at the two points where C_1 and C_2 intersect. We are therefore motivated to introduce the representations of the boundary values given by

$$f(z) = \begin{cases} r(z,\bar{z}) + i\left[a_0 + \sum_{n\geq 1} a_n\zeta^n + \frac{\overline{a_n}}{\zeta^n}\right], & \text{on } C_1, \\ b_0 + \sum_{n\geq 1} b_n\zeta^n + \frac{\overline{b_n}}{\zeta^n}, & \text{on } C_2, \end{cases} \tag{86}$$

where the coefficients $a_0, b_0 \in \mathbb{R}, \{a_n, b_n \in \mathbb{C} | n \geq 1\}$ are to be found; these expressions encode the known boundary conditions (80). On each boundary the complex variable ζ has a different identification with the variable z given by

$$z = \begin{cases} g(\zeta), & \text{for } z \text{ on } C_1, \\ 1 - g(\zeta), & \text{for } z \text{ on } C_2, \end{cases} \tag{87}$$

with

$$g(\zeta) = -\frac{\omega(\zeta,\alpha)\omega(\zeta,1/\alpha)}{\omega(\zeta,\bar{\alpha})\omega(\zeta,1/\bar{\alpha})}, \quad \omega(\zeta,\alpha) = (\zeta - \alpha). \tag{88}$$

where α=is for some $0<s<1$, and where $\zeta=e^{i\theta}$ for $\theta\in[0,\pi]$. For *D* shown in Figure 13 we choose s=0.2679492. $g(\zeta)$ has an interpretation as a conformal slit mapping [6] and, with the value of *s* just stated, (88) maps the unit ζ-disc to the unbounded region exterior to a circular arc slit on the unit *z*-circle for $-\pi/3<\arg[z]<\pi/3$. The upper half semi-circle in the ζ plane, i.e., $\zeta=e^{i\theta}$ for $\theta\in[0,\pi]$ maps in a one-to-one fashion onto the slit $-\pi/3<\arg[z]<\pi/3$. The relevant *s* value is found using Newton's method to ensure that $g(1)=e^{-i\pi/3}$. It is helpful to think of $g(\zeta)$ in (88) as a generalization of the classical Joukowski map to one taking the unit disc to the exterior of a circular arc segment (rather than a straight line segment). We anticipate that $g(\zeta)$ will find great utility in the general numerical implementation of our method.

On substitution of (86) into (85), we find

$$a_0A(k,0)+\sum_{n\geq 1}\{a_nA(k,n)+\overline{a_n}A(k,-n)\}$$
$$+b_0B(k,0)+\sum_{n\geq 1}\{b_nB(k,n)+\overline{b_n}B(k,-n)\}=R(k),\quad k\in -\mathbb{N}. \tag{89}$$

where

$$A(k,n)\equiv i\int_0^{\pi}\frac{\zeta^n g'(\zeta)}{g(\zeta)^{k+1}}\,i\zeta\,d\theta,\quad B(k,n)\equiv -\int_0^{\pi}\frac{\zeta^n g'(\zeta)}{(1-g(\zeta))^{k+1}}\,i\zeta\,d\theta, \tag{90}$$

and

$$R(k)\equiv -\int_0^{\pi}\frac{\zeta^n g'(\zeta)}{g(\zeta)^{k+1}}\,i\zeta\,d\theta. \tag{91}$$

This system can be solved by truncating the infinite sums and choosing sufficiently many values of $k\in-\mathbb{N}$ to form an overdetermined system solvable by a least-squares procedure for the unknown coefficients.

The effectiveness of this scheme is tested by making particular choices of data $r(z,\overline{z})$ for which analytical solutions are known via alternative conformal mapping methods [6]. A composition of the sequence of conformal maps

$$\zeta\mapsto\eta=\frac{1-\zeta}{1+\zeta};\quad \eta\mapsto\xi=e^{i\pi/4}\eta;$$
$$\xi\mapsto\chi=\xi^{4/3};\quad \chi\mapsto z=\frac{1}{2}-\frac{i\sqrt{3}}{2}\left[\frac{1-\chi}{1+\chi}\right] \tag{92}$$

transplants the upper half unit semi-disc in the ζ-plane to the lens-shape domain *D* in the *z*-plane. The exponent 4 / 3 in the third of the sequence of maps is a consequence of the fact, following from spherical trigonometry arguments, that the angle of intersection between C_1 and C_2 at $z=1/2+i\sqrt{3}/2$ is $2\pi/3$. Suppose the inverse of the composite conformal mapping (92) is denoted by $h(z)$ so that

$$\zeta=h(z). \tag{93}$$

Suppose now that we pick, say,

$$f(z) = h(z)^2 = \zeta^2. \tag{94}$$

This function is analytic in D, clearly real on the real ζ axis (and hence on C_2), and it has real part

$$r(z,\bar{z}) = \frac{1}{2}\left[h^2(z) + \frac{1}{h^2(z)}\right] \tag{95}$$

on the upper half unit ζ-disc (and, hence, on C_1). Formula (95) was therefore used as data in the transform solution scheme described above and the resulting solutions checked against $f(z)$ as given in (94). Other choices of $f(z)$ were also tested. The transform method was found to accurately retrieve all such test functions $f(z)$.

As a remark, our use of the function $g(\zeta)$ here is closely related to the use of Chebyshev polynomial expansions although, as already mentioned, not of the standard kind because they have been tailored to a circular arc slit (rather than a straight line interval). The expansions, however, also include square root singularities at the ends of the arcs so they are *not* just expansions in polynomials. As already mentioned, the many approximation theory challenges arising from the numerical implementation of the new transform approach for circular domains presented here await much further investigation (see [16, 19, 21, 22] for a survey of progress in this respect on polygonal domains).

DISCUSSION

This paper has presented several new ideas and results.

First, we have given a new derivation of the representation of analytic functions in convex N-sided simply connected polygons due to Fokas and Kapaev [14]. The derivation is elementary, relying on use of Cauchy's integral formula and a suitable spectral representation of the Cauchy kernel obtained from simple properties of the exponential function. The notion of a *spectral matrix* and the idea of a *fundamental contour* for straight line edges arise naturally in this approach. It was also shown that, by an N-fold change of spectral variables, the N-by-N spectral matrix reduces to the same set of N spectral functions considered by Fokas and Kapaev [14].

Then, by replacing the spectral representation of the Cauchy kernel by the one relevant to circular discs (rather than half-planes), the very same sequence of constructive steps starting from Cauchy's integral formula is used to derive the natural extensions of the transforms to general circular domains, both bounded and unbounded, understood geometrically to be intersections of N circular discs.

More general circular domains with boundaries comprising a mixture of straight lines and circular arcs are treated too (cf. Sect. 6). Geometrically these are intersections of half-planes and circular discs and a hybrid construction clearly leads to the relevant transform pairs.

The reduction of the spectral matrix to a set of N spectral functions appears to be special to the case of polygons and does not extend (at least, in any obvious way) to domains with circular boundaries of non-zero curvature. We have argued that the notions of a spectral matrix, and the fundamental contour, are natural.

Several simple illustrative examples have been given of the scope of the new method in the context of examples where the approach can be checked by alternative means. It is clear that the approach has great flexibility and possible future applications. Already the method has been extended by the author to the case of multiply connected circular domains; full details can be found in [9] where some important physical applications of the method are also discussed.

We also touched on several new approximation theory questions, including the optimal choice of integration contour when performing the transform inversion by numerical quadrature, and the best ways to represent unknown data on circular arc boundaries. Such matters are the subject of ongoing investigation.

Other questions arise: are there special boundary value problems in circular domains, perhaps with sufficient geometrical symmetries, where the analysis of the global relations leads to previously unrecognized explicit solutions using just algebra (as has been the case when the unified transform method has been applied to elliptic PDEs in simple polygons [10, 16])? Can the new transform formulation here lead to fast and accurate new general numerical methods for the solution of harmonic boundary value problems in general circular domains (evidence now exists that this is possible for polygons [11, 16, 19])? A natural advantage of transform methods—as we have seen by our very construction where the singular Cauchy kernel has been replaced by a non-singular spectral integral (cf. (6) and (28))—is that they often lead to the possibility of alternative numerical formulations that do not involve *singular* integral equations (as do most standard boundary integral methods).

We have deliberately steered away from the very mathematical ideas that gave rise to the discovery of (1)–(3) in the first place, i.e., the spectral analysis of Lax pair equations, differential forms, etc. [14]. This was done in the conviction that (1)–(3) is a basic result in analytic function theory and should be derivable as such. But the formulation in terms of the spectral analysis of differential forms is the gateway to generalization to other partial differential equations [16, 18, 25] and it is now of interest to determine how one might reproduce the new transform pairs for circular domains found herein starting from, for example, the spectral analysis of an appropriate differential form [13]. This, in turn, might point the way to generalization of Fokas' unified transform method to other partial differential equations (e.g., the modified Helmholtz equation, or Helmholtz equation) in the more general circular domains of the kind considered here.

Crowdy and Fokas [5] have shown how boundary value problems for the biharmonic equation in polygonal domains in two dimensions can be formulated by appropriate extension of the ideas of Fokas and Kapaev [14] who restricted to harmonic fields. Similarly, the present extended formulation is useful for the solution of biharmonic boundary value problems in general circular domains. There are currently very few alternative techniques to solve biharmonic problems in such domains. Work in this direction is in progress.

The extension of all these ideas to boundary value problems for Laplace's equation in three dimensions—that is, regions interior and exterior to spheres—is an important open problem. In this vein it is worth pointing out that Ashton [2, 3] has recently extended the unified transform approach of Fokas to various elliptic boundary value problems defined in convex polyhedra.

ACKNOWLEDGMENTS

The author is grateful for many useful discussions, over the years, with Thanasis Fokas (whose work has inspired the investigations herein), Euan Spence and Elena Luca. This work is supported by an Established Career Fellowship from the Engineering and Physical Sciences Research Council in the UK. Funds from the Department of Mathematics at Imperial College to support the ACCA-UK network are also acknowledged.

APPENDIX A: ALTERNATIVE DERIVATION OF (31)

Here we show how to rederive the transform pair for the unit disc by combining a simple conformal mapping and the result for a general polygon.

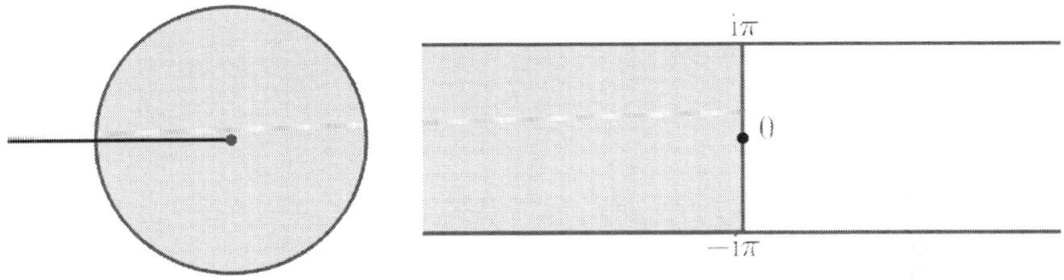

Figure 14: The logarithmic transformation (96) takes the unit z disc to the left semi-strip in an η-plane.

Introduce the (Schwarz–Christoffel) conformal mapping

$$\eta = \log z, \quad z = e^{\eta} \tag{96}$$

and take the branch cut of the logarithm to be along the negative z-axis with the choice $-\pi<\arg[z]<\pi$$-\pi<\arg[z]<\pi$ (say). The map then transplants the unit disc to the left semi-strip in the η-plane (Figure 14):

$$-\infty<\mathrm{Re}[\eta]<0, -\pi<\mathrm{Im}[\eta]<\pi. \tag{97}$$

Now let

$$F(\eta)\equiv f(z(\eta)), \tag{98}$$

where $f(z)$ is analytic in the unit disc. Since $F(\eta)$ inherits the property of analyticity, now in the left semi-strip, the transform expression for a function analytic in this polygon, and vanishing as $|\eta|\to\infty$, is

$$F(\eta) = \frac{1}{2\pi}\int_{\tilde{L}_1} e^{ik\eta}\rho_1(k)\,dk + \frac{1}{2\pi}\int_{\tilde{L}_2} e^{ik\eta}\rho_2(k)\,dk + \frac{1}{2\pi}\int_{\tilde{L}_3} e^{ik\eta}\rho_3(k)\,dk, \tag{99}$$

where $\tilde{L}_1$, $\tilde{L}_2$ and $\tilde{L}_3$ are rays from the origin with arguments $0, -\pi/2$ and $-\pi$. The spectral functions are

$$\rho_1(k) = \int_{-i\pi-\infty}^{-i\pi} F(\eta)e^{-ik\eta}\,d\eta, \quad \rho_2(k) = \int_{-i\pi}^{i\pi} F(\eta)e^{-ik\eta}\,d\eta,$$

$$\rho_3(k) = \int_{i\pi}^{i\pi-\infty} F(\eta)e^{-ik\eta}\,d\eta. \tag{100}$$

We will return later to the restriction to functions vanishing as $|\eta|\to\infty$.

Since $f(z)$ is single-valued in the z-disc, we need to enforce the constraint

$$F(\eta)=F(\eta-2\pi i). \tag{101}$$

On multiplying condition (101) by $e^{-ik\eta}$ and integrating along the upper side of the semi-strip, we find

$$\int_{i\pi}^{i\pi-\infty} e^{-ik\eta} F(\eta)\,d\eta = \int_{i\pi}^{i\pi-\infty} e^{-ik\eta} F(\eta-2\pi i)\,d\eta = e^{2\pi k}\int_{-i\pi}^{-i\pi-\infty} e^{-ik\zeta} F(\zeta)\,d\zeta, \tag{102}$$

where we have used the change of variable $\zeta=\eta-2\pi i$. We conclude that

$$\rho_3(k) = -e^{2\pi k}\rho_1(k). \tag{103}$$

The global relation for the representation of $F(\eta)$ in the left semi-strip is

$$\rho_1(k)+\rho_2(k)+\rho_3(k)=0, \quad \mathrm{Im}[k]\geq 0. \tag{104}$$

Use of (103) in (104) leads to

$$\rho_1(k) = \frac{\rho_2(k)}{e^{2\pi k}-1}, \tag{105}$$

which is valid on $\tilde{L}_1$ and $\tilde{L}_3$. Relation (99) then gives

$$F(\eta)=\frac{1}{2\pi}\int_{\tilde{L}_1} e^{ik\eta}\frac{\rho_2(k)}{e^{2\pi k}-1}\,dk+\frac{1}{2\pi}\int_{\tilde{L}_2} e^{ik\eta}\rho_2(k)\,dk-\frac{1}{2\pi}\int_{\tilde{L}_3} e^{ik\eta}\frac{e^{2\pi k}\rho_2(k)}{e^{2\pi k}-1}\,dk. \tag{106}$$

On redefining the spectral variable to be

$$\lambda=ik, \tag{107}$$

and on use of (96), the transform pair for the unit disc becomes

$$f(z)=\frac{1}{2\pi i}\int_0^{i\infty}\frac{\rho(\lambda)e^{2\pi i\lambda}}{1-e^{2\pi i\lambda}}z^{\lambda}\,d\lambda+\frac{1}{2\pi i}\int_0^{\infty}\rho(\lambda)z^{\lambda}\,d\lambda+\frac{1}{2\pi i}\int_{-i\infty}^{0}\frac{\rho(\lambda)}{1-e^{2\pi i\lambda}}z^{\lambda}\,d\lambda,$$

$$\rho(\lambda)=\oint_{|z'|=1}f(z')[z']^{-\lambda-1}\,dz'. \qquad (108)$$

and the spectral rays $\tilde{L}_1$, $\tilde{L}_2$ and $\tilde{L}_3$ are rotated by angle $\pi/2$ to be almost those defining the fundamental contour in Figure 5.

The global relation can be deduced from (105): since $\rho_1(k)$ must be analytic in the upper half k-plane (105) implies that

$$\rho_2(k)=0, k\in iN, \qquad (109)$$

or, equivalently,

$$\rho(\lambda)=0, \lambda\in -N. \qquad (110)$$

The only difference now between what has just been derived and the formulae (31) is the integration contour. But this is because in (31) we allow for $f(z)$ to be non-zero at z=0 while the derivation of (108) assumes *a priori* that $F(\eta)\to 0$ as $|\eta|\to\infty$. The radius-r circular arc deformations of the above contours $\tilde{L}_j$ to L_j for j=1,2,3 as shown in Figure 5 are included to add in this extra constant term contribution from the residue at k=0.

APPENDIX B: THE CLASSICAL (COMPLEX) MELLIN TRANSFORM PAIR

It is very instructive to see how the classical (complex) Mellin transform pair manifests itself in our much more general transform approach. The reader will then see how our generalized transforms for circular domains are really just generalizations of the Mellin transform.

Let $f(z)$ be analytic in a wedge in a complex z-plane (see Figure 15) defined by

$$-\theta<\arg[z]<0 \qquad (111)$$

for some $0<\theta<2\pi$. We now seek a transform pair for a function $f(z)$ which is analytic in this wedge.

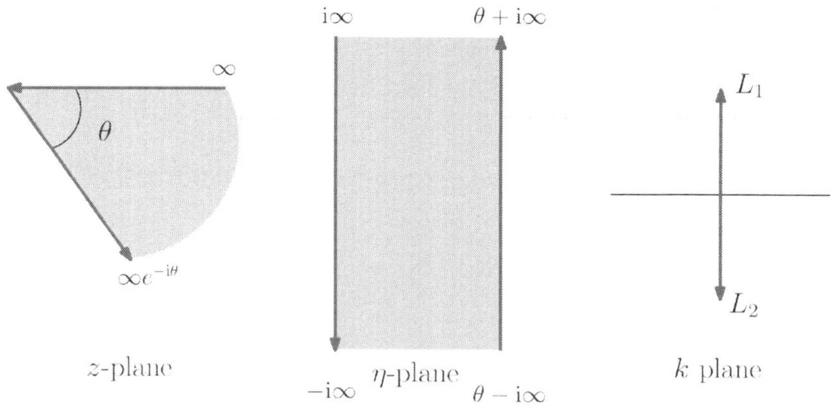

Figure 15: The logarithmic transformation from a wedge to a vertical strip.

In Appendix A, at the end, we had to perform the change of spectral variable (107). To avoid this, instead we modify the original logarithmic change of variable (96) to

$$\eta = \mathrm{i}\log z, \quad z = \mathrm{e}^{-\mathrm{i}\eta}. \tag{112}$$

Under this transformation the wedge in the *z*-plane is transplanted to a vertical strip in the η plane given by

$$0<\mathrm{Re}[\eta]<\theta. \tag{113}$$

Let

$$F(\eta)\equiv f(z(\eta)). \tag{114}$$

Since *f*(*z*) is analytic in the *z*-wedge, and z(η) is analytic in the vertical η-strip, then F(η) is analytic in the vertical η-strip. Hence, from the transform representation of functions analytic in polygons we can write

$$F(\eta) = \frac{1}{2\pi}\left[\int_0^{\mathrm{i}\infty} \rho_1(k)\mathrm{e}^{\mathrm{i}k\eta}\,\mathrm{d}k + \int_0^{-\mathrm{i}\infty} \rho_2(k)\mathrm{e}^{\mathrm{i}k\eta}\,\mathrm{d}k\right], \tag{115}$$

where

$$\rho_1(k) = \int_{\mathrm{i}\infty}^{-\mathrm{i}\infty} F(\eta)\mathrm{e}^{-\mathrm{i}k\eta}\,\mathrm{d}\eta = -\int_0^{\infty} \mathrm{i}f(z)z^{k-1}\,\mathrm{d}z \tag{116}$$

and

$$\rho_2(k) = \int_{\theta-\mathrm{i}\infty}^{\theta+\mathrm{i}\infty} F(z)\mathrm{e}^{-\mathrm{i}k\eta}\,\mathrm{d}\eta = \int_0^{\infty\mathrm{e}^{-\mathrm{i}\theta}} \mathrm{i}f(z)z^{k-1}\,\mathrm{d}z. \tag{117}$$

But the global relation says that

$$\rho_1(k) + \rho_2(k) = 0, \quad k \in \mathrm{i}\mathbb{R}. \tag{118}$$

Hence we can write (115) as

$$F(\eta) = \frac{1}{2\pi}\int_{-\mathrm{i}\infty}^{\mathrm{i}\infty} \rho_1(k)\mathrm{e}^{\mathrm{i}k\eta}\,\mathrm{d}k. \tag{119}$$

If we now define

$$\Phi(k) = \mathrm{i}\rho_1(k) \tag{120}$$

then (116) and (119) imply

$$\begin{cases} \Phi(k) = \int_0^\infty f(z)z^{k-1}\,\mathrm{d}z, \\ f(z) = \frac{1}{2\pi\mathrm{i}}\int_{-\mathrm{i}\infty}^{\mathrm{i}\infty} \Phi(k)z^{-k}\,\mathrm{d}k. \end{cases} \tag{121}$$

This is the classical complex *Mellin transform pair* with the inverse transform taken along the classical Bromwich contour running parallel along the imaginary k-axis. Since the new transforms for circular domains in (42) are built on the results of Appendix A, which themselves are patently "Mellin transforms for a semi-strip", it is natural to call our generalized transform pairs "Fourier–Mellin transforms for circular domains". In the latter, the Bromwich contour is replaced by our fundamental contour in Figure 5.

Fokas and Kapaev [14] studied some boundary value problems for Laplace's equation in a wedge, but without first making the logarithmic transformation (112). The scheme above can be a convenient alternative to that approach, one that connects the approach to more classical Mellin transform techniques.

REFERENCES

1. Ashton, A.C.L.: On the rigorous foundations of the Fokas method for linear elliptic PDEs. Proc. R. Soc. A 468, 1325–1331 (2012)
2. Ashton, A.C.L.: Laplace's equation on convex polyhedra via the unified transform method. Proc. R. Soc. A 20140884, 471 (2015)
3. Ashton, A.C.L.: Elliptic PDEs with constant coefficients on convex polyhedra via the unified method. J. Math. Anal. Appl. 425(1), 160–177 (2014)
4. Charalambopoulos, A., Dassios, G., Fokas, A.S.: Laplace's equation in the exterior of a convex polygon. The equilateral triangle. Q. Appl. Math. 68, 645–660 (2010)
5. Crowdy, D.G., Fokas, A.S.: Explicit integral solutions for the plane elastostatic semi-strip. Proc. R. Soc. A 460, 1285–1310 (2004)
6. Crowdy, D.G.: Conformal slit maps in applied mathematics. ANZIAM J. 53(3), 171–189 (2012)
7. Crowdy, D.G.: Exact solutions for cylindrical "slip-stick" Janus swimmers in Stokes flow. J. Fluid Mech. 719, R2 (2013)
8. Crowdy, D.G., Davis, A.M.J.: Stokes flow singularities in a two-dimensional channel: a novel transform approach with application to microswimming. Proc. R. Soc. A. 469, 20130198 (2013)
9. Crowdy, D.G.: A transform method for Laplace's equation in multiply connected circular domains. IMA J. Appl. Math. (2015). doi:10.1093/imamat/hxv019
10. Dassios, G., Fokas, A.S.: The basic elliptic equations in an equilateral triangle. Proc. R. Soc. A 461, 2721–2748 (2005)
11. Fulton, S.R., Fokas, A.S., Xenophontos, C.: An analytical method for linear elliptic PDEs and its numerical implementation. J. Comput. Appl. Math. 167(2), 465–483 (2004)
12. Fokas, A.S., Gel'fand, I.M.: Integrability of linear and nonlinear evolution equations and the associated nonlinear Fourier transforms. Lett. Math. Phys. 32, 189–210 (1994)
13. Fokas, A.S., Zyskin, M.: The fundamental differential form and boundary value problems. Q. J. Mech. Appl. Math. 55, 457–479 (2002)
14. Fokas, A.S., Kapaev, A.A.: On a transform method for the Laplace equation in a polygon. IMA J. Appl. Math. 68, 355–408 (2003)

15. Fokas, A.S., Pinotsis, D.: The Dbar formalism for certain linear non-homogeneous elliptic PDEs in two dimensions. Eur. J. Appl. Math. 17, 323–346 (2006)
16. Fokas, A.S.: A Unified Approach to Boundary Value Problems. CBMS-NSF Regional Conference Series in Applied Mathematics, vol. 78. SIAM, Philadelphia (2008)
17. Fokas, A.S., Pelloni, B. (eds.): Unified Transform for Boundary Value Problems: Applications and Advances. SIAM, Philadelphia (2015)
18. Fokas, A.S., Spence, E.A.: Synthesis, as opposed to separation, of variables. SIAM Rev. 54(2), 291–324 (2012)
19. Fornberg, B., Flyer, N.: A numerical implementation of Fokas boundary integral approach: Laplace's equation on a polygonal domain. Proc. R. Soc. A 467, 2983–3003 (2011)
20. Shepherd, W.M.: On trigonometric series with mixed conditions. Proc. Lond. Math. Soc. 2(43), 366– 375 (1937)
21. Sifalakis, A.G., Fokas, A.S., Fulton, S.R., Saridakis, Y.G.: The generalized Dirichlet–Neumann map for linear elliptic PDEs and its numerical implementation. J. Comput. Appl. Math. 219, 9–34 (2008)
22. Smitheman, S.A., Spence, E.A., Fokas, A.S.: A spectral collocation method for the Laplace and modified Helmholtz equations in a convex polygon. IMA J. Numer. Anal. 30, 1184–1205 (2010)
23. Sneddon, I.M.: Mixed Boundary Value Problems in Potential Theory. North Holland Publishing Company, Amsterdam (1966)
24. Spence, E.A., Fokas, A.S.: A new transform method I: domain dependent fundamental solutions and integral representations. Proc. R. Soc. A. 466, 2259–2281 (2010)
25. Spence, E.A., Fokas, A.S.: A new transform method II: the global relation and boundary value problems in polar co-ordinates. Proc. R. Soc. A 466, 2283–2307 (2010)
26. Tranter, C.J.: Dual trigonometrical series. Proc. Glasg. Math. Assoc. 4, 4957 (1959)

CHAPTER 5

SINGULAR INTEGRAL EQUATIONS OF CONVOLUTION TYPE WITH HILBERT KERNEL AND A DISCRETE JUMP PROBLEM

Pingrun Li

School of Mathematical Science, Qufu Normal University, Qufu, P.R. China

ABSTRACT

One class of singular integral equations of convolution type with Hilbert kernel is studied in the spacc $L^2[-\pi,\pi]$ in the article. Such equations can be changed into either a system of discrete equations or a discrete jump

CITATION (APA): Li, P. (2017). Singular integral equations of convolution type with Hilbert kernel and a discrete jump problem. Advances in Difference Equations, 2017(1), 360. (13 pages)

problem depending on some parameter via the discrete Laurent transform. We can thus solve the equations with an explicit representation of solutions under certain conditions.

Keywords: singular integral equation, convolution type, Hilbert kernel, a discrete jump problem

INTRODUCTION

It is well known that singular integral equations and boundary value problems for analytic functions are the main branches of complex analysis and have a lot of applications, *e.g.*, in elasticity theory, fluid dynamics, shell theory, underwater acoustics, and quantum mechanics. The theory is well developed by many authors [1, 2, 3, 4, 5, 6, 7]. Integral equations of convolution type are closely related to boundary value problem for an analytic function. There have been many papers studying integral equations with convolution type or singular type, see, for example, Litvinchuc [8], Li [9, 10], De-Bonis [11], Du [12], Jiang [13], among which a series of valuable achievements have been obtained. In recent years, the author [14] discussed some kinds of singular integral equations of convolution type with reflection and translation shifts. Subsequently, the author [15] studied one class of generalized boundary value problems for analytic functions and obtained the general solutions and the conditions of solvability.

The purpose of this article is to extend further the theory to a periodic singular integral equation of convolution type with Hilbert kernel. We remark that integral equations with periodicity have important applications in the elastic theory. Such equations can be changed into either a system of discrete equations or a discrete jump problem (that is, discrete boundary value problems) depending on some parameter via the discrete Laurent transform. We can thus solve the equations with an explicit representation of solutions in $L^2[-\pi,\pi]$ under certain conditions. This paper improves some results for the references [1, 2, 3, 4, 5, 6, 7].

We shall consider the following singular integral equation of convolution type with Hilbert kernel and periodicity:

$$a_1f + b_1Hf + K_1 * f + M_1 * Hf + \xi(a_2f + b_2Hf + K_2 * f + M_2 * Hf)$$
$$= (c_1 + c_2\xi)G, \tag{1.1}$$

where $K_j, M_j, G \in L^2[-\pi,\pi]$ are given functions, $a_j, b_j, c_j \in R$ for any j=1,2, and $f \in L^2[-\pi,\pi]$ is an unknown function. We denote by $*$ the convolution operator and by Hf the Hilbert type singular integral of f, that is,

$$Hf = \frac{1}{\pi}\int_{-\pi}^{\pi} f(\tau)\cot\frac{\tau-\theta}{2}\,d\tau.$$

Assumption A

We shall make the following assumption:

- $a_1^2 + a_2^2 \neq 0,\ b_1^2 + b_2^2 \neq 0.$
- There exist constants $a,b,c,d \in R$ and some $m \in N$ such that

$$\xi(\theta) = \frac{a + b\tan\frac{m}{2}\theta}{c + d\tan\frac{m}{2}\theta}$$

whenever $ad-bc\neq 0$; otherwise we have $\xi(\theta)=C$ (constant). We shall also represent $\xi(\theta)$ as

$$\xi(\theta) = \frac{\alpha^+ + \alpha^- e^{im\theta}}{\beta^+ + \beta^- e^{im\theta}},$$

where

$$\alpha^\pm = a \pm ib, \qquad \beta^\pm = c \pm id.$$

Let $\mathbb{L}, \mathbb{L}^{-1}$ be the discrete Laurent transform and the inverse transform, respectively. We denote

$$S^\pm = (a_2 a + a_1 c) \pm i(a_2 b + a_1 d),$$

$$T^\pm = (b_1 c + b_2 a) \pm i(b_1 d + b_2 b),$$

$$B_k^{(j)} = \mathbb{L}^{-1}K_j, \qquad C_k^{(j)} = \mathbb{L}^{-1}M_j$$

and

$$w_k = \frac{\Delta_{k-m}^-}{\Delta_k^+},$$

where

$$\Delta_k^\pm = S^\pm + \beta^\pm B_k^{(1)} + \alpha^\pm B_k^{(2)} + 2i\,\mathrm{sgn}\,k\left(T^\pm + \beta^\pm C_k^{(1)} + \alpha^\pm C_k^{(2)}\right).$$

Assumption B

To assure the solvability of Eq. (1.1), we need to make the assumption

$$\sum_{k\in\mathbb{Z}}\left(\log\frac{S^+}{S^-}w_k\right)e^{ik\theta_j}=0, \tag{1.2}$$

where

$$\theta_j=\frac{2j}{m}\pi$$

for any j=0,±1,…,±[m/2]. Via using the method of complex analysis, we can take a continuous branch of $\log\left(\frac{S^+}{S^-}w_k\right)$ such that $\{\log\left(\frac{S^+}{S^-}w_k\right)\}_{k\in Z}\in l^2$. There exist constants $\varepsilon_0\in(0,1)$ and $k_0\in N$ such that when $|k|>k_0$ we have

$$\left|\Delta_k^+\right|>\varepsilon_0, \qquad \left|\Delta_{k-m}^-\right|>\varepsilon_0.$$

We can thus introduce the equalities

$$A_k=\frac{1}{S^-\Delta_k^+}\{\beta^-S^+B_{k-m}^{(1)}-\beta^+S^-B_k^{(1)}+\alpha^-S^+B_{k-m}^{(2)}-\alpha^+S^-B_k^{(2)}$$
$$+2i[(\beta^-S^+C_{k-m}^{(1)}+\alpha^-S^+C_{k-m}^{(2)})\operatorname{sgn}(k-m)-(\beta^+S^-C_k^{(1)}+\alpha^+S^-C_k^{(2)})\operatorname{sgn}k]\}. \tag{1.3}$$

We may make η_k such that

$$\eta_k=(1+A_k)\eta_{k-m}. \tag{1.4}$$

Assumption C

To assure the solvability, we need also

$$\sum_{k\in\mathbb{Z}}\frac{H_k}{\eta_k}e^{ik\theta_j'}=0, \tag{1.5}$$

where η_k is determined by (1.4),

$$H_k=\frac{1}{\Delta_k^+}(E^+g_k+E^-g_{k-m}), \quad \{g_k\}_{k\in\mathbb{Z}}=\mathbb{L}^{-1}G$$

and

$$E^{\pm}=(c_1c+c_2a)\pm i(c_2b+c_1d).$$

In order to illustrate that Eq. (1.1) has a solution, at the end of Section 3, we shall present an example and satisfy the above conditions (Assumptions A-C), then we can conclude that a solution set of (1.1) is not empty.

Now we can state our main result, and we will prove it in Section 3.

Theorem 1.1

Under Assumptions A-C, *Eq.* (1.1) *has a solution in* $L^2[-\pi,\pi]$.

- *When* $\Delta_k^+ \neq 0,\ \Delta_{k-m}^- \neq 0$ *for any* $k \in Z$, *the solution of Eq.* (1.1) *is given by*

$$f(\theta) = \sum_{k=-\infty}^{+\infty} f_k e^{ik\theta},$$

and the coefficients f_k *are determined uniquely by the formula*

$$\Delta_k^+ f_k + \Delta_{k-m}^- f_{k-m} = E^+ g_k + E^- g_{k-m}. \tag{1.6}$$

When $\Delta_{n_j}^+ = 0,\ \Delta_{n_j-m}^- = 0$ ($|n_j| \leq k_0$; $j=1,2,\ldots,p_1$), *then*

$$E^+ g_{n_j} + E^- g_{n_j-m} = 0$$

must be satisfied, and f_{nj} *can be taken to be an arbitrary constant* c_{nj} ($1 \leq j \leq p_1$), *then the solution of Eq.* (1.1) *is*

$$f(\theta) = \sum_{j=1}^{p_1} c_{n_j} e^{in_j\theta} + \sum_{k=-\infty, k\neq n_j}^{+\infty} f_k e^{ik\theta}.$$

When $\Delta_{n'_j}^+ = 0,\ \Delta_{n'_j-m}^- \neq 0$ ($|n'_j| \leq k_0$; $j=1,2,\ldots,p_2$), fn'_j *are given by the following formula*:

$$f_{n'_j} = \frac{1}{\Delta_{n'_j}^-}\left(E^+ g_{n'_j+m} + E^- g_{n'_j}\right), \tag{1.7}$$

then the solution of Eq. (1.1) *is*

$$f(\theta) = \sum_{j=1}^{p_2} f_{n'_j} e^{in'_j\theta} + \sum_{k=-\infty, k\neq n'_j}^{+\infty} f_k e^{ik\theta}.$$

- *When* $\Delta_{n''_j}^+ \neq 0,\ \Delta_{n''_j-m}^- = 0$ ($|n''_j| \leq k_0$; $j=1,2,\ldots,p_3$), fn''_j *are given by the following formula*:

$$f_{n''_j} = \frac{1}{\Delta_{n''_j}^+}\left(E^+ g_{n''_j} + E^- g_{n''_j-m}\right), \tag{1.8}$$

then the solution of Eq. (1.1) *is*

$$f(\theta) = \sum_{j=1}^{p_3} f_{n''_j} e^{in''_j\theta} + \sum_{k=-\infty, k\neq n''_j}^{+\infty} f_k e^{ik\theta}. \tag{1.9}$$

In (ii)-(iv), *when* $k\neq n_1,n_2,\ldots,n_{p1},n'_1,n'_2,\ldots,n'_{p2},n''_1,n''_2,\ldots,n''_{p3}$, *the coefficients* f_k *are determined uniquely by* (1.6).

HILBERT TRANSFORM AND ITS DISCRETE LAURENT TRANSFORM

The crucial tool to study the Hilbert transform is to calculate its discrete Laurent transform.

For any $f\in L^2[-\pi,\pi]$, its Hilbert transform is defined as

$$Hf(\theta) = \frac{1}{\pi}\int_{-\pi}^{\pi} f(\tau)\cot\frac{\tau-\theta}{2}\,d\tau.$$

Associated to the operator H is the operator $\tilde{H}$ defined by

$$\tilde{H}f(\theta) = \frac{1}{\pi}\int_{-\pi}^{\pi} f(\tau)\cot\frac{\tau+\theta}{2}\,d\tau,$$

that is,

$$\tilde{H}f(\theta) = Hf(-\theta).$$

It is well known that the Hilbert transform H as well $\tilde{H}$ is a self-map of the space $L^2[-\pi,\pi]$ in virtue of the Riesz theorem (see, *e.g.*, [16]).

Let $l^2(Z)$ be a linear space consisting of sequences $\{f_k\}k\in Z$ for which

$$\sum_{k=-\infty}^{+\infty} |f_k|^2 < +\infty.$$

Definition 2.1

The discrete Laurent transform

$$\mathbb{L} : l^2(\mathbb{Z}) \longrightarrow L^2[-\pi,\pi]$$

is defined by

$$\mathbb{L}\{f_k\}_{k\in\mathbb{Z}} = \sum_{k=-\infty}^{+\infty} f_k e^{ik\theta} =: f(\theta) \tag{2.1}$$

for any $f=\{f_k\}k\in Z\in l^2(Z)$. Its inverse transform is clearly given by

$$\mathbb{L}^{-1}[f]_k = \{f_k\}_{k\in\mathbb{Z}} \tag{2.2}$$

with

$$f_k = \frac{1}{2\pi}\int_{-\pi}^{\pi} f(\theta)e^{-ik\theta}\,d\theta, \quad \forall k \in \mathbb{Z}. \tag{2.3}$$

Now we come to calculate the inverse discrete Laurent transform of a function which is a Hilbert transform of a given function. The result shall be crucial to studying the singular integral equations.

Lemma 2.1

Let $f(\theta)\in L^2[-\pi,\pi]$ *with its discrete Laurent transform*

$$\{f_k\}_{k\in\mathbb{Z}} = \mathbb{L}^{-1}f(\theta).$$

Then we have

$$\mathbb{L}^{-1}(Hf(\theta)) = \{2if_k\,\mathrm{sgn}\,k\}_{k\in\mathbb{Z}},$$

$$\mathbb{L}^{-1}(\tilde{H}f(\theta)) = \{-2if_{-k}\,\mathrm{sgn}\,k\}_{k\in\mathbb{Z}}.$$

Proof

Since

$$\cot\frac{\tau-\theta}{2} = \frac{e^{i\tau}}{e^{i\tau}-e^{i\theta}} - \frac{e^{-i\tau}}{e^{-i\tau}-e^{-i\theta}},$$

by Definition 2.1 we have

$$Hf(\theta) = \frac{i}{\pi}\int_{-\pi}^{\pi} f(\tau)\left(\frac{e^{i\tau}}{e^{i\tau}-e^{i\theta}} - \frac{e^{-i\tau}}{e^{-i\tau}-e^{-i\theta}}\right)d\tau.$$

For any $f(\theta)\in L^2[-\pi,\pi]$, we already know that $Hf(\theta)\in L^2[-\pi,\pi]$ so that $L^{-1}(Hf(\theta))\in l^2(Z)$. We denote

$$\mathbb{L}^{-1}(Hf(\theta)) = \{g_k\}_{k\in\mathbb{Z}}.$$

Then

$$g_k = \frac{1}{2\pi}\int_{-\pi}^{\pi} Hf(\theta)e^{-ik\theta}\,d\theta$$

$$= \frac{i}{2\pi^2}\int_{-\pi}^{\pi}\int_{-\pi}^{\pi} f(\tau)\frac{e^{i\tau}e^{-ik\theta}}{e^{i\tau}-e^{i\theta}}\,d\tau\,d\theta - \frac{i}{2\pi^2}\int_{-\pi}^{\pi}\int_{-\pi}^{\pi} f(\tau)\frac{e^{-i\tau}e^{-ik\theta}}{e^{-i\tau}-e^{-i\theta}}\,d\tau\,d\theta$$

$$=: I_1 - I_2.$$

For the first term, we have

$$I_1 = \frac{i}{2\pi^2}\int_{-\pi}^{\pi}\left(\int_{-\pi}^{\pi}\frac{e^{-ik\theta}}{e^{i\tau}-e^{i\theta}}\,d\theta\right)f(\tau)e^{i\tau}\,d\tau.$$

By a change of variable t=exp(iθ), the inner integral above becomes

$$M(\tau) := \int_{-\pi}^{\pi}\frac{e^{-ik\theta}}{e^{i\tau}-e^{i\theta}}\,d\theta = \int_{|t|=1}\frac{t^{-k}}{(e^{i\tau}-t)it}\,dt.$$

Applying the extended residue theory, we thus obtain

$$M(\tau) = \begin{cases} \frac{\pi}{e^{i(k+1)\tau}}, & k \ge 0, \\ -\frac{\pi}{e^{i(k+1)\tau}}, & k < 0. \end{cases}$$

Therefore, when k≥0, we have

$$I_1 = \frac{i}{2\pi^2}\int_{-\pi}^{\pi}\frac{\pi}{e^{i(k+1)\tau}}f(\tau)e^{i\tau}\,d\tau = \frac{i}{2\pi}\int_{-\pi}^{\pi}f(\tau)e^{-ik\tau}\,d\tau = if_k$$

and similarly

$$I_1 = -if_k, \quad k < 0.$$

In other words,

$$I_1 = if_k \operatorname{sgn} k, \quad \forall k \in \mathbb{Z}.$$

Similarly, we have

$$I_2 = -if_k \operatorname{sgn} k, \quad \forall k \in \mathbb{Z}.$$

As a result,

$$g_k = I_1 - I_2 = 2if_k \operatorname{sgn} k, \quad \forall k \in \mathbb{Z}.$$

The other equality can be proven similarly.

Lemma 2.2

Let f(θ)= $\sum_{k=-\infty}^{\infty} f_k e^{ik\theta}$ *and* f∈l^1, *then* f(θ)∈$L^2[-\pi,\pi]$ *if and only if* f∈l^2.

Proof

Since f(θ)∈$L^2[-\pi,\pi]$ and f∈l^1, then $\sum_{k=-\infty}^{\infty} f_k$ is convergent absolutely and uniformly. It is easy to see that

$$\int_{-\pi}^{\pi} |f(\theta)|^2 d\theta = \int_{-\pi}^{\pi} f(\theta)\overline{f(\theta)}\, d\theta = \int_{-\pi}^{\pi} \sum f_k e^{ik\theta} \sum \bar{f}_j e^{-ij\theta}\, d\theta$$

$$= \int_{-\pi}^{\pi} \sum_{k,j=-\infty}^{\infty} f_k \bar{f}_j e^{i(k-j)\theta}\, d\theta = 2\pi \sum_{k=-\infty}^{\infty} |f_k|^2. \quad (2.4)$$

The proof of Lemma 2.2 is complete.

Finally, we remark that

$$\mathbb{L}^{-1}[f * g(\theta)] = \{f_k g_k\}_{k\in\mathbb{Z}}$$

for any $f,g \in L^2[-\pi,\pi]$ with the discrete Laurent transforms $\{f_k\} k\in Z$ and $\{g_k\}$ $k\in Z$, respectively. Here the convolution in $L^2[-\pi,\pi]$ is defined by

$$(f * g)(\theta) = \frac{1}{2\pi}\int_{-\pi}^{\pi} f(\theta-\tau)g(\tau)\, d\tau. \quad (2.5)$$

PROBLEM PRESENTATION AND SOLUTION

In this section, we study the method of solution for Eq. (1.1). By Euler's formula eimθ=cosmθ+isinmθ, Eq. (1.1) can be written as

$$S^+f + T^+Hf + \beta^+K_1 * f + \alpha^+K_2 * f + \beta^+M_1 * Hf + \alpha^+M_2 * Hf - E^+G$$
$$+ e^{im\theta}\left(S^-f + T^-Hf + \beta^-K_1 * f + \alpha^-K_2 * f + \beta^-M_1 * Hf + \alpha^-M_2 * Hf - E^-G\right)$$
$$= 0. \quad (3.1)$$

In view of Lemma 2.1, by applying L^{-1} to both sides of (3.1), we see that (3.1) is readily reduced to the equation

$$\Delta_k^+ f_k - E^+ g_k + \Delta_{k-m}^- f_{k-m} - E^- g_{k\ m} = 0, \quad (3.2)$$

where

$$\{f_k\}_{k\in\mathbb{Z}} = \mathbb{L}^{-1}f, \qquad \{g_k\}_{k\in\mathbb{Z}} = \mathbb{L}^{-1}G,$$

$$\{B_k^{(j)}\}_{k\in\mathbb{Z}} = \mathbb{L}^{-1}K_j, \qquad \{C_k^{(j)}\}_{k\in\mathbb{Z}} = \mathbb{L}^{-1}M_j$$

with any j=1,2; k∈Z.

Since by assumption, for each j=1,2,

$$\{B_k^{(j)}\}_{k\in\mathbb{Z}} \in l^2(\mathbb{Z}), \qquad \{C_k^{(j)}\}_{k\in\mathbb{Z}} \in l^2(\mathbb{Z}),$$

it follows that

$$\lim_{k\to\infty} B_k^{(j)} = 0, \qquad \lim_{k\to\infty} C_k^{(j)} = 0.$$

Therefore,

$$\lim_{k\to\infty} \Delta_k^{+} = S^{+} \pm 2iT^{+} \neq 0, \qquad \lim_{k\to\infty} \Delta_k^{-} = S^{-} \pm 2iT^{-} \neq 0,$$

which means that for any ε_0 sufficiently small, there exists $k_0>0$ such that when $|k|>k_0$ we have

$$\left|\Delta_k^{+}\right| > \varepsilon_0, \qquad \left|\Delta_{k-m}^{-}\right| > \varepsilon_0.$$

Case 1: $|k|>k_0$.

Since $|\Delta_k^{+}| > \varepsilon_0,\ |\Delta_{k-m}^{-}| > \varepsilon_0$ for $|k|>k_0$, it follows from (1.6) that

$$f_k = -\frac{\Delta_{k-m}^{-}}{\Delta_k^{+}} f_{k-m} + \frac{1}{\Delta_k^{+}}\left(E^{+}g_k + E^{-}g_{k-m}\right), \quad \forall |k| > k_0, \tag{3.3}$$

so that Eq. (3.3) can be rewritten as

$$f_k = -w_k f_{k-m} + H_k, \quad \forall |k| > k_0. \tag{3.4}$$

Denote

$$\rho = \frac{S^{-}}{S^{+}}, \quad w_k = \rho(1 + A_k).$$

When $|k|$ is large enough, we see that sgn(k−m) and sgn*k* equal 1 or −1 simultaneously so that

$$S^{+}T^{-}\operatorname{sgn}(k-m) - S^{-}T^{+}\operatorname{sgn}k = 0.$$

Consequently,

$$A_k = \frac{1}{S^{-}\Delta_k^{+}}\left\{\beta^{-}S^{+}B_{k-m}^{(1)} - \beta^{+}S^{-}B_k^{(1)} + \alpha^{-}S^{+}B_{k-m}^{(2)} - \alpha^{+}S^{-}B_k^{(2)}\right.$$
$$\left. + 2i\left[\left(\beta^{-}S^{+}C_{k-m}^{(1)} + \alpha^{-}S^{+}C_{k-m}^{(2)}\right)\operatorname{sgn}(k-m) - \left(\beta^{+}S^{-}C_k^{(1)} + \alpha^{+}S^{-}C_k^{(2)}\right)\operatorname{sgn}k\right]\right\}. \tag{3.5}$$

Therefore, (3.4) becomes a discrete jump problem

$$f_k = -\rho(1 + A_k)f_{k-m} + H_k, \quad \forall |k| > k_0. \tag{3.6}$$

In order to solve (3.6), we may make η_k such that

$$\eta_k = (1 + A_k)\eta_{k-m} \tag{3.7}$$

with

$$\varepsilon_0 < |\eta_k| < \varepsilon_0^{-1}.$$

First, we need to construct η_k. By taking logarithms on both sides of (3.7) and denoting

$$M_k = \log(1 + A_k), \qquad o_k = \log \eta_k,$$

we get

$$o_k = o_{k-m} + M_k, \tag{3.8}$$

where we have taken a continuous branch of $\log(1+A_k)$ so that $\{\log(1+A_k)\}_{k\in Z} \in l^2$. Taking the Laurent transform L on both sides of (3.8) yields

$$O(\theta) = e^{im\theta} O(\theta) + \tilde{M}(\theta), \tag{3.9}$$

that is,

$$\left(1 - e^{im\theta}\right) O(\theta) = \tilde{M}(\theta),$$

where

$$O(\theta) = \mathbb{L}o, \qquad \tilde{M}(\theta) = \mathbb{L}M, \qquad o = \{o_k\}_{k\in\mathbb{Z}}, \qquad M = \{M_k\}_{k\in\mathbb{Z}}.$$

Notice that when $\tilde{M}(\theta_j) \neq 0$, Eq. (3.8) is not solvable. This means $\tilde{M}(\theta_j)=0$ or, equivalently, Assumption B becomes the solvability conditions of Eq. (3.8), where θ_j are $1+[m/2]$ roots on $[-\pi,\pi]$ for equation $1-e^{im\theta}=0$. Finally, from (3.9) we get

$$\eta_k = \exp o_k, \qquad o_k = \mathbb{L}^{-1} \frac{\tilde{M}(\theta)}{1 - e^{im\theta}}. \tag{3.10}$$

Now we come to solve Eq. (3.6). Denote

$$p_k = \frac{1}{\eta_k} f_k, \qquad q_k = \frac{1}{\eta_k} H_k$$

and rewrite (3.6) as

$$p_k = -\rho p_{k-m} + q_k. \tag{3.11}$$

Due to $\{H_k\}_{k\in\mathbb{Z}} \in l^2(\mathbb{Z})$, $\{\eta_k\}_{k\in\mathbb{Z}} \in l^2(\mathbb{Z})$, we can know that $\{p_k\}_{k\in Z} \in l^2(Z)$, $\{q_k\}_{k\in Z} \in l^2(Z)$. Taking the Laurent transform L on both sides of (3.11), we thus obtain

$$P(\theta) = -\rho e^{im\theta} P(\theta) + Q(\theta), \tag{3.12}$$

where $P(\theta)=Lp$ and $Q(\theta)=Lq$ with $p=\{p_k\}_{k\in Z}$ and $q=\{q_k\}_{k\in Z}$.

Owing to $|\rho|=1$, we know that $1+\rho eim\theta$ has a finite number of zero points, say $\theta'_1,\theta'_2,\ldots,\theta'_n$ in $[-\pi,\pi]$.

The same approach as in the discussion of Eq. (3.8) shows that (3.11) is not solvable if $Q(\theta'_j)\neq 0$. Therefore, $Q(\theta'_j)=0$ $(j=1,2,\dots,n)$, or equivalently, Assumption C becomes the solvability conditions of Eq. (3.11).

Under Assumption C, (3.12) becomes

$$P(\theta) = \frac{Q(\theta)}{1+\rho e^{im\theta}}. \tag{3.13}$$

This determines p_k so does $f_k=p_k\eta_k$.

Case 2: $|k|\leq k_0$.

We split the situation into four cases.

$\Delta_k^+=0$ and $\Delta_{k-m}^-=0$ for some $|k|\leq k_0$. By (3.2), we have

$$E^+g_k = -E^-g_{k-m} \tag{3.14}$$

and fk can be taken to be any constant.

$\Delta_k^+=0$ and $\Delta_{k-m}^-\neq 0$ for some $|k|\leq k_0$. By (3.2), we get

$$f_k = \frac{1}{\Delta_k^-}\left(E^+g_{k+m}+E^-g_k\right). \tag{3.15}$$

$\Delta_k^+\neq 0$ and $\Delta_{k-m}^-=0$ for some $|k|\leq k_0$. By (3.2), we get

$$f_k = \frac{1}{\Delta_k^+}\left(E^+g_k+E^-g_{k-m}\right). \tag{3.16}$$

$\Delta_k^+\neq 0$ and $\Delta_{k-m}^-\neq 0$ for some $|k|\leq k_0$.

In this situation, f_k can be determined as in the proof of Case (i).

In the following, we give the proof of Theorem 1.1.

Proof of Theorem 1.1

From the above discussion, we only need to prove that the function $f(\theta)$ obtained by (1.9) belongs to $L^2[-\pi,\pi]$. Obviously, Eqs. (1.1), (1.6) and (3.1) are equivalent to each other. Since $\{A_k\},\{B_k\},\{C_k\},\{g_k\}\in l^2$, then $\{\Delta^\pm_k\}\in l^2$ and $\Delta^\pm_k\neq 0$. It follows from (3.3) that $\{f_k\}\in l^2$ is a bounded sequence and $\sum_{k=-\infty}^{\infty} f_k e^{ik\theta}$ is convergent. Thus, by Lemma 2.2, Eq. (1.1) has a unique solution $f(\theta)=Lf$, and $f(\theta)\in L^2[-\pi,\pi]$.

Finally, in order to illustrate that Eq. (1.1) has a solution, we shall present an example. For example, suppose that

$$a_1 = b_1 = c_1 = a_2 = b_2 = c_2 = 1, \qquad a = d = 0, \qquad b = c,$$

$$K_1(\theta) = K_2(\theta) = \sin\theta, \qquad M_1(\theta) = M_2(\theta) = 0, \qquad G(\theta) = \sin\theta - \cos\theta,$$

then ξ(θ)=tanm/2θ, and Eq. (1.1) can be transformed into

$$f(\theta) + \frac{1}{\pi}\int_{-\pi}^{\pi} f(t)\cot\frac{t-\theta}{2}\,dt + \frac{1}{\sqrt{2\pi}}\int_{-\pi}^{\pi}\sin(t-\theta)f(t)\,dt$$

$$= \sin\theta - \cos\theta, \quad \theta \in [-\pi,\pi]. \tag{3.17}$$

Equation (3.17) is often used in engineering mechanics. It is easy to verify that Eq. (3.17) satisfies the above conditions (Assumptions A-C). Via using the methods of Section 3, we can obtain the exact solution of Eq. (3.17):

$$f(\theta) = \frac{\sqrt{2}}{2}\left(\frac{\pi-\theta}{\pi+\theta}\right)^{\frac{1}{4}}\sin\theta, \quad \theta \in [-\pi,\pi]. \tag{3.18}$$

As for the solving method of (3.17), we will not elaborate. We can verify that (3.18) is indeed the solution of (3.17). Therefore, we conclude that a solution set of (1.1) is non-empty.

HOMOGENOUS EQUATION AND SOME SPECIFIC EQUATION

In this section we consider the homogenous equation and some specific equation. First we consider the homogenous equation (that is, G(θ)≡0)

$$a_1f + b_1Hf + K_1 * f + M_1 * Hf + \xi(a_2f + b_2Hf + K_2 * f + M_2 * Hf) = 0. \tag{4.1}$$

Via the Laurent transform, it can be reduced to the equation

$$\Delta_k^+ f_k + \Delta_{k-m}^- f_{k-m} = 0. \tag{4.2}$$

That is,

$$f_k = -w_k f_{k-m}, \quad w_k = \frac{\Delta_{k-m}^-}{\Delta_k^+}. \tag{4.3}$$

Again we apply the same approach as the discussion for Eq. (3.8) to deduce that f_k≡0 for all *k* so that f(θ)≡0. As a result, the homogeneous equation (4.1) has only a trivial solution.

Next we consider the specific case that ξ(θ) is a constant. Since ad−bc=0, Eq. (1.1) can be expressed in the form

$$a'f + b'Hf + K * f + M * Hf = G. \tag{4.4}$$

Via transform L^{-1}, (4.4) can be written as

$$\left(a' + 2ib' \operatorname{sgn} k + B_k + 2i \operatorname{sgn} k C_k\right) f_k = g_k, \tag{4.5}$$

where a′, b′ are constants and $\{f_k\}_{k\in Z}=L^{-1} f$, $\{B_k\}_{k\in Z}=L^{-1}K$, $\{C_k\}_{k\in Z}=L^{-1}M$, $\{g_k\}_{k\in Z}=L^{-1}G$.

One can solve out f_k from (4.5) and get the solution $f(\theta)=\sum_{k=-\infty}^{+\infty} f_k e^{ik\theta}$.

CONCLUSIONS

In this paper, we first proposed one class of singular integral equations of convolution type with Hilbert kernel and periodicity. Applying the discrete Laurent transform and its properties, such an equation can be changed into a discrete boundary value problem depending on some parameter, here we call it 'a discrete jump problem'. In this article, our method is different from the ones of the classical boundary value problem, and it is novel and simple. The exact solution, denoted by series, of Eq. (1.1) and the conditions of solvability are obtained. We remark that our approach is also effective to some other classes of equations such as the equations of dual type with periodicity and Hilbert kernel, the Wiener-Hopf type equations, and the equations with periodicity and cosecant kernel. Thus, this paper generalizes the classical theory of boundary value problems and singular integral equations.

One can also consider a similar problem in the setting of Clifford analysis (see, *e.g.*, [17, 18, 19, 20, 21]).

ACKNOWLEDGEMENTS

The author is grateful to the referees for many suggestions to improve the exposition of the paper.

REFERENCES

1. Giang, BT, Tuan, NM: Generalized convolutions and the integral equations of the convolution type. Complex Var. Elliptic Equ. 55(4), 331-345 (2010)
2. Gomez, C, Prado, H, Trofimchuk, S: Separation dichotomy and wavefronts for a nonlinear convolution equation. J. Math. Anal. Appl. 420, 1-19 (2014)
3. Mohankumar, N, Natarajan, A: On the numerical solution of Cauchy singular integral equations in neutron transport. Ann. Nucl. Energy 35, 1800-1804 (2008)
4. Nakazi, T, Yamamoto, T: Normal singular integral operators with Cauchy kernel. Integral Equ. Oper. Theory 78, 233-248 (2014)
5. Wójcik, P, Sheshko, MA, Sheshko, SM: Application of Faber polynomials to the approximate solution of singular integral equations with the Cauchy kernel. Differ. Equ. 49(2), 198-209 (2013)
6. Noor, MA, Mohyud-Din, ST: Variational iteration technique for solving higher order boundary value problems. Appl. Math. Comput. 189, 1929-1942 (2007)
7. Noor, MA, Noor, KI, Mohyud-Din, ST, Shabbir, A: An iterative method with cubic convergence for nonlinear equations. Appl. Math. Comput. 183, 1249-1255 (2006)
8. Litvinchuc, GS: Solvability Theory of Boundary Value Problems and Singular Integral Equations with Shift. Kluwer Academic, London (2004)
9. Li, PR: Two classes of linear equations of discrete convolution type with harmonic singular operators. Complex Var. Elliptic Equ. 61(1), 67-75 (2016)
10. Li, PR, Ren, GB: Some classes of equations of discrete type with harmonic singular operator and convolution. Appl. Math. Comput. 284, 185-194 (2016)
11. De-Bonis, MC, Laurita, C: Numerical solution of systems of Cauchy singular integral equations with constant coefficients. Appl. Math. Comput. 219, 1391-1410 (2012)
12. Du, H, Shen, JH: Reproducing kernel method of solving singular integral equation with cosecant kernel. J. Math. Anal. Appl. 348(1), 308-314 (2008)

13. Jiang, Y, Xu, Y: Fast Fourier-Galerkin methods for solving singular boundary integral equations: numerical integration and precondition. J. Comput. Appl. Math. 234, 2792-2807 (2010)
14. Li, PR: One class of generalized boundary value problem for analytic functions. Bound. Value Probl. 2015, Article ID 40 (2015). doi:10.1186/s13661-015-0301-0
15. Li, PR: Generalized convolution-type singular integral equations. Appl. Math. Comput. 311, 314-323 (2017)
16. Muskhelishvilli, NI: Singular Integral Equations. Nauka, Moscow (2002)
17. Gong, YF, Leong, LT, Qiao, T: Two integral operators in Clifford analysis. J. Math. Anal. Appl. 354, 435-444 (2009)
18. Ren, GB, Che, QH, Cerejeiras, P, Kahler, U: Chirp transforms and Chirp series. J. Math. Anal. Appl. 373, 356-369 (2011)
19. Abreu-Blaya, R, Bory-Reyes, J, Brackx, F, De Schepper, H, Sommen, F: Boundary value problems for the quaternionic Hermitian analysis in R4. Bound. Value Probl. 2012, Article ID 74 (2012). doi:10.1687/1687-2770-2012-74
20. Lu, JK: Boundary Value Problems for Analytic Functions. World Scientific, Singapore (2004)
21. Li, PR: Singular integral equations of convolution type with cosecant kernels and periodic coefficients. Math. Probl. Eng. 2017, Article ID 6148393 (2017). doi:10.1155/2017/6148393

CHAPTER 6

MATRIX FOURIER TRANSFORMS FOR CONSISTENT MATHEMATICAL MODELS

Oleg Yaremko and Natalia Yaremko

Penza State University, Krasnaya Street 40, 440038 Penza, Russia

ABSTRACT

We create a matrix integral transforms method; it allows us to describe analytically the consistent mathematical models. An explicit constructions for direct and inverse Fourier matrix transforms with discontinuous coefficients are established. We introduce special types of Fourier matrix transforms: matrix cosine transforms, matrix sine transforms, and matrix transforms with piecewise trigonometric kernels. The integral transforms of such kinds are used for problems solving of mathematical physics in

CITATION (APA): Yaremko, O., & Yaremko, N. (2016). Matrix Fourier Transforms for Consistent Mathematical Models. Chinese Journal of Mathematics, 2016. (10 pages)

homogeneous and piecewise homogeneous media. Analytical solution of iterated heat conduction equation is obtained. Stress produced in the elastic semi-infinite solid by pressure is obtained in the integral form.

INTRODUCTION

Matrix integral Fourier transforms with sine, cosine, and piecewise trigonometric kernels represent an important branch of mathematical analysis. It is based on the expansion of a function over a set of cosine or sine basis functions. Integral Fourier transforms of such kinds have shown their special applicability in description of consistent mathematical models. To show the versatility of these transforms, we solve the problems of mathematical physics in homogeneous and piecewise homogeneous media. We find analytical solutions of iterated heat conduction equation and solve a problem about stress in the elastic semi-infinite solid.

Given a real function $f(x)$, which is defined over the positive real line $x \geq 0$, for $\lambda \geq 0$, which is piecewise continuous and absolutely integrable over $[0,\infty)$, the Fourier *cosine* transform of $f(x)$ is defined as

$$F[f](\lambda) = \tilde{f}(\lambda) = \int_0^\infty \cos \lambda x f(x)\, dx \tag{1}$$

subject to the existence of the integral. The inverse Fourier cosine transform is given by

$$F^{-1}\left[\tilde{f}\right](\lambda) = f(x) = \frac{2}{\pi}\int_0^\infty \cos \lambda x \tilde{f}(\lambda)\, d\lambda \tag{2}$$

again subject to the existence of the integral used in the definition. The functions (x) and $\tilde{f}(\lambda)$, if they exist, are said to form a Fourier cosine transform pair.

Given a real function $f(x)$, which is defined over the positive real line $x \geq 0$, for $x \geq 0$, which is piecewise continuous and absolutely integrable over $[0,\infty)$, the Fourier *sine* transform of $f(x)$ is defined as

$$F[f](\lambda) = \tilde{f}(\lambda) = \int_0^\infty \sin \lambda x f(x)\, dx \tag{3}$$

subject to the existence of the integral. The inverse Fourier transform is given by

$$F^{-1}\left[\tilde{f}\right](\lambda) = f(x) = \frac{2}{\pi}\int_0^\infty \sin\lambda x \tilde{f}(\lambda)\, d\lambda \tag{4}$$

again subject to the existence of the integral used in the definition. The functions (x) and $\tilde{f}(\lambda)$, if they exist, are said to form a Fourier transform pair.

Given a real function $f(x)$, which is defined over the positive real line $x \geq 0$, for $x \geq 0$, which is piecewise continuous and absolutely integrable over $[0,\infty)$, the Fourier type transform of $f(x)$ is defined as

$$F[f](\lambda) = \tilde{f}(\lambda) = \int_0^\infty (h\sin\lambda x - \lambda\cos\lambda x)\, f(x)\, dx \tag{5}$$

subject to the existence of the integral.

Theorem 1. The unit normalization constant used here provides a definition for the inverse Fourier type transform, given by

$$F^{-1}\left[\tilde{f}\right](\lambda) = \frac{2}{\pi}\int_0^\infty \tilde{f}(\lambda)\, \frac{h\sin\lambda x - \lambda\cos\lambda x}{h^2 + \lambda^2}\, d\lambda \tag{6}$$

again subject to the existence of the integral used in the definition.

Proof. Let function f(x) take the following form:

$$f(x) = \frac{2}{\pi}\int_0^\infty \tilde{f}(\lambda)\, \frac{h\sin\lambda x - \lambda\cos\lambda x}{h^2 + \lambda^2}\, d\lambda. \tag{7}$$

Then,

$$hf(x) + f'(x) = \frac{2}{\pi}\int_0^\infty \tilde{f}(\lambda)\sin\lambda x\, d\lambda. \tag{8}$$

Due to the inverse Fourier cosine transform, we get

$$\begin{aligned}\tilde{f}(\lambda) &= \int_0^\infty \sin\lambda x\left(hf(x) + f'(x)\right) d\lambda \\ &= \sin\lambda x f(x)\Big|_0^\infty \\ &\quad + \int_0^\infty (h\sin\lambda x - \lambda\cos\lambda x)\, f(x)\, dx \\ &= \int_0^\infty (h\sin\lambda x - \lambda\cos\lambda x)\, f(x)\, dx.\end{aligned} \tag{9}$$

The general theory of linear integral transform with some of its applications gave an account of [1–12].

In order to define integral Fourier matrix transforms with piecewise trigonometric kernels, we consider Sturm-Liouville matrix problem:

$$-\frac{d}{dx}\left(A(x)\frac{d}{dx}\varphi(x,\lambda)\right)=\lambda^2\varphi(x,\lambda),\quad x\in I_n^+, \tag{10}$$

where

$$\varphi(x,\lambda)=\sum_{j=1}^{n+1}\left(\theta\left(x-l_{j-1}\right)-\theta\left(x-l_j\right)\right)\varphi_j(x,\lambda), \tag{11}$$

bounded nontrivial unknown matrix function of size $r \times r$ called matrix eigenfunction of Sturm-Liouville problem, $A(x)$ is the matrix-valued function of size $r{\times}r$, and

$$I_n^+=\left\{x:x\in\bigcup_{j=1}^{n+1}\left(l_{j-1},l_j\right),\ l_0=0,\ l_{n+1}=\infty\right\}. \tag{12}$$

In general, Sturm-Liouville matrix problem does not possess an analytical solution. Therefore, we consider the Sturm-Liouville piecewise approximation as follows.

A(x) is piecewise constant; that is,

$$A(x)=\sum_{j=1}^{n+1}A_i^2\left(\theta\left(x-l_{j-1}\right)-\theta\left(x-l_j\right)\right), \tag{13}$$

where $x=l_j$, $j = 1, \ldots, n + 1$, are points of discontinuity in I^+_n, and $\theta(x)$ is the Heaviside step function.

The elements of matrix eigenfunctions (x, λ) of a Sturm-Liouville matrix problem are piecewise trigonometric functions. The explicit expression of spectral matrix-valued function $\varphi(x, \lambda)$ allows for defining direct integral Fourier matrix transform with piecewise trigonometric kernels. The explicit solution of dual Sturm-Liouville matrix problem serves as a kernel for an inverse integral Fourier matrix transform.

Integral transforms arise in a natural way through the principle of linear superposition in constructing integral representations of linear differential equations solutions [13– 15]. The theory of integral Fourier transforms with piecewise trigonometric kernels in a scalar case was studied by Ufljand

[16], Najda [17], Procenko and Solov'jov [18], and Lenjuk [19]. The matrix version is adapted for the problems solving in piecewise homogeneous medium and has been developed by Yaremko in [10, 20]. The necessary proofs by method of contour integration were conducted in [11, 21]. It is clear that this method is effective to obtain the exact solution of boundary value problems for piecewise homogeneous media.

MATRIX FOURIER TRANSFORMS WITH PIECEWISE TRIGONOMETRIC KERNELS

The Sturm-Liouville matrix problem [1] is to find the nontrivial solution y bounded on the set I^{+}_{n}

$$y(x,\lambda) = \sum_{k=1}^{n+1} \theta(x - l_{k-1})\,\theta(l_k - x)\, y_k(x,\lambda), \tag{14}$$

to a system of an ordinary differential equations with constant matrix coefficients

$$\left(A_m^2 \frac{d^2}{dx^2} + \lambda^2 E\right) y_m = 0, \quad m = \overline{1, n+1} \tag{15}$$

with the boundary conditions at the points l_0 and l_{n+1},

$$\left(\left(\alpha_{11}^0 + \lambda^2 \delta_{11}^0\right) \frac{d}{dx} + \left(\beta_{11}^0 + \lambda^2 \gamma_{11}^0\right)\right) y_1 = 0, \tag{16}$$

$$\left\| y_{n+1} \right\| \Big|_{x=\infty} < \infty, \tag{17}$$

and internal boundary conditions at the points $x=l_k$

$$\left(\left(\alpha_{j1}^k + \lambda^2 \delta_{j1}^k\right) \frac{d}{dx} + \left(\beta_{j1}^k + \lambda^2 \gamma_{j1}^k\right)\right) y_k$$

$$= \left(\left(\alpha_{j2}^k + \lambda^2 \delta_{j2}^k\right) \frac{d}{dx} + \left(\beta_{j2}^k + \lambda^2 \gamma_{j2}^k\right)\right) y_{k+1}, \tag{18}$$

where

$$\alpha_{ji}^k, \beta_{ji}^k, \gamma_{ji}^k, \delta_{ji}^k, A_j \tag{19}$$

are matrices of size $r \times r$:

$$y_m(x,\lambda) = \begin{pmatrix} y_{1m}(x,\lambda) \\ \vdots \\ y_{rm}(x,\lambda) \end{pmatrix},$$

$$\|y_m\| = \sqrt{y_{1m}^2 + \cdots + y_{rm}^2}. \tag{20}$$

Let y be nontrivial solution to boundary value problems (52)– (69) for some λ. The number λ is called an eigenvalue, and the corresponding solution $y(x, \lambda)$ is called matrix-valued eigenfunction.

We will required invertible conditions

$$\det M_{mk} \neq 0, \quad \lambda \in [0,\infty) \tag{21}$$

for matrices

$$M_{mk} \equiv \begin{pmatrix} \beta_{1m}^k + \lambda^2\gamma_{1m}^k & \alpha_{1m}^k + \lambda^2\delta_{1m}^k \\ \beta_{2m}^k + \lambda^2\gamma_{2m}^k & \alpha_{2m}^k + \lambda^2\delta_{2m}^k \end{pmatrix}. \tag{22}$$

The matrices $A^2{}_m$ are positive-defined [4]. We denote

$$\Phi_{n+1}(x) = e^{q_{n+1}xi};$$

$$\Psi_{n+1}(x) = e^{-q_{n+1}xi},$$

$$q_{n+1}^2 = \lambda^2 A_{n+1}^{-2}. \tag{23}$$

We define the other n pairs of matrix-value functions $(\Phi_k, \Psi_k),\ k = \overline{1,n}$ by the induction relations:

$$\left[\left(\alpha_{j1}^k + \lambda^2\delta_{j1}^k\right)\frac{d}{dx} + \left(\beta_{j1}^k + \lambda^2\gamma_{j1}^k\right)\right](\Phi_k \Psi_k)$$

$$= \left[\left(\alpha_{j2}^k + \lambda^2\delta_{j2}^k\right)\frac{d}{dx} + \left(\beta_{j2}^k + \lambda^2\gamma_{j2}^k\right)\right](\Phi_{k+1}\Psi_{k+1}). \tag{24}$$

Introduce the following notations:

$$\Phi_{0,1}(\lambda) = \left(\alpha_{11}^{0} + \lambda^2 \delta_{11}^{0}\right) \Phi_1'(l_0, \lambda)$$

$$+ \left(\beta_{11}^{0} + \lambda^2 \gamma_{11}^{0}\right) \Phi_1(l_0, \lambda),$$

$$\Psi_{0,1}(\lambda) = \left(\alpha_{11}^{0} + \lambda^2 \delta_{11}^{0}\right) \Psi_1'(l_0, \lambda)$$

$$+ \left(\beta_{11}^{0} + \lambda^2 \gamma_{11}^{0}\right) \Psi_1(l_0, \lambda),$$

$$\Omega_k = \begin{pmatrix} \Phi_k & \Psi_k \\ \Phi_k' & \Psi_k' \end{pmatrix}. \tag{25}$$

Lemma 2. The following identity

$$\left(\Omega_k^{-1}(x, \lambda)\right)' = \begin{pmatrix} 0 & E \\ -q_k^2 & 0 \end{pmatrix} \Omega_k,$$

$$\left(\Omega_k^{-1}(x, \lambda)\right)'' = -\Omega_k^{-1} \lambda^2 A_k^{-2} \tag{26}$$

holds true for $k = 1, \ldots, n$.

Proof. Using formula [20]

$$\left(\Omega_k^{-1}\right)' = -\Omega_k^{-1} \Omega_k' \Omega_k^{-1} \tag{27}$$

and identity

$$(\Omega_k)' = \begin{pmatrix} \Phi_k' & \Psi_k' \\ \Phi_k'' & \Psi_k'' \end{pmatrix} = \begin{pmatrix} 0 & E \\ -q_k^2 & 0 \end{pmatrix} \Omega_k, \tag{28}$$

we can conclude

$$\left(\Omega_k^{-1}\right)' = -\Omega_k^{-1} \begin{pmatrix} 0 & \mathrm{E} \\ -q_k^2 & 0 \end{pmatrix},$$

$$\left(\Omega_k^{-1}\right)'' = \Omega_k^{-1} \begin{pmatrix} 0 & \mathrm{E} \\ -q_k^2 & 0 \end{pmatrix} \begin{pmatrix} 0 & \mathrm{E} \\ -q_k^2 & 0 \end{pmatrix}$$

$$= \Omega_k^{-1} \begin{pmatrix} -q_k^2 & 0 \\ 0 & -q_k^2 \end{pmatrix}. \tag{29}$$

Lemma 3. If the following inequality

$$\det \begin{pmatrix} \beta_{11}^{0} + \lambda^2 \gamma_{11}^{0} & \alpha_{11}^{0} + \lambda^2 \delta_{11}^{0} \\ 0 & E \end{pmatrix} \neq 0, \quad \lambda \in [0, \infty) \tag{30}$$

holds true, then

$\det \Phi_{0,1}(\lambda) \neq 0,$

$\det \Psi_{0,1}(\lambda) \neq 0,$

$$\lambda \in [0, \infty). \tag{31}$$

Proof. Matrices $\Phi_{0,1}$ and $\Psi_{0,1}$ are nonsingular due to the following identity:

$$\begin{pmatrix} \Phi_{0,1} & \Psi_{0,1} \\ \Phi_1' & \Psi_1' \end{pmatrix} = \begin{pmatrix} \beta_{11}^0 + \lambda^2 \gamma_{11}^0 & \alpha_{11}^0 + \lambda^2 \delta_{11}^0 \\ 0 & E \end{pmatrix} \begin{pmatrix} \Phi_1 & \Psi_1 \\ \Phi_1' & \Psi_1' \end{pmatrix}. \tag{32}$$

Theorem 4. The spectrum of problems (15), (16), and (18) is continuous and fills semiaxis $(0,\infty)$. Sturm-Liouville problem is r time singular. Exactly r linearly independent matrix-valued functions correspond to each eigenvalue λ. It is possible to take r columns of matrix-value functions:

$$v(x,\lambda) = \sum_{k=1}^{n+1} \theta(x - l_{k-1})\,\theta(l_k - x)\, v_k(x,\lambda),$$

$$v_j(x,\lambda) = \Phi_j(x,\lambda)\,\Phi_{0,1}^{-1}(\lambda) - \Psi_j(x,\lambda)\,\Psi_{0,1}^{-1}(\lambda). \tag{33}$$

That is

$$y_m(x,\lambda) = \begin{pmatrix} v_{1m}(x,\lambda) \\ \vdots \\ v_{rm}(x,\lambda) \end{pmatrix}. \tag{34}$$

Theorem 4 follows from Lemmas 2 and 3.

Now we consider the dual matrix Sturm-Liouville problem. We find the nontrivial solution y^* of a system of ordinary differential equations with constant matrix coefficients

$$\left(A_m^2 \frac{d^2}{dx^2} + \lambda^2 \mathrm{E} \right) y_m^* = 0, \quad m = \overline{1, n+1}, \tag{35}$$

with boundary conditions at the points $x=l_0$ and $x=l_{n+1}$

$$y_1^* \left(\alpha_{11}^0 + \lambda^2 \delta_{11}^0\right)^{-1} + \frac{d}{dx} y_1^* \left(\beta_{11}^0 + \lambda^2 \gamma_{11}^0\right)^{-1} = 0, \tag{36}$$

$$\left\| y_{n+1}^* \right\| < \infty, \tag{37}$$

and internal boundary conditions at the points $x=l_k$,

$$
\left(-(y_k^*)'\ y_k^*\right)\begin{pmatrix}\beta_{11}^k+\lambda^2\gamma_{11}^k & \alpha_{11}^k+\lambda^2\delta_{11}^k\\ \beta_{21}^k+\lambda^2\gamma_{21}^k & \alpha_{21}^k+\lambda^2\delta_{21}^k\end{pmatrix}^{-1}
= \left(-(y_{k+1}^*)'\ y_{k+1}^*\right)\begin{pmatrix}\beta_{12}^k+\lambda^2\gamma_{12}^k & \alpha_{12}^k+\lambda^2\delta_{12}^k\\ \beta_{22}^k+\lambda^2\gamma_{22}^k & \alpha_{22}^k+\lambda^2\delta_{22}^k\end{pmatrix}^{-1}. \tag{38}
$$

We write the solution of the boundary value problem in the following form:

$$
\begin{aligned}
y^*(\xi,\lambda) &= \sum_{k=1}^{n+1}\theta(\xi-l_{k-1})\,\theta(l_k-\xi)\,y_k^*(\xi,\lambda),\\
y_m^*(\xi,\lambda) &= \left(y_{m1}^*(\xi,\lambda)\ \cdots\ y_{mr}^*(\xi,\lambda)\right),\\
\|y_m^*\| &= \sqrt{(y_{1m}^*)^2+\cdots+(y_{rm}^*)^2}.
\end{aligned} \tag{39}
$$

Theorem 5. The spectrum of problems (35), (36), and (38) is continuous and fills semiaxis $(0,\infty)$. Sturm-Liouville problem is r time singular. Exactly r linearly independent matrix-valued functions correspond to each eigenvalue λ. It is possible to take r rows of matrix-value functions:

$$
\begin{aligned}
v^*(x,\lambda) &= \sum_{k=1}^{n+1}\theta(x-l_{k-1})\,\theta(l_k-x)\,v_k^*(x,\lambda),\\
v_j^*(x,\beta) &= (\Phi_{0,1}\Psi_{0,1})\,\Omega_j^{-1}(x,\beta)\begin{pmatrix}0\\E\end{pmatrix}A_j^{-2}.
\end{aligned} \tag{40}
$$

That is,

$$
y^{*j}(\xi,\lambda) = \left(v_{j1}^*(\xi,\lambda)\ \cdots\ v_{jr}^*(\xi,\lambda)\right),\quad j=\overline{1,r}. \tag{41}
$$

Theorem 5 follows from Lemmas 2 and 3.

The explicit expression of spectral matrix-valued function $u(x,\lambda)$ and the dual spectral function $u^*(x,\lambda)$ allow for writing the decomposition theorem on the set I_n^+.

Theorem 6. Let vector-valued function (x) be defined over I_n^+, continuous, absolutely integrable, and has a bounded variation. Then, for any $x\in I+$, the decomposition formula

$$
\begin{aligned}
f(x) = -\frac{1}{\pi i}\int_0^\infty v(x,\lambda)\Bigg(&\int_{l_0}^\infty v^*(\xi,\lambda)\,f(\xi)\,d\xi+\gamma_{11}^0 f_1(l_0)+\delta_{11}^0 f_1'(l_0)\\
&+\sum_{k=1}^{n}(\Phi_{0,1}\Psi_{0,1})\,\Omega_k^{-1}(l_k,\lambda)\,M_{k1}^{-1}(\lambda)\cdot\left(\begin{pmatrix}\gamma_{21}^k & \delta_{21}^k\\ \gamma_{22}^k & \delta_{22}^k\end{pmatrix}\begin{pmatrix}f_{k+1}(l_k)\\ f_{k+1}'(l_k)\end{pmatrix}-\begin{pmatrix}\gamma_{11}^k & \delta_{11}^k\\ \gamma_{12}^k & \delta_{12}^k\end{pmatrix}\begin{pmatrix}f_k(l_k)\\ f_k'(l_k)\end{pmatrix}\right)\Bigg)\lambda\,d\lambda
\end{aligned} \tag{42}
$$

holds true.

This theorem can be proved by method of contour integration [17].

We define the direct and inverse matrix integral Fourier transforms on the real semiline with piecewise trigonometric kernels according to Theorem 6.

The direct transform is

$$F_{n+}[f](\lambda) = \int_{l_0}^{\infty} v^*(\xi,\lambda) f(\xi)\, d\xi + \gamma_{11}^0 f_1(l_0)$$
$$+ \delta_{11}^0 f_1'(l_0) + \sum_{k=1}^{n} (\Phi_{0,1} \Psi_{0,1}) \Omega_k^{-1}(l_k,\lambda) M_{k1}^{-1}(\lambda)$$
$$\cdot \left(\begin{pmatrix} \gamma_{21}^k & \delta_{21}^k \\ \gamma_{22}^k & \delta_{22}^k \end{pmatrix} \begin{pmatrix} f_{k+1}(l_k) \\ f_{k+1}'(l_k) \end{pmatrix} \right.$$
$$\left. - \begin{pmatrix} \gamma_{11}^k & \delta_{11}^k \\ \gamma_{12}^k & \delta_{12}^k \end{pmatrix} \begin{pmatrix} f_k(l_k) \\ f_k'(l_k) \end{pmatrix} \right) \tag{43}$$

and the inverse transform is

$$F_{n+}^{-1}\left[\tilde{f}\right](x) = -\frac{1}{\pi i} \int_0^{\infty} \lambda v(x,\lambda) \tilde{f}(\lambda)\, d\lambda \equiv f(x), \tag{44}$$

when

$$f(x) = \sum_{k=1}^{n+1} \theta(l_k - x)\, \theta(x - l_{k-1})\, f_k(x). \tag{45}$$

Now we will get the result of the basic identity of matrix integral transforms with piecewise trigonometric kernels for differential operator:

$$B = \sum_{j=1}^{n+1} \theta(x - l_{j-1})\, \theta(l_j - x)\, A_j^2 \frac{d^2}{dx^2}. \tag{46}$$

Theorem 7. If vector-valued function

$$f(x) = \sum_{k=1}^{n+1} \theta(x - l_{k-1})\, \theta(l_k - x)\, f_k(x) \tag{47}$$

is three times continuously differentiable over set I_{n+}, has the limit values of its third-order derivatives at the points $x=l_k$, and satisfies boundary condition on infinity

$$\lim_{x\to\infty} \left(u^*(x,\lambda) \frac{d}{dx} f(x) - \frac{d}{dx} u^*(x,\lambda) f(x) \right) = 0 \tag{48}$$

and internal boundary conditions at the points $x=l_k$, then the basic identity

$$F_{n+}\left[B(f)\right](\lambda) = -\lambda^2 \tilde{f}(\lambda) - \left\{\left(\beta_{11}^0 f_1(l_0) + \alpha_{11}^0 f_1'(l_0)\right) - \left(\gamma_{11}^0 A_1^2 f_1''(l_0) + \delta_{11}^0 A_1^2 f_1'''(l_0)\right)\right\}$$

$$- \sum_{k=1}^{n} \left(\Phi_{0,1} \Psi_{0,1}\right) \Omega_k^{-1}(l_k, \lambda) M_{k1}^{-1}(\lambda) \tag{49}$$

$$\cdot \left\{\left(\begin{pmatrix} \beta_{21}^k & \alpha_{21}^k \\ \beta_{22}^k & \alpha_{22}^k \end{pmatrix} \begin{pmatrix} f_{k+1}(l_k) \\ f_{k+1}'(l_k) \end{pmatrix} - \begin{pmatrix} \gamma_{21}^k & \delta_{21}^k \\ \gamma_{22}^k & \delta_{22}^k \end{pmatrix} \begin{pmatrix} A_{k+1}^2 f_{k+1}''(l_k) \\ A_{k+1}^2 f_{k+1}'''(l_k) \end{pmatrix}\right) - \left(\begin{pmatrix} \beta_{11}^k & \alpha_{11}^k \\ \beta_{12}^k & \alpha_{12}^k \end{pmatrix} \begin{pmatrix} f_k(l_k) \\ f_k'(l_k) \end{pmatrix} - \begin{pmatrix} \gamma_{11}^k & \delta_{11}^k \\ \gamma_{12}^k & \delta_{12}^k \end{pmatrix} \begin{pmatrix} A_k^2 f_k'(l_k) \\ A_k^2 f_k'''(l_k) \end{pmatrix}\right)\right\}$$

holds true.

This theorem can be proved by method of integration by parts.

SPECIAL TYPES OF MATRIX INTEGRAL FOURIER TRANSFORMS

Theorem 8. The matrix-valued Sturm-Liouville problem with Dirichlet boundary condition

$$-A^2 \frac{d^2}{dx^2} \varphi(x, \lambda) = \lambda^2 \varphi(x, \lambda), \quad 0 < x < \infty,$$

$$\varphi(0, \lambda) = 0,$$

$$\left\|\varphi(x, \lambda)\right\| < \infty \tag{50}$$

provides the direct and inverse matrix integral sine transforms on the real semiline:

$$F_{+}[f](\lambda) \equiv \tilde{f}(\lambda) = \int_0^{\infty} \sin\left(A^{-1}\lambda\xi\right) f(\xi)\, d\xi,$$

$$f(x) = \frac{2A^{-1}}{\pi} \int_0^{\infty} \sin\left(A^{-1}\lambda x\right) \tilde{f}(\lambda)\, d\lambda. \tag{51}$$

Proof. Performing calculations in formulas (33) and (41), we get

$$\Phi(x,\lambda) = \exp\left(iA^{-1}\lambda x\right),$$

$$\Psi(x,\lambda) = \exp\left(-iA^{-1}\lambda x\right),$$

$$\Phi_{0,1} = \mathrm{E},$$

$$\Psi_{0,1} = \mathrm{E},$$

$$v(x,\lambda) = \Phi(x,\lambda) - \Psi(x,\lambda) = 2i\sin\left(A^{-1}\lambda x\right), \tag{52}$$

$$v^{*}(x,\beta) = (\mathrm{EE})\begin{pmatrix} \Phi & \Psi \\ \Phi' & \Psi' \end{pmatrix}^{-1} \begin{pmatrix} 0 \\ \mathrm{E} \end{pmatrix} A^{-2}$$

$$= -\frac{A}{2i}(\mathrm{EE})\begin{pmatrix} -iA^{-1}\lambda x e^{-iA^{-1}\lambda x} & -e^{-iA^{-1}\lambda x} \\ -iA^{-1}\lambda x e^{iA^{-1}\lambda x} & e^{iA^{-1}\lambda x} \end{pmatrix}$$

$$\cdot \begin{pmatrix} 0 \\ \mathrm{E} \end{pmatrix} A^{-2} = A^{-1}\sin\left(A^{-1}\lambda x\right).$$

Theorem 9. The matrix-valued Sturm-Liouville problem with Neumann boundary condition

$$-A^2 \frac{d^2}{dx^2} \varphi(x,\lambda) = \lambda^2 \varphi(x,\lambda), \quad 0 < x < \infty,$$

$$\varphi'(0,\lambda) = 0,$$

$$\|\varphi(x,\lambda)\| < \infty \tag{53}$$

provides the direct and inverse matrix integral cosine transforms on the real semiline:

$$F_+[f](\lambda) \equiv \tilde{f}(\lambda) = \int_0^\infty \cos\left(A^{-1}\lambda\xi\right) f(\xi)\, d\xi,$$

$$f(x) = \frac{2A^{-1}}{\pi} \int_0^\infty \cos\left(A^{-1}\lambda x\right) \tilde{f}(\lambda)\, d\lambda. \tag{54}$$

Theorem 10. The matrix-valued Sturm-Liouville problem with Robin boundary condition

$$-A^2 \frac{d^2\varphi}{dx^2} = \lambda^2 \varphi(x,\lambda), \quad x > 0,$$

$$H\varphi + \frac{d\varphi}{dx} = 0, \quad x = 0,$$

$$\|\varphi(x,\lambda)\| < \infty, \tag{55}$$

where H is the square matrix with negative cigenvalues, provides the direct and inverse matrix Fourier type transforms on the real semiline:

$$F_+[f](\lambda) \equiv \tilde{f}(\lambda)$$
$$= \int_0^\infty \left(H \sin A^{-1}\lambda\xi - A^{-1}\lambda \cos A^{-1}\lambda\xi\right) A f(\xi)\, d\xi,$$

$$f(x)$$
$$= \frac{2}{\pi} \int_0^\infty A\left(H \sin A^{-1}\lambda\xi - A^{-1}\lambda \cos A^{-1}\lambda\xi\right)$$
$$\cdot A\left((HA)^2 + \lambda^2 \mathrm{E}\right)^{-1} \tilde{f}(\lambda)\, d\lambda. \tag{56}$$

Proof. Substitute

$$\Phi(x,\lambda) = \exp\left(iA^{-1}\lambda x\right),$$

$$\Psi(x,\lambda) = \exp\left(-iA^{-1}\lambda x\right) \tag{57}$$

into (33) and (41); then,

$$\Phi_{0,1} = H + i\lambda A^{-1},$$

$$\Psi_{0,1} = H - i\lambda A^{-1},$$

$$\begin{aligned} &v(x,\lambda) \\ &= \Phi(x,\lambda)\left(H + i\lambda A^{-1}\right)^{-1} \\ &\quad - \Psi(x,\lambda)\left(H - i\lambda A^{-1}\right)^{-1} \\ &= A\left(\Phi(x,\lambda)(HA - i\lambda \mathrm{E}) - \Psi(x,\lambda)(HA + i\lambda \mathrm{E})\right) \\ &\quad \cdot \left((HA)^2 + \lambda^2 \mathrm{E}\right)^{-2} \\ &= 2iA\left(\sin\left(\lambda A^{-1}x\right)H - A^{-1}\lambda\cos\left(A^{-1}\lambda x\right)\right)A; \end{aligned}$$

$$\begin{aligned} v^*(x,\beta) &= (\mathrm{EE})\begin{pmatrix}\Phi & \Psi \\ \Phi' & \Psi'\end{pmatrix}^{-1}\begin{pmatrix}0 \\ \mathrm{E}\end{pmatrix}A^{-2} \\ &= -\frac{\mathrm{E}}{2i}\left(H + i\lambda A^{-1} H - i\lambda A^{-1}\right) \\ &\quad \cdot \begin{pmatrix}-e^{-iA^{-1}\lambda x} \\ e^{iA^{-1}\lambda x}\end{pmatrix}\begin{pmatrix}0 \\ \mathrm{E}\end{pmatrix}A^{-1} \\ &= \left(-H\sin\left(\lambda A^{-1}x\right) + A^{-1}\lambda\cos\left(A^{-1}\lambda x\right)\right)A. \end{aligned} \tag{58}$$

Theorem 11. The matrix-valued Sturm-Liouville problem with Dirichlet boundary condition on the composite semiline $I_{1+} = (0, l) \cup (l,\infty)$

$$-A_1^2\frac{d^2\varphi_1}{dx^2} = \lambda^2\varphi_1(x,\lambda), \quad x \in (0,l);$$

$$A_2^2\frac{d^2\varphi_2}{dx^2} = \lambda^2\varphi_2(x,\lambda), \quad x \in (l,\infty)$$

$$\varphi_1 = 0, \; x = 0,$$

$$\varphi_1 = \varphi_2,$$

$$\lambda_1 \partial_x \varphi_1 = \lambda_2 \partial_x \varphi_2, \quad x = l,$$

$$\|\varphi_2(x, \lambda)\| < \infty \tag{59}$$

provides the direct and inverse matrix sine type integral transforms on the composite real semiline:

$$F_+[f](\lambda) = \tilde{f}(\lambda) = \int_0^l \sin q_1 \xi \cdot A_1^{-1} f_1(\xi)\, d\xi$$

$$+ \int_l^\infty (\cos q_1 l \cdot \sin q_2 (\xi - l)$$

$$+ \sin q_1 l \chi \cos q_2 (\xi - l))\, A_2^{-1} f_2(\xi)\, d\xi,$$

$$f_1(x) = \frac{2}{\pi} \int_0^\infty \sin q_1 x \tilde{f}(\lambda)\, d\lambda, \tag{60}$$

$$f_2(x) = \frac{2}{\pi} \int_0^\infty \Big(\cos q_2 (x - l) \sin q_1 l$$

$$+ \sin q_2 (x - l)\, \chi^{-1} \cos q_1 l\Big)\, \Omega \tilde{f}(\lambda)\, d\lambda,$$

where

$$\Omega = \left(\sin q_1 l \chi \sin q_1 l + \cos q_1 l \chi^{-1} \cos q_1 l\right)^{-1},$$

$$q_1 = A_1^{-1} \lambda_1,$$

$$q_2 = A_2^{-1} \lambda_1 \cdot \chi = \lambda_2 A_2^{-1} A_1 \lambda_1^{-1}. \tag{61}$$

Proof. Performing calculations in formulas (33) and (41), we get

$$\Phi_2 = \cos q_2 (x-l) + i \sin q_2 (x-l), \quad x > l,$$

$$\Psi_2 = \cos q_2 (x-l) - i \sin q_2 (x-l), \quad x > l,$$

$$\Phi_1 = \cos q_1 (x-l) + i \sin q_1 (x-l) \cdot \chi, \quad 0 < x < l,$$

$$\Psi_1 = \cos q_1 (x-l) - i \sin q_1 (x-l) \cdot \chi, \quad 0 < x < l,$$

$$\Phi_{0,1} = \cos q_1 l - i \sin q_1 l \cdot \chi,$$

$$\Psi_{0,1} = \cos q_1 l + i \sin q_1 l \cdot \chi. \tag{62}$$

Then, matrix eigenfunctions of Sturm-Liouville problem have the following form:

$$\begin{aligned} v_2 &= \Phi_2 (x,\lambda) \Phi_{0,1}^{-1} (\lambda) - \Psi_2 (x,\lambda) \Psi_{0,1}^{-1} (\lambda) \\ &= 2i \left(\cos q_2 (x-l) \quad \sin q_2 (x-l) \right) \\ &\cdot \begin{pmatrix} \cos q_1 l & \sin q_1 l \cdot \chi \\ -\sin q_1 l \cdot \chi & \cos q_1 l \end{pmatrix}^{-1} \begin{pmatrix} 0 \\ \mathrm{E} \end{pmatrix} = 2i \\ &\cdot \left(\cos q_2 (x-l) \sin q_1 l + \sin q_2 (x-l) \chi^{-1} \cos q_1 l \right) \\ &\cdot \Omega, \\ v_1 &= \Phi_1 (x,\lambda) \Phi_{0,1}^{-1} (\lambda) - \Psi_1 (x,\lambda) \Psi_{0,1}^{-1} (\lambda) \\ &= 2i \left(\cos q_1 (x-l) \quad \sin q_1 (x-l) \chi^{-1} \right) \\ &\cdot \begin{pmatrix} \cos q_1 l & \sin q_1 l \cdot \chi \\ -\sin q_1 l \cdot \chi & \cos q_1 l \end{pmatrix}^{-1} \begin{pmatrix} 0 \\ \mathrm{E} \end{pmatrix} = 2i \\ &\cdot \left(\cos q_2 (x-l) \sin q_1 l + \sin q_2 (x-l) \cos q_1 l \right) \Omega \\ &= 2i \sin q_1 x \cdot \Omega. \end{aligned} \tag{63}$$

And dual matrix eigenfunctions of Sturm-Liouville problem have the following form:

$$v_1^* (x, \beta) = (\Phi_{0,1} \Psi_{0,1}) \Omega_1^{-1} (x, \beta) \begin{pmatrix} 0 \\ \mathrm{E} \end{pmatrix} A_1^{-2}$$

$$= (\Phi_{0,1} \Psi_{0,1})$$

$$\cdot \begin{pmatrix} \cos q_1 (x - l) & \sin q_1 (x - l) \chi \\ -q_1 \sin q_1 (x - l) \chi & q_1 \cos q_1 (x - l) \end{pmatrix}^{-1}$$

$$\cdot \begin{pmatrix} 0 \\ \mathrm{E} \end{pmatrix} A_1^{-2} = (\cos q_1 l \;\; \sin q_1 l) \tag{64}$$

$$\cdot \begin{pmatrix} \cos q_1 (x - l) & \sin q_1 (x - l) \cdot \chi \\ -q_1 \sin q_1 (x - l) \cdot \chi & q_1 \cos q_1 (x - l) \end{pmatrix}^{-1}$$

$$\cdot \begin{pmatrix} 0 \\ \mathrm{E} \end{pmatrix} A_1^{-2} = -\frac{A_1^{-1}}{\lambda} \sin q_1 x,$$

$$v_2^* (x, \beta) = (\Phi_{0,1} \Psi_{0,1}) \Omega_2^{-1} (x, \beta) \begin{pmatrix} 0 \\ \mathrm{E} \end{pmatrix} A_2^{-2}$$

$$= (\Phi_{0,1} \Psi_{0,1})$$

$$\cdot \begin{pmatrix} \cos q_2 (x - l) & \sin q_2 (x - l) \chi \\ -q_2 \sin q_2 (x - l) \chi & q_2 \cos q_2 (x - l) \end{pmatrix}^{-1}$$

$$\cdot \begin{pmatrix} 0 \\ \mathrm{E} \end{pmatrix} A_2^{-2} = (\cos q_1 l \;\; \sin q_1 l) \begin{pmatrix} 1 & 0 \\ 0 & -\chi \end{pmatrix}$$

$$\cdot \begin{pmatrix} \cos q_2 (x - l) & \sin q_2 (x - l) \\ -q_1 \sin q_2 (x - l) & q_1 \cos q_2 (x - l) \end{pmatrix}^{-1} \tag{65}$$

$$\cdot \begin{pmatrix} 0 \\ \mathrm{E} \end{pmatrix} A_2^{-2} = -(\cos q_1 l \sin q_2 (x - l) + \sin q_1 l$$

$$\cdot \chi \cos q_2 (x - l)) \frac{A_2^{-1}}{\lambda}.$$

Now we can use the matrix sine type integral transforms on the composite real semiline (43) and (44) to describe analytically the consistent mathematical models.

ANALYTICAL SOLUTION OF ITERATED HEAT CONDUCTION EQUATION

In this section, we can solve a mixed boundary value problem for iterated heat conduction equation [22].

Let

$$u = \sum_{k=1}^{n+1} \theta\left(x - l_{k-1}\right)\theta\left(l_k - x\right)u_k\left(t, x\right) \tag{66}$$

be a solution of system of differential equations

$$\left(\partial_t - a_k^2\partial_{xx}\right)^2 u_k = 0,$$

$$x \in \left(l_{k-1}, l_k\right),\ k = 1, \ldots, n+1 \tag{67}$$

with initial conditions

$$u_k\left(0, x\right) = f_k\left(x\right),$$

$$\partial_t u_k\left(0, x\right) = 0 \tag{68}$$

with boundary conditions

$$h_1 u_1 + \partial_x u_1 = 0,$$

$$h_2\partial_{xx}u_1 + \partial_{xxx}u_1 = 0,$$

$$\text{if } x = 0 \tag{69}$$

and internal boundary conditions at the points $x=l_k$

$$u_k = u_{k+1},$$

$$\lambda_k\partial_x u_k = \lambda_{k+1}\partial_x u_{k+1},$$

$$\mu_k\partial_{xx}u_k = \mu_k\partial_{xx}u_{k+1},$$

$$\nu_k\partial_{xxx}u_k = \nu_{k+1}\partial_{xxx}u_{k+1}. \tag{70}$$

At the beginning, we will solve an auxiliary vector mixed boundary value problem. Let

$$\begin{pmatrix} u \\ v \end{pmatrix} = \sum_{k=1}^{n+1} \theta\left(x - l_{k-1}\right)\theta\left(l_k - x\right)\begin{pmatrix} u_k \\ v_k \end{pmatrix} \tag{71}$$

be a solution to the system of the differential equations

$$\partial_t \begin{pmatrix} u_k \\ v_k \end{pmatrix} = \begin{pmatrix} a_k^2 & -a_k^2 \\ 0 & a_k^2 \end{pmatrix} \partial_{xx} \begin{pmatrix} u_k \\ v_k \end{pmatrix}, \tag{72}$$

with initial conditions

$$\begin{pmatrix} u_k \\ v_k \end{pmatrix} = \begin{pmatrix} f_k(x) \\ f_k(x) \end{pmatrix}, \quad \text{if } t = 0 \tag{73}$$

with boundary conditions at the point $x=0$

$$\begin{pmatrix} h_1 & 0 \\ h_2 - h_1 & h_2 \end{pmatrix}\begin{pmatrix} u_1 \\ v_1 \end{pmatrix} + \partial_x \begin{pmatrix} u_1 \\ v_1 \end{pmatrix} = \begin{pmatrix} 0 \\ 0 \end{pmatrix} \tag{74}$$

and internal boundary conditions at the points $x=l_k$

$$\begin{pmatrix} 1 & 0 \\ \frac{\mu_k}{a_k^2} & \frac{\mu_k}{a_k^2} \end{pmatrix}\begin{pmatrix} u_k \\ v_k \end{pmatrix} = \begin{pmatrix} 1 & 0 \\ \frac{\mu_{k+1}}{a_{k+1}^2} & \frac{\mu_{k+1}}{a_{k+1}^2} \end{pmatrix}\begin{pmatrix} u_{k+1} \\ v_{k+1} \end{pmatrix}, \tag{75}$$

$$\begin{pmatrix} \lambda_1 & 0 \\ \frac{\nu_k}{a_k^2} & \frac{\nu_k}{a_k^2} \end{pmatrix}\partial_x \begin{pmatrix} u_k \\ v_k \end{pmatrix} = \begin{pmatrix} \lambda_2 & 0 \\ \frac{\nu_{k+1}}{a_{k+1}^2} & \frac{\nu_{k+1}}{a_{k+1}^2} \end{pmatrix}\partial_x \begin{pmatrix} u_{k+1} \\ v_{k+1} \end{pmatrix}. \tag{76}$$

Lemma 12. The solution of problems (72)–(76) has the following form:

$$\begin{pmatrix} u \\ v \end{pmatrix} = F_{n+}^{-1}\left[e^{-\lambda^2 t}\tilde{f}(\lambda)\right], \tag{77}$$

where

$$\tilde{f}(\lambda) = \int_0^{\infty} v^*(\xi, \lambda)\begin{pmatrix} f(\xi) \\ f(\xi) \end{pmatrix} d\xi. \tag{78}$$

Proof. This lemma can be proved by Fourier integral transform method with piecewise trigonometric kernels (43)-(44). Let the vector-function

$$\begin{pmatrix} \tilde{u} \\ \tilde{v} \end{pmatrix} \equiv F_{n+} \begin{pmatrix} u \\ v \end{pmatrix} = \int_0^{\infty} v^* (\xi, \lambda) \begin{pmatrix} u(\xi) \\ v(\xi) \end{pmatrix} d\xi \tag{79}$$

be the Fourier transform. Then, from Theorem 7, vector-function $\begin{pmatrix} \tilde{u} \\ \tilde{v} \end{pmatrix}$ will be a solution of the Cauchy problem:

$$\frac{d}{dt} \begin{pmatrix} \tilde{u} \\ \tilde{v} \end{pmatrix} = -\lambda^2 \begin{pmatrix} \tilde{u} \\ \tilde{v} \end{pmatrix},$$

$$\begin{pmatrix} \tilde{u}(0) \\ \tilde{v}(0) \end{pmatrix} = F_{n+} \left[\begin{pmatrix} f(x) \\ f(x) \end{pmatrix} \right]. \tag{80}$$

The solution has the following form:

$$\begin{pmatrix} \tilde{u} \\ \tilde{v} \end{pmatrix} = e^{-\lambda^2 t} F_{n+} \left[\begin{pmatrix} f(x) \\ f(x) \end{pmatrix} \right]. \tag{81}$$

To complete the proof, we apply inverse Fourier transform F_{n+}^{-1}:

$$\begin{pmatrix} u \\ v \end{pmatrix} = F_{n+}^{-1} \left[e^{-\lambda^2 t} \tilde{f}(\lambda) \right]. \tag{82}$$

Theorem 13. Let (u/v) be a solution of vector problems (72)–(76); then,

$$u_k = \begin{pmatrix} 1 & 0 \end{pmatrix} \begin{pmatrix} u_k \\ v_k \end{pmatrix}, \quad k = 1, \ldots, n+1 \tag{83}$$

is a solution to mixed boundary value problems (67)–(70) for iterated heat conduction equation (67).

Proof. In accordance with (72), the function u is a solution of iterated heat conduction equation (67):

$$\left(\partial_t - a^2 \partial_{xx}\right)^2 u = a^2 \partial_t \left(\partial_t - a^2 \partial_{xx}\right) v = 0. \tag{84}$$

Due to (76), we get the initial condition:

$$\partial_t u(0,x) = a^2 \partial_{xx} u(0,x) - a^2 \partial_{xx} v(0,x)$$
$$= a^2 f''(x) - a^2 f''(x) = 0. \quad (85)$$

On the basis of (74), the Robin boundary conditions have the following form: the first condition is as follows:

$$(h_2 - h_1) u + h_2 v + \partial_x v = 0, \quad \text{if } x = 0, \quad (86)$$

and the second condition is as follows: if x=0, then

$$h_2 \partial_{xx} u + \partial_{xxx} u = h_2 \frac{\partial_t u + \partial_t v}{a^2} + \partial_x \frac{\partial_t u + \partial_t v}{a^2}$$
$$= \frac{h_2 \partial_t u + \partial_{tx} u + (h_2 \partial_t v + \partial_{tx} v)}{a^2}$$
$$= \frac{h_2 \partial_t u + \partial_{tx} u + (h_1 - h_2) \partial_t u}{a^2}$$
$$= \partial_t \frac{h_1 u + \partial_x u}{a^2} = 0. \quad (87)$$

It follows from (93)-(94) that

$$\partial_t \frac{\mu_k (u_k + v_k)}{a_k^2} = \partial_t \frac{\mu_{k+1} (u_{k+1} + v_{k+1})}{a_{k+1}^2};$$
$$\partial_x \partial_t \frac{\nu_k (u_k + v_k)}{a_k^2} = \partial_x \partial_t \frac{\nu_{k+1} (u_{k+1} + v_{k+1})}{a_{k+1}^2}$$
$$\partial_t \frac{u_i + v_i}{a_i^2} = \partial_{xx} u_i. \quad (88)$$

Then, the internal boundary conditions at the points $x = l_k$

$$\mu_k \partial_{xx} u_k = \mu_{k+1} \partial_{xx} u_{k+1},$$
$$\nu_k \partial_{xxx} u_k = \nu_{k+1} \partial_{xxx} u_{k+1} \quad (89)$$

hold true.

Corollary 14. The solution of problems (67)–(70) has the following form:

$$u = (1\ \ 0)\begin{pmatrix} u \\ v \end{pmatrix} = (1\ \ 0)\, F_{n+}^{-1}\left[e^{-\lambda^2 t}\tilde{f}(\lambda)\right]$$

$$= (1\ \ 0)\int_0^{\infty} v_k^*(\xi,\lambda)\begin{pmatrix} 1 \\ 1 \end{pmatrix} f_k(\xi)\, d\xi. \tag{90}$$

STRESS PRODUCED IN THE ELASTIC SEMI-INFINITE SOLID BY PRESSURE

Let us consider a problem about distribution of tension in an n+1-layer elastic semi-infinite solid:

$$I_n^+ \times R = \{(x,y) : x \in I_n^+, y \in R\}. \tag{91}$$

In the case of plane, the strain vector of displacement $\overline{u}_i$ has components u_i, V_i, 0. Introduce Airy stress function [10]

$$p = \sum_{k=1}^{n+1} \theta(x - l_{k-1})\,\theta(l_k - x)\, p_k(y,x) \tag{92}$$

as a solution to system of differential equations

$$\left(\partial_{yy} + \partial_{xx}\right)^2 p_k = 0, \quad x \in (l_{k-1}, l_k),\ k = 1,\ldots,n+1 \tag{93}$$

with boundary conditions

$$\sigma_{1x} = -p''(y),$$

$$\tau_{1x} = 0,$$

$$\text{if } x = 0 \tag{94}$$

and internal boundary conditions at the points $x=l_k$

$$\sigma_{kx} = \sigma_{k+1x},$$

$$\tau_{kxy} = \tau_{k+1xy},$$

$$u_k = u_{k+1},$$

$$v_k = v_{k+1}, \tag{95}$$

where σ_{kx} is the normal stress and t_{kxy} is the shearing stresses. Fourier transform of Airy stress function with respect to y

$$\overline{p} = \sum_{k=1}^{n+1} \theta(x - l_{k-1})\theta(l_k - x)\overline{p}_k(y, x) \tag{96}$$

is the solution to system of differential equations

$$(-\xi^2 + d_{xx})^2 \overline{p}_k = 0,$$

$$x \in (l_{k-1}, l_k), \ k = 1, \ldots, n+1 \tag{97}$$

with boundary conditions

$$\overline{\sigma}_{1x} = \xi^2 \overline{p}(\xi),$$

$$\overline{\tau}_{1x} = 0,$$

$$\text{if } x = 0 \tag{98}$$

and internal boundary conditions at the points $x=l_k$

$$\overline{\sigma}_{kx} = \overline{\sigma}_{k+1x},$$

$$\overline{\tau}_{kxy} = \overline{\tau}_{k+1xy},$$

$$\overline{u}_k = \overline{u}_{k+1},$$

$$\overline{v}_k = \overline{v}_{k+1}. \tag{99}$$

At the beginning, we will solve an auxiliary vector mixed boundary value problem. Let

$$\begin{pmatrix} \overline{p} \\ \overline{q} \end{pmatrix} = \sum_{k=1}^{n+1} \theta(x - l_{k-1})\theta(l_k - x) \begin{pmatrix} \overline{p}_k \\ \overline{q}_k \end{pmatrix} \tag{100}$$

be a solution to the system of differential equations

$$-\xi^2 \begin{pmatrix} \overline{p}_k \\ \overline{q}_k \end{pmatrix} + \begin{pmatrix} 1 & -1 \\ 0 & 1 \end{pmatrix} d_{xx} \begin{pmatrix} \overline{p}_k \\ \overline{q}_k \end{pmatrix} = 0, \tag{101}$$

with boundary conditions

$$\overline{p}_1 = -\overline{p}(\xi),$$

$$\partial_x \overline{p}_1 = 0 \tag{102}$$

and internal boundary conditions at the points $x=l_k$

$$\begin{pmatrix} 1 & 0 \\ \dfrac{1+\sigma_k}{E_k} & -\dfrac{1-\sigma_k^2}{E_k} \end{pmatrix} \begin{pmatrix} \overline{p}_k \\ \overline{q}_k \end{pmatrix} = \begin{pmatrix} 1 & 0 \\ \dfrac{1+\sigma_{k+1}}{E_{k+1}} & -\dfrac{1-\sigma_{k+1}^2}{E_{k+1}} \end{pmatrix} \begin{pmatrix} \overline{p}_{k+1} \\ \overline{q}_{k+1} \end{pmatrix}, \tag{103}$$

$$\begin{pmatrix} 1 & 0 \\ \dfrac{1+\sigma_k}{E_k} & \dfrac{1-\sigma_k^2}{E_k} \end{pmatrix} \partial_x \begin{pmatrix} \overline{p}_k \\ \overline{q}_k \end{pmatrix} = \begin{pmatrix} 1 & 0 \\ \dfrac{1+\sigma_{k+1}}{E_{k+1}} & \dfrac{1-\sigma_{k+1}^2}{E_{k+1}} \end{pmatrix} \partial_x \begin{pmatrix} \overline{p}_{k+1} \\ \overline{q}_{k+1} \end{pmatrix}. \tag{104}$$

Lemma 15. The solution of problems (101)–(104) in the Fourier images takes the form

$$\begin{pmatrix} \overline{p} \\ \overline{q} \end{pmatrix} = -F_{n+}^{-1}\left[\frac{1}{\xi^2+\lambda^2}\right]\begin{pmatrix} \overline{p}(\xi) \\ 0 \end{pmatrix} \tag{105}$$

and in the Fourier originals has the form

$$\begin{pmatrix} p \\ q \end{pmatrix} = -\frac{1}{2\pi}\int_0^{\infty} F_{n+}^{-1}\left[e^{-\lambda|y-\eta|}\right]\begin{pmatrix} p(\eta) \\ 0 \end{pmatrix} d\eta. \tag{106}$$

Inverse Fourier transform F_{n+}^{-1} is constructed in accordance with (44).

Theorem 16. Let $\begin{pmatrix} \overline{p} \\ \overline{q} \end{pmatrix}$ be a solution of vector problems (101)–(104); then,

$$\overline{p}_k = (1 \;\; 0)\begin{pmatrix} \overline{p}_k \\ \overline{q}_k \end{pmatrix}, \quad k = 1, \ldots, n+1 \tag{107}$$

is the solution of scalar problems (75)-(76).

Proof. In accordance with (101), the function p is a solution of iterated Laplace equation (93):

$$\left(-\xi^2 + d_{xx}\right)^2 p_k = d_{xx}\left(-\xi^2 + d_{xx}\right) q_k = 0. \tag{108}$$

Calculating the components $\overline{u}_k$ and $\overline{v}_k$ of strain vector of displacement on the basis of [10], we get

$$\overline{v}_k = \frac{1+\sigma_k}{2\pi E}\left(\left(1-\sigma_k\right)\overline{p}_k'' + \sigma_k \xi^2 \overline{p}_k\right)$$

$$= \frac{1+\sigma_k}{2\pi E}\left(\left(1-\sigma_k\right)\left(\xi^2\overline{p}_k - \xi^2\overline{q}_k\right) + \sigma_k \xi^2 \overline{p}_k\right)$$

$$= \frac{1+\sigma_k}{2\pi E}\left(\xi^2\overline{p}_k - \left(1-\xi^2\right)\overline{q}_k\right), \tag{109}$$

$$\overline{u}_k = \frac{1+\sigma_k}{2\pi E}\left(\left(1-\sigma_k\right)\overline{p}_k''' - \left(2-\sigma_k\right)\xi^2\overline{p}_k'\right)$$

$$= \frac{1+\sigma_k}{2\pi E}\left(\left(1-\sigma_k\right)\left(\xi^2\overline{p}_k' - \xi^2\overline{q}_k'\right) - \left(2-\sigma_k\right)\xi^2\overline{p}_k'\right)$$

$$= \frac{1+\sigma_k}{2\pi E}\left(-\xi^2\overline{p}_k - \left(1-\xi^2\right)\overline{q}_k\right).$$

As a result, the internal boundary conditions (103)-(104) hold.

Corollary 17. The solution of problems (93)–(95) has the form

$$p = \begin{pmatrix}1 & 0\end{pmatrix}\begin{pmatrix}p \\ q\end{pmatrix}$$

$$= -\begin{pmatrix}1 & 0\end{pmatrix}\frac{1}{2\pi}\int_0^{\infty} F_{n+}^{-1}\left[e^{-\lambda|y-\eta|}\right]\begin{pmatrix}p(\eta) \\ 0\end{pmatrix} d\eta. \tag{110}$$

CONCLUSION

Usage of the integral Fourier matrix transforms with piecewise trigonometric kernels method allows us to solve internal boundary conditions problems. Internal boundary conditions problems arise in mathematical modeling of heat conduction and stress produced in the piecewise homogeneous media.

REFERENCES

1. R. Bracewell, The Fourier Transform and Its Applications, McGraw-Hill College, 3rd edition, 1999.
2. J. F. James, A Student's Guide to Fourier Transforms, Cambridge University Press, Cambridge, UK, 1995.
3. J. T. W. Keorner, Fourier Analysis, Cambridge University Press, Cambridge, UK, 1988.
4. L. Grafakos, Classical and Modern Fourier Analysis, Prentice-Hall, Upper Saddle River, NJ, USA, 2004.
5. B. Osgood, The Fourier Transform and Its Applications, Stanford University, 2009.
6. M. A. Pinsky, Introduction to Fourier Analysis and Wavelets, Brooks/Cole, 2002.
7. M. A. Pinsky, "Fourier inversion for piecewise smooth functions in several variables," Proceedings of the American Mathematical Society, vol. 118, no. 3, pp. 903–910, 1993.
8. A. D. Polyanin and A. V. Manzhirov, Handbook of Integral Equations, CRC Press, Boca Raton, Fla, USA, 1998.
9. M. S. Salih, Fourier Transform Applications, InTech, Rijeka, Croatia, 2012.
10. I. N. Sneddon, Fourier Transforms, McGraw-Hill, 1951.
11. E. Stein and R. Shakarchi, Fourier Analysis: An Introduction, Princeton University Press, Princeton, NJ, USA, 2003.
12. E. C. Titchmarsh, Introduction to the Theory of Fourier Integrals, Clarendon, Oxford, UK, 1948.
13. G.-C. Kuo, Y.-H. Hu, W.-L. Liaw, K.-J. Wang, and K.-Y. Kung, "Transient temperature solutions of a cylindrical fin," WSEAS Transactions on Mathematics, vol. 10, no. 2, pp. 47–55, 2011.
14. K.-T. Chiang, G. C. Kuo, K.-J. Wang, Y. F. Hsiao, and K.-Y. Kung, "Transient temperature analysis of a cylindrical heat equation," WSEAS Transactions on Mathematics, vol. 7, no. 8, pp. 309–319, 2009.
15. P.-Y. Wang, G.-C. Kuo, Y.-H. Hu, and W.-L. Liaw, "Transient temperature solutions of a cylindrical fin with lateral heat loss," WSEAS Transactions on Mathematics, vol. 11, no. 10, pp. 918–952, 2012.

16. J. S. Ufljand, Integral'nye preobrazovanija v zadachah teorii uprugosti, Nauka, Leningrad, Russia, 1967.
17. L. S. Najda, "Gibridnye integral'nye preobrazovanija tipa Hankelja-Lezhandra," Matematicheskie Metody Analiza Dinamicheskih Sistem, Harkov, vol. 8, pp. 132–135, 1984.
18. V. S. Procenko and A. I. Solov'jov, "Gibridnye integral'nye preobrazovanija i ih prilozhenija v teorii uprugosti neodnorodnyh sred," Prikladnaja Mehanika, vol. 1, no. 13, pp. 62–67, 1982.
19. M. P. Lenjuk, "Integral'noe preobrazovanie Fur'e na kusochno-odnorodnoj poluprjamoj," Izvestiya Vysshikh Uchebnykh Zavedenii. Matematika, vol. 4, pp. 14–18, 1989.
20. F. R. Gantmacher, The Theory of Matrices, Reprinted by American Mathematical Society, AMS Chelsea, 2000.
21. O. Yaremko, V. Selutin, and N. Yaremko, "The fourier transform with piecewise trigonometric kernels and its applications," WSEAS Transactions on Mathematics, vol. 13, pp. 615–625, 2014.
22. F. M. Mors and G. Fishbah, Methods of Theoretical Physics, part 1, McGraw-Hill, New York, NY, USA, 1953.

CHAPTER 7

APPLICATIONS OF THE FUZZY SUMUDU TRANSFORM FOR THE SOLUTION OF FIRST ORDER FUZZY DIFFERENTIAL EQUATIONS

Norazrizal Aswad Abdul Rahman and Muhammad Zaini Ahmad

Institute of Engineering Mathematics, Universiti Malaysia Perlis, Pauh Putra Main Campus, 02600 Arau, Perlis, Malaysia

ABSTRACT

In this paper, we study the classical Sumudu transform in fuzzy environment, referred to as the fuzzy Sumudu transform (FST). We also propose some results on the properties of the FST, such as linearity, preserving, fuzzy derivative, shifting and convolution theorem. In order to show the capability

CITATION (APA): Rahman, N. A. A., & Ahmad, M. Z. (2015). Applications of the fuzzy Sumudu transform for the solution of first order fuzzy differential equations. Entropy, 17(7), 4582-4601. (20 pages)

of the FST, we provide a detailed procedure to solve fuzzy differential equations (FDEs). A numerical example is provided to illustrate the usage of the FST.

Keywords: fuzzy number; fuzzy Sumudu transform; generalized differentiability; fuzzy differential equation; Sumudu transform; integral transform

INTRODUCTION

Integral transforms constitute fundamental tools in operational calculus. They are mathematical operators that have been used widely in solving many practical problems in applied mathematics, physics and engineering [1–4]. The precursor of the integral transforms is the Fourier transform, which is used to express functions in a finite interval. Thenceforward, there are a number of works on the theories and applications of integral transforms, some of which are Laplace, Mellin and Hankel transforms [5–7]. Subsequently, the concept of integral transforms was expanded to remove the necessity of finite intervals. Watugala [8,9] has proposed a new integral transform called the Sumudu transform. The Sumudu transform has been used to solve ordinary differential equations in control engineering problems. Weerakon [10,11] has extended the Sumudu transform on partial differential equations. The work was then continued by Asiru [12], who studied the convolution theorem of the Sumudu transform, which can be expressed in terms of polynomial and convergent infinite series. Next, Belgacem *et al.* [13] have emphasized the Laplace–Sumudu duality, which is a vital step in establishing new results on the Sumudu transform. For instance, the duality property has been used to invoke a complex inverse of the Sumudu transform, as a Bromwich contour integral formula [14]. Furthermore, in [15], the applications of the Sumudu transform on Bessel functions and equations have been explored. The theories and applications of the Sumudu transform have been examined and explored by many authors (see in [16–21]).

In general, the Sumudu transform is considered a popular integral transform for solving differential equations. This is due to its unity property, which eases the process of finding solutions. It is also more powerful compared to other integral transforms, as the function transformed is a similitude of the resulting function.

Many real-world problems are modeled by differential equations. However, we cannot be sure that the models is perfect. For example, the initial value of the models might not be known accurately. The initial value may contain some uncertainty quantities, such as "less than x_0", "about x_0" or "more than x_0". If this is the case, the classical differential equations cannot be used to handle this situation. Therefore, it is necessary to study other theories in order to overcome this problem. One of the most popular theories for describing this situation is the fuzzy set theory [22]. By incorporating fuzziness into classical mathematics, many authors studied fuzzy derivatives [23–28], fuzzy differential equations (FDEs) [29–31] and fuzzy fractional differential equations (FFDEs) [32–34].

Recently, Allahviranloo and Ahmadi [35] have proposed the fuzzy Laplace transform (FLT) and showed its applications to solve FDEs. The FLT is then used to solve second order linear FDEs and the state-space description of fuzzy linear continuous time systems [36]. This work has motivated many authors to expand the theories and applications of FLT in the mathematics and engineering fields [37–41]. The work has also motivated a few researchers to study the classical Sumudu transform in the fuzzy setting. The first effort was initiated by Ahmad and Abdul Rahman [42] and further studied by Alam Khan *et al.* [43]. In this paper, we add some new results on the Sumudu transform in the fuzzy setting, especially on the linearity and preserving properties. Some other results may parallel the ones proposed in [43]. However, our definition of the Sumudu transform in the fuzzy setting is quite general, and the results are presented in different ways.

This paper is organized as follows. In Section 2, we recall several basic definitions and concepts of fuzzy numbers. In Section 3, we provide a general definition of the fuzzy Sumudu transform (FST) and investigate the duality property between FST and FLT. We also provide some theorems and properties regarding the FST. In Section 4, we construct detailed procedures to solve FDEs. Later, in Section 6, we give conclusions.

PRELIMINARIES

In this section, we recall some definitions and theorems needed in order to understand the contribution in this paper. The definition of a fuzzy number is as follows.

Definition 1. [44] *By* $\mathbb{R}$, *we denote the set of all real numbers. A fuzzy number is a mapping* $U : \mathbb{R} \to [0, 1]$ *with the following properties:*

- *U is upper semi-continuous,*
- *U is fuzzy convex, i.e.,* $U(\lambda x + (1 - \lambda)y) \geq \min\{U(x), U(y)\}$ *for all* $x, y \in \mathbb{R}, \lambda \in [0, 1]$,
- *U is normal, i.e.,* $\exists x_0 \in \mathbb{R}$ *for which* $U(x_0) = 1$,
- $\mathrm{supp}\, U = \{x \in \mathbb{R} | U(x) > 0\}$ *is the support of the U, and its closure, i.e. cl(supp U) is compact.*

The following definition is the α-level set of fuzzy numbers.

Definition 2. [44] *Let* $F\mathscr{F}(\mathbb{R})$ *be the set of all fuzzy numbers on* $\mathbb{R}$. *The* α*-level set of a fuzzy number* $U \in F\mathscr{F}(\mathbb{R})$, $\alpha \in [0, 1]$, *denoted by* U_α, *is defined as*:

$$U_\alpha = \begin{cases} \{x \in \mathbb{R} | U(x) \geq \alpha\}, & \text{if } 0 \leq \alpha \leq 1, \\ cl(\mathrm{supp}\ U), & \text{if } \alpha = 0. \end{cases}$$

It is clear that the α-level set of a fuzzy number is a closed and bounded interval, *i.e.*, $[\underline{u}_\alpha, \overline{u}_\alpha]$, where $\underline{u}_\alpha$ and $\overline{u}_\alpha$ denote the lower bound and the upper bound of U, respectively. Since each $y \in \mathbb{R}$ can be regarded as a fuzzy number $\tilde{y}$ defined by:

$$\tilde{y}(t) = \begin{cases} 1, & \text{if } t = y, \\ 0, & \text{if } t \neq y, \end{cases}$$

$\mathbb{R}$ can be embedded in $F\mathscr{F}(\mathbb{R})$.

Remark. [45] *Let X be the Cartesian product of universes* $X = X_1 \times \ldots \times X_n$ *and* $A_1, \ldots, A_n$ *be n fuzzy numbers in* $X_1, \ldots, X_n$, *respectively. A fuzzy function f maps from X to a universe Y,* $y = f(x_1, \ldots, x_n)$. *Then, the extension principle allows us to define a fuzzy set B in Y by:*

$$B = \{(y, U_B(y)) | y = f(x_1, \ldots, x_n), (x_1, \ldots, x_n) \in X\},$$

where:

$$U_B(y) = \begin{cases} \sup_{(x_1, \ldots, x_n) \in f^{-1}(y)}, & \\ \min\{U_{A_1}(x_1), \ldots, U_{A_n}(x_n)\}, & \text{if } f^{-1} \neq 0, \\ 0, & \text{if otherwise,} \end{cases}$$

where f^{-1} *is the inverse of f.*

For $n = 1$, *the extension principle is reduced to:*

$$B = \{(y, U_B(y)) | y = f(x), x \in X\},$$

where:

$$U_B(y) = \begin{cases} \sup_{x \in f^{-1}(y)} U_A(x), & \text{if } f^{-1} \neq 0, \\ 0, & \text{if otherwise.} \end{cases}$$

Referring to the extension principle, addition on F$\mathscr{F}(\mathbb{R})$ can be defined by:

$$(U \oplus V)(x) = \sup_{y \in \mathbb{R}} \min\{U(y), V(x-y)\}, x \in \mathbb{R},$$

and scalar multiplication is defined by:

$$(k \odot U)(x) = \begin{cases} U(x/k) & k > 0, \\ \widetilde{0}, & k = 0, \end{cases}$$

where $\widetilde{0} \in F(R)$.

Furthermore, for all α-levels,

$$[U_\alpha \oplus V_\alpha] = U_\alpha + V_\alpha,$$

and:

$$[k \odot U]_\alpha = kU_\alpha,$$

is true.

In this paper, the notation U_α represents the α-level set of a fuzzy number. We may conclude that the fuzzy number is determined by the endpoints of the intervals U_α. This leads to other representation of a fuzzy number, which will be defined by two endpoint functions $\underline{u}_\alpha$ and $\overline{u}_\alpha$. Friedman *et al.* [46] and Ma *et al.* [47] defined the representation as:

Definition 3. *A fuzzy number U in parametric form is a pair* $[\underline{u}_\alpha, \overline{u}_\alpha]$ *of functions* $\underline{u}_\alpha$ and $\overline{u}_\alpha$, $\alpha \in [0, 1]$, *which satisfy the following requirements:*

- $\underline{u}_\alpha$ *is a bounded non-decreasing left continuous function in* (0, 1] *and right continuous at zero,*
- $\overline{u}_\alpha$ *is a bounded non-increasing left continuous function in* (0, 1] *and right continuous at zero,*
- $\underline{u}_\alpha \leq \overline{u}_\alpha$.

A fuzzy number can be represented as a fuzzy membership function. One of the most commonly-used fuzzy membership functions in the literature is the fuzzy triangular membership function. It is defined as follows.

Definition 4. *Let* $U \in$ F$\mathscr{F}(\mathbb{R})$. *U is called a triangular fuzzy number if its membership function has the following form:*

$$U(x) = \begin{cases} 0, & \text{if } x < a, \\ \dfrac{x-a}{b-a}, & \text{if } a \le x < b, \\ \dfrac{c-x}{c-b}, & \text{if } b \le x \le c, \\ 0, & \text{if } x > c, \end{cases}$$

and its α-level is simply $U_\alpha = [a + (b - c)\alpha, c - (c - b)\alpha]$, *for any* $\alpha \in [0, 1]$.

Definition 5. [46] *For arbitrary* $U_\alpha = [\underline{u}_\alpha, \overline{u}_\alpha]$, $V_\alpha = [\underline{v}_\alpha, \overline{v}_\alpha]$ *and* $k > 0$, *we define addition, subtraction and multiplication by k for* U_α *and* V_α *as:*

- *addition,*

$$U_\alpha \oplus V_\alpha = [\underline{u}_\alpha + \underline{v}_\alpha, \overline{u}_\alpha + \overline{v}_\alpha],$$

- *subtraction,*

$$U_\alpha \ominus V_\alpha = [\underline{u}_\alpha - \overline{v}_\alpha, \overline{u}_\alpha - \underline{v}_\alpha],$$

- *scalar multiplication,*

$$k \odot U_\alpha = \begin{cases} [k\underline{u}_\alpha, k\overline{u}_\alpha], & k \ge 0, \\ [k\overline{u}_\alpha, k\underline{u}_\alpha], & k < 0. \end{cases}$$

If $k = -1$, then $k \odot U_\alpha = -U_\alpha$.

Definition 6. [48] *The distance* $D(U, V)$ *between two fuzzy intervals U and V is defined as:*

$$D(U, V) = \sup_{\alpha \in [0,1]} d_H(U_\alpha, V_\alpha),$$

where:

$$d_H(U_\alpha, V_\alpha) = \max\{|\underline{u}_\alpha - \underline{v}_\alpha|, |\overline{u}_\alpha - \overline{v}_\alpha|\}$$

is the Hausdorff distance between U_α *and* V_α.

Thus, we can conclude that D is a metric space and has the following properties:

1. $D(U \oplus W, V \oplus W) = D(U, V), \forall U, V, W \in F\mathscr{F}(\mathbb{R})$,
2. $D(k \odot U, k \odot V) = |k| D(U, V), \forall k \in \mathbb{R}, U, V \in F\mathscr{F}(\mathbb{R})$,
3. $D(U \oplus V, W \oplus E) \le D(U, W) + D(V, E), \forall U, V, W, E \in F\mathscr{F}(\mathbb{R})$,
4. $(D, F\mathscr{F}(\mathbb{R}))$ is a complete metric space.

Definition 7. [49] *Let* $f : \mathbb{R} \to F\mathscr{F}(\mathbb{R})$. *The function f is called continuous if for every* $x_0 \in \mathbb{R}$ *and every* $\epsilon > 0$, *there exists* $\delta > 0$, *such that if* $|x - x_0| < \delta$, *then* $D(f(x), f(x_0)) < \epsilon$.

Theorem 1. [50] *Let* $f: \mathbb{R} \rightarrow F\mathscr{F}(\mathbb{R})$, *and it is represented by* $[\underline{f}_\alpha(x), \overline{f}_\alpha(x)]$. *For any fixed* $\alpha \in [0, 1]$, *assume that* $\underline{f}_\alpha(x)$ *and* $\overline{f}_\alpha(x)$ *are Riemann-integrable on* $[a, b]$ *for every* $b \geq a$, *and assume that there are two positive* $\underline{M}_\alpha$ *and* $\overline{M}_\alpha$, *such that* $\int_a^b |\underline{f}_\alpha(x)|dx \leq \underline{M}_\alpha$ *and* $\int_a^b |\overline{f}_\alpha(x)|dx \leq \overline{M}_\alpha$ *for every* $b \geq a$. *Then,* $f(x)$ *is improper fuzzy Riemann-integrable on* $[a, \infty)$, *and the improper fuzzy Riemann-integrable is a fuzzy number. Furthermore, we have:*

$$\int_a^\infty f(x)dx = \left[\int_a^\infty \underline{f}_\alpha(x)dx, \int_a^\infty \overline{f}_\alpha(x)dx\right].$$

Proposition 1. [51] *If each of* $f(x)$ *and* $g(x)$ *is a fuzzy-valued function and fuzzy Riemann-integrable on* $[a, \infty)$, *then* $f(x) \oplus g(x)$ *is fuzzy Riemann-integrable on* $[a, \infty)$. *Moreover, we have:*

$$\int_I (f(x) \oplus g(x))dx = \int_I f(x)dx \oplus \int_I g(x)dx.$$

The next definition is Hukuhara's differentiability, also known as H-derivatives. The definition is about H-differences of sets, and it is introduced as follows.

Definition 8. [27] *Let* $x, y \in F\mathscr{F}(\mathbb{R})$. *If there exists* $z \in F\mathscr{F}(\mathbb{R})$, *such that* $x = y \oplus z$, *then z is called the H-difference of x and y, and it is denoted by* $x -^H y$.

Definition 9. [52,53] *Let* $f: (a, b) \rightarrow F\mathscr{F}(\mathbb{R})$ *and* $x_0 \in (a, b)$. *We say that f is strongly generalized differentiable at* x_0, *if there exists an element* $f'(x_0) \in F\mathscr{F}(\mathbb{R})$, *such that:*

- *for all* $h > 0$ *sufficiently small, there exist* $f(x_0 + h) -^H f(x_0)$, $f(x_0) -^H f(x_0 - h)$ *and the limits (in the metric D):*

$$\lim_{h\to 0} \frac{f(x_0+h) -^H f(x_0)}{h} = \lim_{h\to 0} \frac{f(x_0) -^H f(x_0-h)}{h} = f'(x_0),$$

or

- *for all* $h > 0$ *sufficiently small, there exist* $f(x_0) -^H f(x_0 + h)$, $f(x_0 - h) -^H f(x_0)$ *and the limits (in the metric D):*

$$\lim_{h\to 0} \frac{f(x_0) -^H f(x_0+h)}{-h} = \lim_{h\to 0} \frac{f(x_0-h) -^H f(x_0)}{-h} = f'(x_0).$$

(h and −h in the denominators mean $1/\mathrm{h}$ *and* $-1/\mathrm{h}$, *respectively.)*

Note. *In this paper, we only consider Cases (1) and (2) in the strongly generalized differentiability proposed by Bede and Gal* [52]. *Chalco-Cano* *and Roman-Florés* [54] *stated that Cases (1) and (2) are more important, since Cases (3) and (4) occur only on a discrete set of points.*

Theorem 2. [54] *Let* $f : \mathbb{R} \rightarrow \mathrm{F}\mathscr{F}(\mathbb{R})$ *be a continuous fuzzy-valued function and denote* $[\underline{f}_\alpha(x), \overline{f}_\alpha(x)]$, *for each* $\alpha \in [0, 1]$. *Then:*

- *if f is (1)-differentiable, then* $\underline{f}_\alpha(x)$ *and* $\overline{f}_\alpha(x)$ *are differentiable functions and* $[\underline{f}'_\alpha(x), \overline{f}'_\alpha(x)]$,
- *if f is (2)-differentiable, then* $\underline{f}_\alpha(x)$ *and* $\overline{f}_\alpha(x)$ *are differentiable functions and* $[\overline{f}'_\alpha(x), \underline{f}'_\alpha(x)]$.

FUZZY SUMUDU TRANSFORM

In order to establish results, some definitions are needed. $G(u)$ and S[f(x)] will be used as the notations for the fuzzy Sumudu transform throughout this paper.

Definition 10. *Let* $f : \mathbb{R} \rightarrow \mathrm{F}\mathscr{F}(\mathbb{R})$ *be a continuous fuzzy-valued function. Suppose that* $f(ux) \odot e^{-x}$ *is improper fuzzy Riemann-integrable on* $[0, \infty)$, *then* $\int_0^\infty f(ux) \odot e^{-x} dx$ *is called the fuzzy Sumudu transform and is denoted by:*

$$G(u) = \mathcal{S}[f(x)] = \int_0^\infty f(ux) \odot e^{-x} dx, \qquad (u \in [-\tau_1, \tau_2]), \tag{1}$$

where the variable u is used to factor the variable x in the argument of the fuzzy-valued function.

From Theorem 1, we obtain:

$$\int_0^\infty f(ux) \odot e^{-x} dx = \left[\int_0^\infty \underline{f}_\alpha(ux) e^{-x} dx, \int_0^\infty \overline{f}_\alpha(ux) e^{-x} dx \right].$$

From the classical Sumudu transform, we have:

$$s[\underline{f}_\alpha(x)] = \int_0^\infty \underline{f}_\alpha(ux) e^{-x} dx,$$

and:

$$s[\overline{f}_\alpha(x)] = \int_0^\infty \overline{f}_\alpha(ux) e^{-x} dx.$$

Finally, we have:

$$\mathcal{S}[f(x)] = [s[\underline{f}_{\alpha}(x)], s[\overline{f}_{\alpha}(x)]].$$

Duality Properties of the Fuzzy Laplace and Fuzzy Sumudu Transform

FLT has a close relationship with FST. It is necessary for us to be able to link between the two transforms in order to prove theorems and properties of the FST. The definition for FLT is given as follows.

Definition 11. [35] *Let $f(x)$ be a continuous fuzzy-valued function. Suppose that $f(x) \odot e^{-px}$ is improper fuzzy Riemann-integrable on* $[0, \infty)$, *then* $\int_0^{\infty} f(ux) \odot e^{-x}dx$ *is called the fuzzy Laplace transform and is denoted by:*

$$F(p) = \mathcal{L}[f(x)] = \int_0^{\infty} f(x) \odot e^{-px}dx, \qquad (p > 0). \tag{2}$$

Theorem 3. *Let $f(x)$ be a continuous fuzzy-valued function. If F is the fuzzy Laplace transform of $f(x)$ and G is the fuzzy Sumudu transform of $f(x)$, then:*

$$G(u) = \frac{F(1/u)}{u}. \tag{3}$$

Proof. Let $f(x) \in F\mathscr{F}(\mathbb{R})$, then for $-\tau_1 < u < \tau_2$,

$$G(u) = \left[\int_0^{\infty} \underline{f}_{\alpha}(ux)e^{-x}dx, \int_0^{\infty} \overline{f}_{\alpha}(ux)e^{-x}dx\right].$$

Substituting $w = ux$ or $x = w / u$, then we have:

$$\begin{aligned} G(u) &= \left[\int_0^{\infty} \underline{f}_{\alpha}(w)e^{-w/u}\frac{dw}{u}, \int_0^{\infty} \overline{f}_{\alpha}(w)e^{-w/u}\frac{dw}{u}\right], \\ &= \frac{1}{u}\left[\int_0^{\infty} \underline{f}_{\alpha}(w)e^{-w/u}dw, \int_0^{\infty} \overline{f}_{\alpha}(w)e^{-w/u}dw\right], \\ &= \frac{1}{u}\int_0^{\infty} f(w) \odot e^{-w/u}dw. \end{aligned}$$

It is clear that the $\frac{1}{u}\int_0^{\infty} f(w) \odot e^{-w/u}dw$ is $\frac{F(1/u)}{u}$.

In the following corollary, we show that the roles of F and G in Equation (3) can be interchanged.

Corollary 1. *Let f(x) ∈ F$\mathscr{F}$($\mathbb{R}$), having F and G for the fuzzy Laplace transform and fuzzy Sumudu transform, respectively. Then:*

$$F(p) = \frac{G(1/p)}{p}. \tag{4}$$

Proof. The proof of Equation (4) can be obtained by changing *u* to 1 / p in Equation (3).

Equations (3) and (4) form the fuzzy Laplace–Sumudu duality and serve as a mean of changing between those two transforms when needed.

Fundamental Theorems and Properties of the Fuzzy Sumudu Transform

In this section, we provide some theorems and properties associated with FST. Please note that the theorems and properties proposed in this section are extension of the classical Sumudu transform, as studied in [13,14].

Theorem 4. *Let f, g : $\mathbb{R} \to$ F$\mathscr{F}$($\mathbb{R}$) be two continuous fuzzy-valued functions. Suppose that c_1 and c_2 are arbitrary constants, then:*

$$\mathcal{S}[(c_1 \odot f(x)) \oplus (c_2 \odot g(x))] = (c_1 \odot \mathcal{S}[f(x)]) \oplus (c_2 \odot \mathcal{S}[g(x)]).$$

Proof. Assume that $f(x) = [\underline{f}_\alpha(x), \overline{f}_\alpha(x)]$ and $g(x) = [\underline{g}_\alpha(x), \overline{g}_\alpha(x)]$. First, we proof for the lower bound of *f*(*x*) and g(*x*).

$$\begin{aligned} s[(c_1\underline{f}_\alpha(x)) + (c_2\underline{g}_\alpha(x))] &= \int_0^\infty ((c_1\underline{f}_\alpha(ux)) + (c_2\underline{g}_\alpha(ux)))e^{-x}dx, \\ &= \int_0^\infty (c_1\underline{f}_\alpha(ux))e^{-x}dx + \int_0^\infty (c_2\underline{g}_\alpha(ux))e^{-x}dx, \\ &= c_1\int_0^\infty \underline{f}_\alpha(ux)e^{-x}dx + c_2\int_0^\infty \underline{g}_\alpha(ux)e^{-x}dx, \\ &= c_1 s[\underline{f}_\alpha(x)] + c_2 s[\underline{g}_\alpha(x)]. \end{aligned}$$

Secondly, we proof for the upper bound of *f*(*x*) and g(*x*).

$$\begin{aligned} s[(c_1\overline{f}_\alpha(x)) + (c_2\overline{g}_\alpha(x))] &= \int_0^\infty ((c_1\overline{f}_\alpha(ux)) + (c_2\overline{g}_\alpha(ux)))e^{-x}dx, \\ &= \int_0^\infty (c_1\overline{f}_\alpha(ux))e^{-x}dx + \int_0^\infty (c_2\overline{g}_\alpha(ux))e^{-x}dx, \\ &= c_1\int_0^\infty \overline{f}_\alpha(ux)e^{-x}dx + c_2\int_0^\infty \overline{g}_\alpha(ux)e^{-x}dx, \\ &= c_1 s[\overline{f}_\alpha(x)] + c_2 s[\overline{g}_\alpha(x)]. \end{aligned}$$

Finally, we conclude that:

$$\mathcal{S}[(c_1 \odot f(x)) \oplus (c_2 \odot g(x))] = (c_1 \odot \mathcal{S}[f(x)]) \oplus (c_2 \odot \mathcal{S}[g(x)]).$$

The proof is complete.

In the following theorem, we provide the first preserving theorem.

Theorem 5. *Let* $f : \mathbb{R} \rightarrow F\mathscr{F}(\mathbb{R})$ *be a continuous fuzzy-valued function and a an arbitrary constant, then:*

S[f(ax)]=G(au).

Proof. From Definition 10,

$$\begin{aligned}\mathcal{S}[f(ax)] &= [s[\underline{f}_\alpha(ax)], s[\overline{f}_\alpha(ax)]],\\ &= \left[\int_0^\infty \underline{f}_\alpha(aux)e^{-x}dx, \int_0^\infty \overline{f}_\alpha(aux)e^{-x}dx\right],\\ &= G(au).\end{aligned}$$

The theorem is proven to be true.

Next, we provide the second preserving theorem.

Theorem 6. *Let* $f : \mathbb{R} \rightarrow F\mathscr{F}(\mathbb{R})$ *be a continuous fuzzy-valued function, then:*

$$\mathcal{S}\left[x \odot \frac{df(x)}{dx}\right] = u\frac{dG(u)}{du}.$$

Proof. From the definition of FST, G(u)= $\int_0^\infty f(ux) \odot e^{-x}dx$. Then, for Case 1 in Theorem 2,

$$\begin{aligned}\frac{dG(u)}{du} &= \frac{d}{du}\left[\int_0^\infty \underline{f}_\alpha(ux)e^{-x}dx, \int_0^\infty \overline{f}_\alpha(ux)e^{-x}dx\right],\\ &= \left[\int_0^\infty \frac{d}{du}\underline{f}_\alpha(ux)e^{-x}dx, \int_0^\infty \frac{d}{du}\overline{f}_\alpha(ux)e^{-x}dx\right],\\ &= \left[\int_0^\infty xe^{-x}\frac{d\underline{f}_\alpha(ux)}{dx}dx, \int_0^\infty xe^{-x}\frac{d\overline{f}_\alpha(ux)}{dx}dx\right].\end{aligned}$$

Hence,

$$\begin{aligned}\frac{dG(u)}{du} &= \frac{1}{u}\left[\int_0^\infty (ux)\underline{f}'_\alpha(ux)e^{-x}dx, \int_0^\infty (ux)\overline{f}'_\alpha(ux)e^{-x}dx\right],\\ &= \frac{1}{u}\mathcal{S}\left[x \odot \frac{df(x)}{dx}\right].\end{aligned}$$

Multiplying both sides with u, we obtain:

$$\mathcal{S}\left[x \odot \frac{df(x)}{d(x)}\right] = u\frac{dG(u)}{du}.$$

For Case 2 from Theorem 2,

$$\frac{dG(u)}{du} = \frac{d}{du}\left[\int_0^\infty \overline{f}_\alpha(ux)e^{-x}dx, \int_0^\infty \underline{f}_\alpha(ux)e^{-x}dx\right].$$
$$= \left[\int_0^\infty \frac{d}{du}\overline{f}_\alpha(ux)e^{-x}dx, \int_0^\infty \frac{d}{du}\underline{f}_\alpha(ux)e^{-x}dx\right].$$
$$= \left[\int_0^\infty xe^{-x}\frac{d\overline{f}_\alpha(ux)}{dx}dx, \int_0^\infty xe^{-x}\frac{d\underline{f}_\alpha(ux)}{dx}dx\right].$$

Hence,

$$\frac{dG(u)}{du} = \frac{1}{u}\left[\int_0^\infty (ux)\overline{f}'_\alpha(ux)e^{-x}dx, \int_0^\infty (ux)\underline{f}'_\alpha(ux)e^{-x}dx\right],$$
$$= \frac{1}{u}\mathcal{S}\left[x \odot \frac{df(x)}{dx}\right].$$

Multiplying both sides with u, we obtain:

$$\mathcal{S}\left[x \odot \frac{df(x)}{d(x)}\right] = u\frac{dG(u)}{du}.$$

Therefore, we can conclude that:

$$\mathcal{S}\left[x \odot \frac{df(x)}{d(x)}\right] = u\frac{dG(u)}{du}$$

for both cases from Theorem 2.

Next, we provide the theorem for the first degree derivative.

Theorem 7. *Let* $f : \mathbb{R} \to F\mathscr{F}(\mathbb{R})$ *be a continuous fuzzy-valued function and f the primitive of* f' *on* $[0, \infty)$. *Then:*

$$\mathcal{S}[f'(x)] = \frac{G(u) -^H f(0)}{u} \quad \text{where } f \text{ is (1)-differentiable.}$$

or:

$$\mathcal{S}[f'(x)] = \frac{(-f(0)) -^H (-G(u))}{u} \quad \text{where } f \text{ is (2)-differentiable.}$$

Proof. First, we assume f is (1)-differentiable. Therefore,

$$\frac{G(u) -^H f(0)}{u} = \left[\frac{s[\underline{f}_\alpha(x)] - \underline{f}_\alpha(0)}{u}, \frac{s[\overline{f}_\alpha(x)] - \overline{f}_\alpha(0)}{u}\right].$$

Since:

$$s[\underline{f}'_\alpha(x)] = \frac{s[\underline{f}_\alpha(x)] - \underline{f}_\alpha(0)}{u},$$

and:

$$s[\overline{f}'_\alpha(x)] = \frac{s[\overline{f}_\alpha(x)] - \overline{f}_\alpha(0)}{u},$$

then:

$$\frac{G(u) -^H f(0)}{u} = [s[\underline{f}'_\alpha(x)], s[\overline{f}'_\alpha(x)]].$$

Since f is (1)-differentiable,

$$\frac{G(u) -^H f(0)}{u} = \mathcal{S}[f'(x)].$$

Now, we assume that f is (2)-differentiable. Therefore,

$$\frac{(-f(0)) -^H (-G(u))}{u} = \left[\frac{-(\overline{f}_\alpha(0)) - (-s[\overline{f}_\alpha(x)])}{u}, \frac{-(\underline{f}_\alpha(0)) - (-s[\underline{f}_\alpha(x)])}{u}\right],$$

equivalent to:

$$\frac{(-f(0)) -^H (-G(u))}{u} = \left[\frac{s[\overline{f}_\alpha(x)] - \overline{f}_\alpha(0)}{u}, \frac{s[\underline{f}_\alpha(x)] - \underline{f}_\alpha(0)}{u}\right].$$

Since:

$$s[\overline{f}'_\alpha(x)] = \frac{s[\overline{f}_\alpha(x)] - \overline{f}_\alpha(0)}{u},$$

and:

$$s[\underline{f}'_\alpha(x)] = \frac{s[\underline{f}_\alpha(x)] - \underline{f}_\alpha(0)}{u},$$

then:

$$\frac{(-f(0)) -^H (-G(u))}{u} = [s[\overline{f}'_\alpha(x)], s[\underline{f}'_\alpha(x)]].$$

Since f is (2)-differentiable, it follows that:

$$\frac{(-f(0)) -^{H} (-G(u))}{u} = \mathcal{S}[f'(x)].$$

The proof is now complete.

In the next theorem, we provide the first shifting theorem.

Theorem 8. *Let $f : \mathbb{R} \to F\mathcal{F}(\mathbb{R})$ be a continuous fuzzy-valued function and a an arbitrary constant, then:*

$$\mathcal{S}[e^{ax} \odot f(x)] = \frac{1}{1-au} G\left(\frac{u}{1-au}\right).$$

Proof. From Definition 10,

$$\mathcal{S}[e^{ax} \odot f(x)] = \left[\int_0^\infty \underline{f}_\alpha(ux)e^{-(1-au)x}dx, \int_0^\infty \overline{f}_\alpha(ux)e^{-(1-au)x}dx\right].$$

By using substitution $w = (1 - au)x$, we then obtain:

$$\begin{aligned}\mathcal{S}[e^{ax} \odot f(x)] &= \left[\frac{1}{1-au}\int_0^\infty \underline{f}_\alpha\left(\frac{uw}{1-au}\right)e^{-w}dw, \frac{1}{1-au}\int_0^\infty \overline{f}_\alpha\left(\frac{uw}{1-au}\right)e^{-w}dw\right],\\ &= \frac{1}{1-au}\int_0^\infty f\left(\frac{uw}{1-au}\right)e^{-w}dw,\\ &= \frac{1}{1-au}G\left(\frac{u}{1-au}\right).\end{aligned}$$

The convolution theorem is provided below.

Theorem 9. *Let $f, g : \mathbb{R} \to F\mathcal{F}(\mathbb{R})$ be two continuous fuzzy-valued functions. Let $F(p)$ and $G(p)$ be fuzzy Laplace transforms, and let $M(u)$ and $N(u)$ be fuzzy Sumudu transforms for f and g, respectively. Then, the Sumudu transform of the convolution of f and g,*

$$(f * g)(x) = \int_0^t f(\tau)g(t-\tau)d\tau,$$

is given by:

$$\mathcal{S}[(f * g)(x)] = uM(u)N(u).$$

Proof. The FLT for $(f * g)$ as in [55] is given by:

$$\mathcal{L}[(\underline{f}_\alpha * \underline{g}_\alpha)(x), (\overline{f}_\alpha * \overline{g}_\alpha)(x)] = F(p)G(p).$$

By the fuzzy Laplace–Sumudu duality relation,

$$\mathcal{S}[(\underline{f}_\alpha * \underline{g}_\alpha)(x), (\overline{f}_\alpha * \overline{g}_\alpha)(x)] = \frac{1}{u}\mathcal{L}[(\underline{f}_\alpha * \underline{g}_\alpha)(x), (\overline{f}_\alpha * \overline{g}_\alpha)(x)],$$

and since:

$$M(u) = \frac{F(1/u)}{u}, \qquad N(u) = \frac{G(1/u)}{u},$$

the FST for $(f * g)$ is as follows:

$$\begin{aligned} \mathcal{S}[(\underline{f}_\alpha * \underline{g}_\alpha)(x), (\overline{f}_\alpha * \overline{g}_\alpha)(x)] &= \frac{F(1/u)G(1/u)}{u}, \\ &= u\frac{F(1/u)}{u}\frac{G(1/u)}{u}, \\ &= uM(u)N(u). \end{aligned}$$

PROCEDURE FOR SOLVING FUZZY DIFFERENTIAL EQUATIONS

We consider a crisp differential equation given by:

$$y'(t) = f(t, y(t)), \quad y(t_0) = y_0, \tag{5}$$

where f: $[t_0, T] \times \mathbb{R} \to \mathbb{R}$. Suppose that the initial value in Equation (5) is not precisely known and modeled with a fuzzy number, we have the following fuzzy initial value problem [56]:

$$\begin{cases} Y'(t) = f(t, Y(t)), & 0 \le t \le T, \\ Y(t_0) = [\underline{y}_\alpha(0), \overline{y}_\alpha(0)], & 0 < \alpha \leqslant 1, \end{cases} \tag{6}$$

where f : $[t_0, T] \times F\mathscr{F}(\mathbb{R}) \to F\mathscr{F}(\mathbb{R})$ is a continuous fuzzy mapping. By referring to Kaleva [57], we observe that Theorem 2 provides a procedure to solve the Equation (6). As a matter of fact, $Y_\alpha(t) = [\underline{y}_\alpha(t), \overline{y}_\alpha(t)]$.

By using FST on Equation (6), we have the following equation.

$$\mathcal{S}[Y'(t)] = \mathcal{S}[f(t, Y(t))]. \tag{7}$$

Case 1: If we consider $Y'(t)$ by using a (1)-differentiable function, then from Theorem 2, we get $Y'(t) = [\underline{y}'_\alpha(t), \overline{y}'_\alpha(t)]$. Now, we obtain the following FDE to be solved.

$$\begin{cases} \underline{y}'_\alpha(t) = \underline{f}_\alpha(t, Y(t)), & \underline{y}_\alpha(t_0) = \underline{y}_\alpha(0), \\ \overline{y}'_\alpha(t) = \overline{f}_\alpha(t, Y(t)), & \overline{y}_\alpha(t_0) = \overline{y}_\alpha(0). \end{cases} \tag{8}$$

From Theorem 7, for Case 1,

$$\mathcal{S}[Y'(t)] = \frac{\mathcal{S}[Y(t)] -^{H} Y(t_0)}{u}.$$

Therefore,

$$\begin{cases} s[\underline{f}_{\alpha}(t, Y(t))] = \dfrac{s[\underline{y}_{\alpha}(t)] - \underline{y}_{\alpha}(0)}{u}, \\ \\ s[\overline{f}_{\alpha}(t, Y(t))] = \dfrac{s[\overline{y}_{\alpha}(t)] - \overline{y}_{\alpha}(0)}{u}. \end{cases} \quad (9)$$

To solve Equation (9), first we assume that:

$s[\underline{y}_{\alpha}(t)] = L_{\alpha}^{1}(u),$

$s[\overline{y}_{\alpha}(t)] = U_{\alpha}^{1}(u),$

where $L_{\alpha}^{1}(u)$ and $U_{\alpha}^{1}(u)$ are solutions of Equation (9). By using the inverse of FST, we then compute $\underline{y}_{\alpha}(t)$ and $\overline{y}_{\alpha}(t)$ as follows.

$\underline{y}_{\alpha}(t) = s^{-1}[L_{\alpha}^{1}(u)],$

$\overline{y}_{\alpha}(t) = s^{-1}[U_{\alpha}^{1}(u)].$

Case 2: If we consider *Y'*(*t*) by using a (2)-differentiable function, then from Theorem 2, we get $Y'(t) = [\overline{y}'_{\alpha}(t), \underline{y}'_{\alpha}(t)]$. Now, we obtain the following FDE to be solved.

$$\begin{cases} \underline{y}'_{\alpha}(t) = \overline{f}_{\alpha}(t, Y(t)), & \underline{y}_{\alpha}(t_0) = \underline{y}_{\alpha}(0), \\ \overline{y}'_{\alpha}(t) = \underline{f}_{\alpha}(t, Y(t)), & \overline{y}_{\alpha}(t_0) = \overline{y}_{\alpha}(0). \end{cases} \quad (10)$$

From Theorem 7, for Case 2,

$$\mathcal{S}[Y'(t)] = \frac{-(Y(t_0)) -^{H} (-\mathcal{S}[Y(t)])}{u}.$$

Therefore,

$$\begin{cases} s[\underline{f}_{\alpha}(t, Y(t))] = \dfrac{s[\underline{y}_{\alpha}(t)] - \underline{y}_{\alpha}(0)}{u}, \\ \\ s[\overline{f}_{\alpha}(t, Y(t))] = \dfrac{s[\overline{y}_{\alpha}(t)] - \overline{y}_{\alpha}(0)}{u}. \end{cases} \quad (11)$$

To solve Equation (11), first we assume that:

$s[\underline{y}_\alpha(t)] = L_\alpha^2(u),$
$s[\overline{y}_\alpha(t)] = U_\alpha^2(u),$

where $L_\alpha^2(u)$ and $U_\alpha^2(u)$ are solutions of Equation (11). By using the inverse of the FST, we then compute $\underline{y}_\alpha(t)$ and $\overline{y}_\alpha(t)$ as follows.

$\underline{y}_\alpha(t) = s^{-1}[L_\alpha^2(u)],$

$\overline{y}_\alpha(t) = s^{-1}[U_\alpha^2(u)].$

A NUMERICAL EXAMPLE

In this section, we provide a numerical example of solving an FDE using the FST.

Example 1. *Consider the following initial value problem:*

$$\begin{cases} Y'(t) = -Y(t), & 0 \leq t \leq T, \\ Y(t_0) = [\underline{y}_\alpha(0), \overline{y}_\alpha(0)], & 0 < \alpha \leqslant 1. \end{cases} \tag{12}$$

By using FST, we have:

$$\mathcal{S}[Y'(t)] = \mathcal{S}[-Y(t)]$$

and:

$$\mathcal{S}[Y'(t)] = \int_0^\infty Y'(ut) \odot e^{-t} dt.$$

First, we consider the condition where $Y'(t)$ is (1)-differentiable. Therefore, from Theorem 2,

$$\begin{cases} \underline{y}'_\alpha(t) = -\underline{y}_\alpha(t), & \underline{y}_\alpha(t_0) = \underline{y}_\alpha(0), \\ \overline{y}'_\alpha(t) = -\overline{y}_\alpha(t), & \overline{y}_\alpha(t_0) = \overline{y}_\alpha(0). \end{cases}$$

By Theorem 7,

$$\mathcal{S}[Y'(t)] = \frac{\mathcal{S}[Y(t)] -^H Y(t_0)}{u}.$$

Therefore,

$$\mathcal{S}[-Y(t)] = \frac{\mathcal{S}[Y(t)] -^{H} Y(t_0)}{u},$$

$$-\mathcal{S}[Y(t)] = \frac{\mathcal{S}[Y(t)] -^{H} Y(t_0)}{u}.$$

From Equation (9), we have:

$$-s[\overline{y}_\alpha(t)] = \frac{s[\underline{y}_\alpha(t)] - (\underline{y}_\alpha(0))}{u},$$

and:

$$-s[\underline{y}_\alpha(t)] = \frac{s[\overline{y}_\alpha(t)] - (\overline{y}_\alpha(0))}{u}.$$

Then, we obtain

$$s[\underline{y}_\alpha(t)] = \underline{y}_\alpha(0)\left(\frac{1}{1-u^2}\right) - \overline{y}_\alpha(0)\left(\frac{u}{1-u^2}\right),$$

and:

$$s[\overline{y}_\alpha(t)] = \overline{y}_\alpha(0)\left(\frac{1}{1-u^2}\right) - \underline{y}_\alpha(0)\left(\frac{u}{1-u^2}\right).$$

Thus,

$$\underline{y}_\alpha(t) = \underline{y}_\alpha(0)s^{-1}\left(\frac{1}{1-u^2}\right) - \overline{y}_\alpha(0)s^{-1}\left(\frac{u}{1-u^2}\right),$$

and:

$$\overline{y}_\alpha(t) = \overline{y}_\alpha(0)s^{-1}\left(\frac{1}{1-u^2}\right) - \underline{y}_\alpha(0)s^{-1}\left(\frac{u}{1-u^2}\right).$$

Therefore,

$$\underline{y}_\alpha(t) = e^{t}\left(\frac{\underline{y}_\alpha(0) - \overline{y}_\alpha(0)}{2}\right) + e^{-t}\left(\frac{\underline{y}_\alpha(0) + \overline{y}_\alpha(0)}{2}\right),$$

and:

$$\overline{y}_\alpha(t) = e^{t}\left(\frac{\overline{y}_\alpha(0) - \underline{y}_\alpha(0)}{2}\right) + e^{-t}\left(\frac{\overline{y}_\alpha(0) + \underline{y}_\alpha(0)}{2}\right).$$

Next, we consider the condition where $Y'(t)$ is (2)-differentiable. Therefore, from Theorem 2,

$$\begin{cases} \underline{y}'_{\alpha}(t) = -\overline{y}_{\alpha}(t), & \underline{y}_{\alpha}(t_0) = \underline{y}_{\alpha}(0), \\ \overline{y}'_{\alpha}(t) = -\underline{y}_{\alpha}(t), & \overline{y}_{\alpha}(t_0) = \overline{y}_{\alpha}(0). \end{cases}$$

By Theorem 7,

$$\mathcal{S}[Y'(t)] = \frac{-(Y(t_0)) -^{H} (-\mathcal{S}[Y(t)])}{u}.$$

Therefore,

$$\mathcal{S}[-Y(t)] = \frac{-(Y(t_0)) -^{H} (-\mathcal{S}[Y(t)])}{u},$$

$$-\mathcal{S}[Y(t)] = \frac{-(Y(t_0)) -^{H} (-\mathcal{S}[Y(t)])}{u}.$$

From Equation (11), we have:

$$-s[\underline{y}_{\alpha}(t)] = \frac{s[\underline{y}_{\alpha}(t)] - (\underline{y}_{\alpha}(0))}{u},$$

and:

$$-s[\overline{y}_{\alpha}(t)] = \frac{s[\overline{y}_{\alpha}(t)] - (\overline{y}_{\alpha}(0))}{u}.$$

Then, we obtain:

$$s[\underline{y}_{\alpha}(t)] = \underline{y}_{\alpha}(0)\left(\frac{1}{1+u}\right);$$

and:

$$s[\overline{y}_{\alpha}(t)] = \overline{y}_{\alpha}(0)\left(\frac{1}{1+u}\right).$$

Therefore,

$$\underline{y}_{\alpha}(t) = \underline{y}_{\alpha}(0)e^{-t},$$

and:

$$\overline{y}_{\alpha}(t) = \overline{y}_{\alpha}(0)e^{-t}.$$

Assume that $[\underline{y}_{\alpha}(t_0), \overline{y}_{\alpha}(t_0)]$=[α−a,a−α]. *Then, the solutions of Equation (12) for Case (1) and Case (2) are as follows.*

Case 1:

$$\underline{y}_{\alpha}(t) = (\alpha - a)e^{t},$$
$$\overline{y}_{\alpha}(t) = (a - \alpha)e^{t}.$$

Case 2:

$$\underline{y}_{\alpha}(t) = (\alpha - a)e^{-t},$$
$$\overline{y}_{\alpha}(t) = (a - \alpha)e^{-t}.$$

The results obtained using the FST for both cases proposed in this paper are shown in Figures 1 and 2, respectively. We can see that for Case 1, the result diverges as *t* increases; while for Case 2, the result implicates that the solution converges as *t* increases.

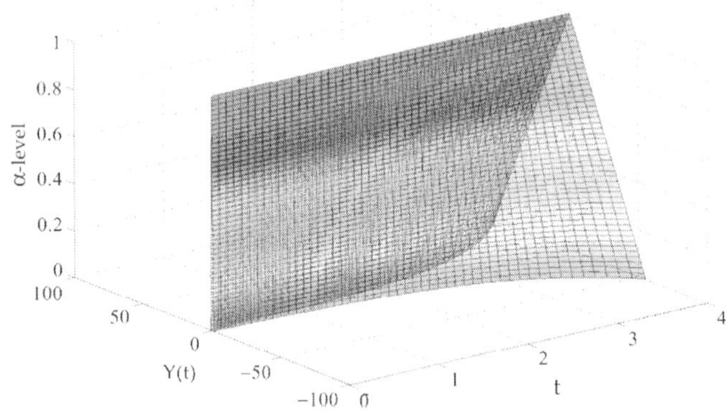

Figure 1. The solution of Equation (12) for Case 1 when $Y(t_0) = (-1, 0, 1)$.

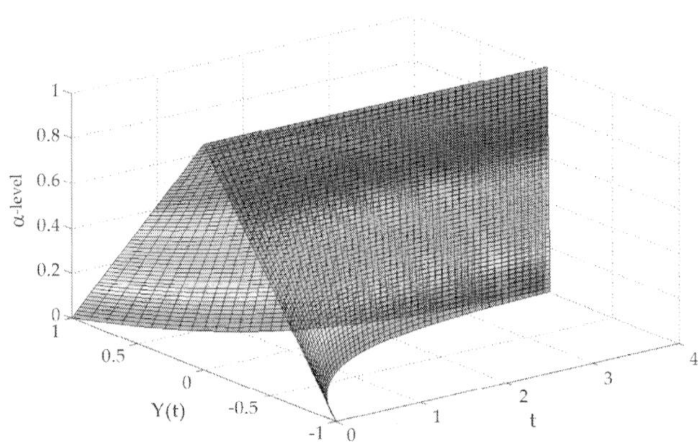

Figure 2. The solution of Equation (12) for Case 2 when $Y(t_0) = (-1, 0, 1)$.

Remark. *From the example, we notice that the solutions depend on the differential equation we chose. For Case 1, the solution has the property that the diameter, i.e. diam(supp $y(t)$) = $2ae^{t}$, which is unbounded as t approaches infinity. Comparing to Case 2, the diam(supp $y(t)$) = $2ae^{-t} \to 0$ as t approaches infinity, which leads to much more intuitive results.*

CONCLUSIONS

In this paper, we have studied the classical Sumudu transform in the fuzzy setting. We have also proposed detailed procedures to solve FDEs. In the last part, we have conducted a numerical example of solving first order linear FDE using the FST.

ACKNOWLEDGMENTS

This research was supported by the Fundamental Research Grant Scheme of Ministry of Science, Technology and Innovation, Malaysia, under Project Code 9003-00417.

AUTHOR CONTRIBUTIONS

Both authors have read and agreed on the submission of this paper. Furthermore, both authors have contributed to every part of this paper. Both authors have read and approved the final manuscript.

REFERENCES

1. Davis, J.A.; McNamara, D.E.; Cottrell, D.M.; Campos, J. Image processing with the radial Hilbert transform: Theory and experiments. *Opt. Lett.* **2000**, *25*, 99–101.
2. Namias, V. The fractional order Fourier transform and its application to quantum mechanics. *IMA J. Appl. Math.* **1980**, *25*, 241–265.
3. Saitoh, S. The Weierstrass transform and an isometry in the heat equation. *Appl. Anal.* **1983**, *16*, 1–6.
4. Ghaemi, F.; Yunus, R.; Ahmadian, A.; Salahshour, S.; Suleiman, M.; Saleh, S.F. Application of Fuzzy Fractional Kinetic Equations to Modelling of the Acid Hydrolysis Reaction. *Abstr. Appl. Anal.* **2013**, *2013*, 610314.
5. Spinelli, R. Numerical inversion of a Laplace transform. *SIAM J. Numer. Anal.***1966**, *3*, 636–649.
6. Layman, J.W. The Hankel transform and some of its properties. *J. Integer Seq.***2001**, *4*, 1–11.
7. Tranter, C. The use of the Mellin transform in finding the stress distribution in an infinite wedge. *Q. J. Mech. Appl. Math.* **1948**, *1*, 125–130.
8. Watugala, G.K. Sumudu transforms—A new integral transform to solve differential equations and control engineering problems. *Int. J. Math. Educ. Sci. Technol.* **1993**, *24*, 35–43.
9. Watugala, G.K. Sumudu transforms-a new integral transform to solve differential equations and control engineering problems. *Math. Eng. Ind.* **1998**, *6*, 319–329.
10. Weerakoon, S. Application of Sumudu transform to partial differential equations. *Int. J. Math. Educ. Sci. Technol.* **1994**, *25*, 277–283.
11. Weerakoon, S. Complex inversion formula for Sumudu transform. *Int. J. Math. Educ. Sci. Technol.* **1998**, *29*, 618–621.
12. Asiru, M.A. Sumudu transform and the solution of integral equations of convolution type. *Int. J. Math. Educ. Sci. Technol.* **2001**, *32*, 906–910.
13. Belgacem, F.B.M.; Karaballi, A.A.; Kalla, S.L. Analytical investigations of the Sumudu transform and applications to integral production equations. *Math. Probl. Eng.* **2003**, *2003*, 103–118.

14. Belgacem, F.B.M.; Karaballi, A.A. Sumudu transform fundamental properties investigations and applications. *Int. J. Stoch. Anal.* **2006**, *2006*.
15. Belgacem, F.B.M. Sumudu transform applications to Bessel functions and equations. *Appl. Math. Sci.* **2010**, *4*, 3665–3686.
16. Agwa, H.A.; Ali, F.M.; Kılıçman, A. A new integral transform on time scales and its application. *Adv. Differ. Equ.* **2012**, *2012*.
17. Kılıçman, A.; Eltayeb, H.; Ismail, M.R. A note on integral transforms and differential equations. *Malays. J. Math. Sci.* **2012**, *6*, 1–18.
18. Asiru, M.A. Further properties of the Sumudu transform and its application. *Int. J. Math. Educ. Sci. Technol.* **2002**, *33*, 441–449.
19. Eltayeb, H.; Kılıçman, A. A note on the Sumudu transforms and differential equations. *Appl. Math. Sci.* **2010**, *4*, 1089–1098.
20. Belgacem, F.B.M. Introducing and analyzing deeper Sumudu properties. *Nonlinear Stud.* **2006**, *13*, 23–41.
21. Rathore, S.; Kumar, D.; Singh, J.; Gupta, S. Homotopy analysis Sumudu transform method for nonlinear equations. *Int. J. Ind. Math.* **2012**, *4*, 301–314.
22. Zadeh, L.A. Fuzzy sets. *Inf. Control.* **1965**, *8*, 338–353.
23. Chang, S.S.L.; Zadeh, L.A. On fuzzy mapping and control. *IEEE Trans. Syst. Man Cybern.* **1972**, *SMC-2*, 30–34.
24. Dubois, D.; Prade, H. Towards fuzzy differential calculus part 1: Integration of fuzzy mappings. *Fuzzy Sets Syst.* **1982**, *8*, 1–17.
25. Dubois, D.; Prade, H. Towards fuzzy differential calculus part 2: Integration on fuzzy intervals. *Fuzzy Sets Syst.* **1982**, *8*, 105–116.
26. Dubois, D.; Prade, H. Towards fuzzy differential calculus part 3: Differentiation. *Fuzzy Sets Syst.* **1982**, *8*, 225–233.
27. Puri, M.L.; Ralescu, D.A. Differentials of fuzzy functions. *J. Math. Anal. Appl.***1983**, *91*, 552–558.
28. Goetschel, R., Jr.; Voxman, W. Elementary fuzzy calculus. *Fuzzy Sets Syst.***1986**, *18*, 31–43.
29. Kaleva, O. Fuzzy differential equations. *Fuzzy Sets Syst.* **1987**, *24*, 301–317.
30. Ding, Z.; Ma, M.; Kandel, A. Existence of the solutions of fuzzy differential equations with parameters. *Inf. Sci.* **1997**, *99*, 205–217.

31. Salahshour, S. Nth-order fuzzy differential equations under generalized differentiability. *J. Fuzzy Set Valued Anal.* **2011**, *2011*.
32. Shahriyar, M.; Ismail, F.; Aghabeigi, S.; Ahmadian, A.; Salahshour, S. An eigenvalue-eigenvector method for solving a system of fractional differential equations with uncertainty. *Math. Probl. Eng.* **2013**, *2013*, 579761.
33. Arshad, S.; Lupulescu, V. On the fractional differential equations with uncertainty. *Nonlinear Anal. Theory Methods Appl.* **2011**, *74*, 3685–3693.
34. Allahviranloo, T.; Salahshour, S.; Abbasbandy, S. Explicit solutions of fractional differential equations with uncertainty. *Soft Comput.* **2012**, *16*, 297–302.
35. Allahviranloo, T.; Ahmadi, M.B. Fuzzy Laplace transforms. *Soft Comput.* **2010**, *14*, 235–243.
36. Salahshour, S.; Allahviranloo, T. Applications of fuzzy Laplace transforms. *Soft Comput.* **2013**, *17*, 145–158.
37. Jafarian, A.; Golmankhaneh, A.K.; Baleanu, D. On fuzzy fractional Laplace transformation. *Adv. Math. Phys.* **2014**, *2014*.
38. Salahshour, S.; Allahviranloo, T.; Abbasbandy, S. Solving fuzzy fractional differential equations by fuzzy Laplace transforms. *Commun. Nonlinear Sci. Numer. Simul.* **2012**, *17*, 1372–1381.
39. Salahshour, S.; Khezerloo, M.; Hajighasemi, S.; Khorasany, M. Solving fuzzy integral equations of the second kind by fuzzy Laplace transform method. *Int. J. Ind. Math.* **2012**, *4*, 21–29.
40. Muhammad Ali, H.F.; Haydar, A.K. On fuzzy Laplace transforms for fuzzy differential equations of the third order. *J. Kerbala Univ.* **2013**, *11*, 251–256.
41. Salahshour, S.; Haghi, E. Solving fuzzy heat equation by fuzzy Laplace transforms, Information Processing and Management of Uncertainty in Knowledge-Based Systems; Springer: Berlin/Heidelberg, Germany, 2010; pp. 512–521.
42. Ahmad, M.Z.; Abdul Rahman, N.A. Explicit solution of fuzzy differential equations by mean of fuzzy Sumudu transform. *Int. J. Appl. Phys. Math.* **2015**, *5*, 86–93.
43. Alam Khan, N.; Abdul Razzaq, O.; Ayyaz, M. On the solution of fuzzy differential equations by fuzzy Sumudu transform. *Nonlinear Eng.* **2015**, *4*, 49–60.

44. Xu, J.; Liao, Z.; Hu, Z. A class of linear differential dynamical systems with fuzzy initial condition. *Fuzzy Sets Syst.* **2007**, *158*, 2339–2358.
45. Zimmerman, H.J. *Fuzzy Set Theory and Its Applications*; Kluwer Academic Publisher and Dordrecht: Dordrecht, The Netherlands, 1991.
46. Friedman, M.; Ma, M.; Kandel, A. Numerical solutions of fuzzy differential and integral equations. *Fuzzy Sets Syst.* **1999**, *106*, 35–48.
47. Ma, M.; Friedman, M.; Kandel, A. Numerical solutions of fuzzy differential equations. *Fuzzy Sets Syst.* **1999**, *105*, 133–138.
48. Ahmad, M.Z.; de Baets, B. A Predator-Prey Model with Fuzzy Initial Populations. Proceedings of the Joint 2009 International Fuzzy Systems Association World Congress and 2009 European Society of Fuzzy Logic and Technology Conference, Lisbon, Portugal, 20–24 July 2009; pp. 1311–1314.
49. Dubois, D.J. *Fuzzy Sets and Systems: Theory and Applications*; Academic Press: Waltham, MA, USA, 1980.
50. Wu, H.C. The improper fuzzy Riemann integral and its numerical integration. *Inf. Sci.* **1998**, *111*, 109–137.
51. Wu, H.C. The fuzzy Riemann integral and its numerical integration. *Fuzzy Sets Syst.* **2000**, *110*, 1–25.
52. Bede, B.; Gal, S.G. Generalizations of the differentiability of fuzzy-number-valued functions with applications to fuzzy differential equations. *Fuzzy Sets Syst.* **2005**, *151*, 581–599.
53. Bede, B.; Rudas, I.J.; Bencsik, A.L. First order linear fuzzy differential equations under generalized differentiability. *Inf. Sci.* **2007**, *177*, 1648–1662.
54. Chalco-Cano, Y.; Román-Flores, H. On new solutions of fuzzy differential equations. *Chaos Solitons Fractals.* **2008**, *38*, 112–119.
55. Das, M.; Talukdar, D. Method For Solving Fuzzy Integro-Differential Equation By Using Fuzzy Laplace Transformation. *Int. J. Sci. Technol. Res.* **2014**, *3*, 291–295.
56. Chalco-Cano, Y.; Román-Flores, H. Comparation between some approaches to solve fuzzy differential equations. *Fuzzy Sets Syst.* **2009**, *160*, 1517–1527.
57. Kaleva, O. A note on fuzzy differential equations. *Nonlinear Anal. Theory Methods Appl.* **2006**, *64*, 895–900.

CHAPTER 8

THE MEHLER-FOCK TRANSFORM IN SIGNAL PROCESSING

Reiner Lenz [1,2]

[1]Department of Electrical Engineering, Linköping University, SE-581 83 Linköping, Sweden

[2]Department of Science and Technology, Linköping University, SE-60174 Norrköping, Sweden

ABSTRACT

Many signals can be described as functions on the unit disk (ball). In the framework of group representations it is well-known how to construct Hilbert-spaces containing these functions that have the groups SU(1,N) as their symmetry groups. One illustration of this construction is three-dimensional color spaces in which chroma properties are described by points on the unit disk. A combination of principal component analysis and the Perron-Frobenius theorem can be used to show that perspective projections

CITATION (APA): Lenz, R. (2017). The Mehler-Fock Transform in Signal Processing. Entropy, 19(6), 289. (12 pages)

map positive signals (i.e., functions with positive values) to a product of the positive half-axis and the unit ball. The representation theory (harmonic analysis) of the group SU(1,1) leads to an integral transform, the Mehler-Fock-transform (MFT), that decomposes functions, depending on the radial coordinate only, into combinations of associated Legendre functions. This transformation is applied to kernel density estimators of probability distributions on the unit disk. It is shown that the transform separates the influence of the data and the measured data. The application of the transform is illustrated by studying the statistical distribution of RGB vectors obtained from a common set of object points under different illuminants.

Keywords: Mehler-Fock transform, kernel density estimator, signal processing, harmonic analysis, SU(1,1), positive signals

INTRODUCTION

Fitting statistical models to empirical data is one of the most fundamental techniques in signal processing. Usually, one distinguishes between parametric and non-parametric models. Parametric models require, by definition, information about the type of the underlying statistical distribution. Non-parametric methods are more data-driven. Two important classes of non-parametric techniques are histograms and kernel-density estimators. Histograms are computationally very efficient but the specification of the bins used can be very difficult. Kernel density estimators, on the other hand, require only the definition of the kernel. One of the drawbacks of a straight-forward implementation of a kernel density estimator is the fact that a new data point to be included into the analysis can interact with a large number of already analysed data points.

In cases where the underlying domain has a group theoretical structure methods from harmonic analysis can be used to separate the contributions from the kernel and the data. In the following this will be illustrated in the case where the domain on which the data is defined is the unit disk and where the symmetry group of the data is the group SU(1,1) (to be introduced later). In the harmonic analysis framework the simplest groups are the commutative groups which lead to ordinary Fourier transforms. Next come the non-commutative but compact groups like the rotation groups which are related to finite-dimensional invariant subspaces and corresponding series expansions (like the spherical harmonics). The group SU(1,1) is the simplest example of a non-commutative and non-compact group and the corresponding Fourier transform is an integral transform, the Mehler-Fock transform to be described in the following sections.

The group SU(1,1) is well-known in mathematics and theoretical physics but there are only a few applications in signal processing. One reason for this is certainly that the usage of the unit disk domain is not as obvious as those of other domains. The natural domain for time-sequences is the real line with the shift-operators as symmetry group. The plane or three-dimensional space are domains for spatial processes with spatial shifts of the origin of the space as group elements. The rotation group is related to unit spheres and orientations. In the next section it will be shown that stochastic processes with positive values only have a natural conical structure with SU(1,1) as the natural symmetry group. Then the main results regarding the Mehler-Fock transform will be presented and a few properties of the transform will be illustrated with the help of an application from color signal processing.

POSITIVE SIGNALS

The raw output values of many measurement processes are non-negative numbers. An extreme example are binary measurements indicating the absence/presence of a response. This is typical for biological and binary technical systems. Other common examples are size, weight, age/duration or the photon counts of a camera. For simplicity only the case of strictly positive measurements will be considered. This is natural in many cases. In imaging science, for example, measurements are often representing counts such as the number of interactions of photons with the sensor in a given time interval. Sensors are never perfect and the cases where zero counts are registered should be very rare. The measurement values are therefore almost always positive numbers.

In the following functions (signals) that are outcomes of stochastic processes are considered. It is assumed that they are elements of a vector (or Hilbert) space and that they assume positive values only. In this space one can use principal component analysis to introduce a basis where the first eigenvector has only positive components. This follows from the Perron-Frobenius (Krein-Milman) theorem. In the following only processes will be considered in which the first eigenvector is (approximately) proportional to the mean vector. It is also assumed that all relevant signals lie in a conical subset which is a product of the positive half-axis and the n-dimensional solid unit ball. In the following only the case R+×D where D denotes the unit disk is considered. Generalizations to higher dimensions are straightforward.

This is formulated in the following definition:

Definition 1. *A positive stochastic process is a function* $f(\omega,x)>0$ *where* ω *is the stochastic variable and* $x \in X$ *denotes a set on which the function is defined. For every* ω *the function* $f(\omega,x)$ *is an element in a Hilbert space with scalar product* $\langle,\rangle$.

Typical examples of X are:

- a set of points xk on the plane which describe the location of the k–th sensor element (pixel)
- a set of points xk in a time-interval denoting the time when the measurement was obtained
- a set of indices xk describing the spectral sensitivity of color sensor number *k*.

For a positive stochastic process *f* the expectation and the correlation function are defined as usual:

Definition 2. *The expectation is defined as* $E_f(x)=\int f(\omega,x)d\omega$.

The correlation function is defined as $C_f(x_1,x_2)=\int f(\omega,x_1)f(\omega,x_2)d\omega$.

The correlation function Cf(x1,x2) *defines an integral operator* If *defined as:*

$$\left(I_f h\right)(\mathbf{x}_2) = \int_X C_f(\mathbf{x}_1,\mathbf{x}_2)h(\mathbf{x}_1)\ d\mathbf{x}_1$$

The correlation function $C_f(x_1,x_2)$ is symmetric, i.e., $C_f(x_1,x_2)=C_f(x_2,x_1)$ and for all positive stochastic processes the expectation and the correlation function have positive values everywhere. For a finite set X the expectation is a vector and the correlation function is a symmetric positive matrix (having only positive elements) which is also positive semi-definite.

The operator I_f is a positive and positive-definite operator and the Krein-Rutman theorem (Perron-Frobenius theorem in the case of matrices) shows that the first eigenfunction of the operator is a positive function. All eigenvalues are non-negative. Denote the eigenfunctions of If (sorted by decreasing eigenvalues) as $b_0,b_1,\ldots$. Then $\langle f,b_0\rangle>0$.

The following example, discussing a model of an RGB camera, should give an idea of a situation in which the unit disk model arises in signal processing. A more general and detailed application is discussed later when the properties of the Mehler-Fock transform are illustrated. In a first approximation it is assumed that such a camera acts as a linear system and that the measured RGB values are linearly related to the incoming light. In practice this implies using the raw data from the camera sensor, excluding processing steps like white-balancing and gamma mapping. Also excluded are, in this first step, pixels outside the linear response region of the camera avoiding critical effects such as a saturation of the sensor. This will be discussed later in connection with the experiments. Under these conditions the doubling of the exposure time will thus result in a multiplication of the RGB vector by a factor of two.

Under this condition the group R^+ of positive real numbers acts on an RGB vector f(x) as $(c,f(x)) \mapsto cf(x)$. The elements of the group R+ are *intensity transformations* and $\|f(x)\|_1$ is defined as the *intensity* of the RGB vector f(x). The result *z* of the projection $f(x) \mapsto z = f(x)/\|f(x)\|_1$ defines the *chroma* of the RGB vector x.

The main observation now is the fact that chroma vectors are located on a disk with a finite radius. After a scaling operation it can therefore be assumed that chroma vectors are all located on the unit disk. An investigation of the general case of multispectral color distributions can be found in [1,2] and will be studied in Section 5.

THE UNIT DISK AND THE GROUP SU(1,1)

Points on the unit disk are in the following described by complex variables z. The group SU(1,1) is defined as follows:

Definition 3. *The group* SU(1,1) *consists of all* 2×2 *matrices with complex elements:*

$$\mathrm{SU}(1,1) = \{\mathbf{M} = \begin{pmatrix} a & b \\ \overline{b} & \overline{a} \end{pmatrix}, a, b \in \mathbb{C}, |a|^2 - |b|^2 = 1\} \tag{1}$$

The group operation is the usual matrix multiplication.

The group acts as a transformation group on the open unit disk D *(consisting of all points* $z \in \mathbb{C}$ *with* $|z|<1$*) as the Möbius transforms:*

$$(\mathbf{M}, z) = \left(\begin{pmatrix} a & b \\ \overline{b} & \overline{a} \end{pmatrix}, z\right) \mapsto \mathbf{M}\langle z \rangle = \frac{az+b}{\overline{b}z+\overline{a}} \tag{2}$$

with $(M_1M_2)\langle z\rangle = M_1\langle M_2\langle z\rangle\rangle$ *for all matrices and all points.*

In the following the notation M is used when the group elements are the matrices. In situations where they represent elements of the abstract group or if the parametrization of the Möbius transforms is essential the symbol g is used. This will become clearer when the general transform is decomposed into a product of three special transforms. This corresponds to the well-known property of the group of ordinary three-dimensional rotations that every rotation can be written as a product of three rotations around the coordinate axes. A similar decomposition holds also for the group SU(1,1). In this case these three parameters are denoted by ψ,τ,φ and the decomposition is

$$g(\varphi,\tau,\psi) = g(\varphi,0,0)g(0,\tau,0)g(0,0,\psi) = \begin{pmatrix} \cosh\frac{\tau}{2}e^{i(\varphi+\psi)/2} & \sinh\frac{\tau}{2}e^{i(\varphi-\psi)/2} \\ \sinh\frac{\tau}{2}e^{-i(\varphi-\psi)/2} & \cosh\frac{\tau}{2}e^{-i(\varphi+\psi)/2} \end{pmatrix} \quad (3)$$

This motivates the introduction of two special types of transformations/ matrices defined as:

$$K = \left\{g(\varphi,0,0) = \begin{pmatrix} e^{i\varphi/2} & 0 \\ 0 & e^{-i\varphi/2} \end{pmatrix} : -2\pi \le \varphi < 2\pi\right\} \quad (4)$$

$$A = \left\{g(0,\tau,0) = \begin{pmatrix} \cosh\frac{\tau}{2} & \sinh\frac{\tau}{2} \\ \sinh\frac{\tau}{2} & \cosh\frac{\tau}{2} \end{pmatrix} : \tau \in \mathbb{R}\right\}. \quad (5)$$

This shows that $g(\varphi,0,0), g(0,0,\psi) \in K$ and $g(0,\tau,0) \in A$. This decomposition is known as Cartan's KAK decomposition of the group and ψ,τ,φ are known as the KAK-coordinates. The following formula for the inverse: $g(\varphi,\tau,\psi)^{-1} = g(-\psi,-\tau,-\varphi)$ follows directly by using the product decomposition and observing that the inverse of a transformation in *K* and *A* is given by the transformation with the negative parameter value.

From the definition follows that matrices in *K* are rotations that leave the origin fixed $0 = g(\varphi,0,0)\langle 0\rangle = g(0,0,\psi)\langle 0\rangle$. For a general element in SU(1,1) the following parametrization of the unit disk follows directly from the KAK decomposition:

$$g(\varphi,\tau,\psi)\langle 0\rangle = g(\varphi,0,0)g(0,\tau,0)g(0,0,\psi)\langle 0\rangle = \tanh\frac{\tau}{2}e^{i\varphi}.$$

Two transformations are defined as equivalent if their first two parameters are identical. From the previous calculations it can be seen that all equivalent transformations map the origin to the same point on the unit disk which defines a correspondence between the points on the unit disk and the equivalence classes of transformations. This is expressed as D=SU(1,1)/K.

It also follows that functions on the unit disk are functions on the group that are independent of the last argument of the KAK decomposition.

KERNEL DENSITY ESTIMATORS AND MEHLER-FOCK TRANSFORM

Both interpretations are useful in the motivation of the the basic idea behind kernel-density estimators on the unit disk. Consider first two points $w,z \in D$. Here $z$denotes the measurement as the actual outcome of a stochastic process. Since it is a stochastic process it is also possible that the outcome could have been a neighboring point w. This leads to the definition of the kernel function $k(w,z)$ as the probability of obtaining a measurement w given that the actual measurement resulted in the point z. In cases where no extra knowledge is available it is natural to treat all point pairs (w,z) on equal footing. This leads to the requirement $k(M\langle w\rangle,M\langle z\rangle)=k(w,z)$for all $M \in SU(1,1)$. In the case where we can expect that the data is concentrated in a few separate clusters other methods like mixture-models (see [3] for an introduction and [4] for an example where the data points are covariance matrices) might be considered.

The Möbius transforms define an invariant metric and the (hyperbolic) distance between points z and w in this metric is denoted by $d(w,z)$. In the following only kernel functions of the form $k(d(w,z))$ will be used. It follows that it is sufficient to consider the special case where one point is the origin and the other is located on the real axis ($w=0$ and $z=\tanh(\tau)$) since every pair (w,z) can be transformed to this form by applying a Möbius transform to both points. For these point configurations the distance is given by $d(0,\tanh\tau)=2\tau$. This shows that the border of the unit disk had to be excluded since its points are infinitely far away from the origin. The general formula for the hyperbolic distance between two points is:

$$d(w,z) = 2\mathrm{arctanh}\left(\left|\frac{z-w}{1-\bar{z}w}\right|\right).$$

In the following this construction will be generalized to the case of general group elements. Thus the measured point z is replaced by the group element g_l and the neighboring point w by the general group element g. The fact that all pairs of group elements are treated equally motivates the following condition on the kernels:

$$\mathrm{k}(g,g_l) = \mathrm{k}(hg,hg_l) \text{ for all } g,g_l,h \in \mathrm{SU}(1,1).$$

Such functions are known as (left-) isotropic kernels. This leads to the following construction of the kernel density estimator (KDE):

Definition 4. *Given a stochastic process defined on the group SU(1,1) with observed group elements* g_l,l=1,…,L. *The kernel density estimator(KDE) of the probability density function p is given by:*

$$p(g) = \sum_{l=1}^{L} \mathrm{k}(g, g_l) \tag{6}$$

Only isotropic kernels will be used in the following and it follows that $k(g,g_l)=k(g^{-1}lg,e)=k(g^{-1}{}_l g)$ where e is the identity element. Using KAK coordinates for two points g=g(φ0,τ0,0). and $g_l=g(\varphi_l,\tau_l,0)$ on the unit disk gives the following relation between them and the difference $g^{-1}{}_l g$ with its own decomposition g−1lg=g(φ,τ,0):

$$g(\varphi, \tau, 0) = g(\varphi_l, \tau_l, 0)^{-1} g(\varphi_0, \tau_0, 0) = g(0, -\tau_l, 0) g(\varphi_0 - \varphi_l, \tau_0, 0).$$

In ([5], Vol. 1, page 271), the relation between the parameters of the three group elements can be found as follows:

$$\cosh \tau = \cosh \tau_l \cosh \tau_0 + \sinh \tau_l \sinh \tau_0 \cos(\varphi_0 - \varphi_l) \tag{7}$$

As before in the case of the hyperbolic distance only kernel functions of the form

$$\begin{aligned} \mathrm{k}(g, g_l) &= \mathrm{k}\left(g_l^{\ 1} g\right) = \mathrm{k}(\cosh \tau) \\ &= \mathrm{k}(\cosh \tau_l \cosh \tau_0 + \sinh \tau_l \sinh \tau_0 \cos(\varphi_0 - \varphi_l)) \end{aligned} \tag{8}$$

are considered in the following.

In general one has to compute the function values $k(g,g_l)$ for every pair (g,g_l). Of special interest are functions which separate these factors. They are the associated Legendre functions (ALFs). They are also known as zonal or Mehler functions ([5], page 324) of order *m* and degree α=−1/2+iκ and defined as (see Equation (9)):

$$\mathfrak{P}_\alpha^m(\cosh \tau) = \frac{1}{2\pi} \frac{\Gamma(\alpha + m + 1)}{\Gamma(\alpha + 1)} \int_0^{2\pi} (\sinh \tau \cos \theta + \cosh \tau)^\alpha \, e^{im\theta} \, d\theta \tag{9}$$

For these functions the following addition formula ([5], page 327) holds:

$$\mathfrak{P}_\alpha(\cosh \tau_l \cosh \tau_0 + \sinh \tau_l \sinh \tau_0 \cos \theta) = \sum_{m \in \mathbb{Z}} \mathfrak{P}_\alpha^{-m}(\cosh \tau_l) \mathfrak{P}_\alpha^m(\cosh \tau_0) \, e^{-im\theta} \tag{10}$$

Using $P_{-1/2+i\kappa}(\cosh\tau)$ in the previous Equations (7) and (8) gives:

$$\begin{aligned} \mathrm{k}(\cosh\tau) &= \mathfrak{P}_{-1/2+i\kappa}(\cosh\tau) \\ &= \textstyle\sum_{m\in\mathbb{Z}} \mathfrak{P}^{-m}_{-1/2+i\kappa}(\cosh\tau_l)\mathrm{e}^{im\varphi_l}\mathfrak{P}^{m}_{-1/2+i\kappa}(\cosh\tau_0)\mathrm{e}^{-im\varphi_0} \end{aligned} \tag{11}$$

separating the influence of the data dependent terms, originating in g_l with parameters τ_l,φ_l and the general part depending on g with τ_0,φ_0.

The choice $\mathrm{k}(\cosh\tau)=P_{-1/2+i\kappa}(\cosh\tau)$ can be generalized by the Mehler-Fock ttransform, that shows that a large class of functions are combinations of ALFs:

Theorem 1 (Mehler-Fock Transform; MFT).

For a function k *defined on the interval* $[1,\infty)$ *define its transform c as:*

$$c(\kappa) = \int_0^\infty \mathrm{k}(\cosh\tau)\mathfrak{P}_{-1/2+i\kappa}(\cosh\tau)\sinh\tau\, d\tau \tag{12}$$

Then k *can be recovered by the inverse transform:*

$$\mathrm{k}(\cosh\tau) = \int_0^\infty \kappa\tanh(\pi\kappa)\mathfrak{P}_{-1/2+i\kappa}(\cosh\tau)c(\kappa)\, d\kappa \tag{13}$$

The MFT also preserves the scalar product (Parseval relation): If $c_n(\kappa)=\int_1^\infty f_n(x)P_{1/2+i\kappa}(x)dx$ *then*

$$\int_0^\infty c_1(\kappa)c_2(\kappa)\kappa\tanh(\pi\kappa)\, d\kappa = \int_1^\infty f_1(x)f_2(x)\, dx \tag{14}$$

This, and many details about the transform, special cases and its applications can be found in ([6], Section 7.6) and [5,7,8,9]. The description of the Parseval relation can be found in ([6], Section 7.6.16) and ([6], Section 7.7.1).

The integral in Equation (12) is essentially the scalar product of the function k and the ALF defined by the invariant Haar integral on the group. Applying the integral transform to the shifted kernel and using the group invariance of the transform and the addition formula for the ALFs results in the following equations:

$$\begin{aligned} c_{\kappa l} &= \int_0^\infty \mathrm{k}(\cosh\tau)\mathfrak{P}_{-1/2+i\kappa}(\cosh\tau_l\cosh\tau+\sinh\tau_l\sinh\tau\cos\theta)\sinh\tau\, d\tau \\ &= \sum_{m\in\mathbb{Z}} \mathrm{e}^{-im\theta}\mathfrak{P}^{-m}_{-1/2+i\kappa}(\cosh\tau_l)\int_0^\infty \mathrm{k}(\cosh\tau)\mathfrak{P}^{m}_{-1/2+i\kappa}(\cosh\tau)\sinh\tau\, d\tau \\ &= \sum_{m\in\mathbb{Z}} \mathrm{e}^{im\varphi_l}\mathfrak{P}^{-m}_{-1/2+i\kappa}(\cosh\tau_l)\, w^{(\mathrm{k})}_{\kappa m} = \sum_{m\in\mathbb{Z}} \gamma_{\kappa m l} w^{(\mathrm{k})}_{\kappa m} \end{aligned} \tag{15}$$

Here the coefficients $\gamma_{\kappa ml}$ are computed from the measurements and the weights $w^{(k)}_{\kappa m}$ are independent of the data. Equation (16) shows how harmonic analysis separates the effects of the data in the γ and the kernel in the w.

ILLUSTRATION AND IMPLEMENTATION

In the following the usage of the MFT will be illustrated with the help of a computational model of a color camera. The Mehler-Fock transform is an integral transform in which the key components are integrals over the positive half-axis. This means that the implementation has to take into account problems such as numerical accuracy and quantization. The main goal of this paper is a first feasibility study of these tools and we will therefore comment on the relevant decisions in connection with the particular application under investigation.

The illustration studied in the following is similar to the motivating example introduced earlier but here a multispectral approach based on the following equation is used:

$$f_l(\mathbf{x}) = \tanh\left(\gamma_l \int_{380}^{780} s(\lambda, l) \cdot r(\lambda, \mathbf{x}) \cdot l(\lambda, T)\, d\lambda\right) \tag{16}$$

where fl(x) is the measured value by channel l at position x, λ denotes the wavelength variable in the relevant interval which is here given by the lower and upper bounds 380 nm and 780 nm, l(λ,T) is the spectral distribution of the black body radiator at temperature T, r(λ,x) is the reflection spectrum at object point x, s(λ,l)describes the spectral characteristics of sensor channel l and γl are constants used for white point correction. More information about spectral models of color imaging and black body radiators can be found in [10].

The reflectance spectra used are from a database consisting of 2782 measurements from the combined Munsell (1269 chips) (see [11]) and NCS (1513 chips) (measured by NCS Colour AB, http://ncscolour.com) color atlas. The spectra are measured in the spectral range from 380 nm to 780 nm in 5 nm steps. The spectral characteristics of the camera channels are described in [12] and the numerical values were downloaded from (https://spectralestimation.wordpress.com/data). From the reflection spectra the CIELAB values where computed first (using the constant spectrum as a whitepoint reference) and then these CIELAB value where converted to RGB using the lab2rgb function in Matlab. The normalization factors γ_l were computed in two steps: first three constants were computed such that the RGB vectors computed from the CIELAB conversion and the RGB vector computed from the camera model had the same mean vectors. Then a global normalization factor was computed such that the median value of the green channel was mapped to 0.5 by the tanh-mapping.

The non-linear tanh map is more realistic than the usual unrestricted usage of the integrated values since all realizable sensors must have a saturation limit related to the maximum capacity of the sensor. Geometrically this has a profound influence on the space of the RGB vectors. It was shown earlier that the positivity of the functions involved result in a conical structure of the measurement space. Increasing the intensity of the illumination source results in a half-line in measurement space when no saturation limits are enforced. With the limitation (for example in the form of the tanh mapping) all curves generated by intensity chances originate in the black point as in the unrestricted case but the saturation limit leads to a termination of all such curves in the white-point. Instead of an open-ended cone the saturation limited systems leads therefore to a double-cone structure ending in a black- and a white-point.

Using a constant illumination spectrum ($l(\lambda)=1$) the three eigenvectors b0,b1,b2were computed from the simulated RGB-camera responses. The simulated RGB vector computed from color chip number *m* under the constant (white) illumination spectrum is denoted by f(m,w). This was first mapped to the PCA coordinates $c_k(m,w)=\langle f(m,w),b_k\rangle$ and then the corresponding 'chroma' value z(m,w) is computed as

$$z(m,w) = \sigma\left(\frac{c_1(m,w)}{c_0(m,w)}, \frac{c_2(m,w)}{c_0(m,w)}\right)$$

where σ is a global constant which ensures that all chroma values are located on the unit disk. The value of σ has to be chosen so that the maximum length of the projected vectors *z* is less than one to ensure that the chroma points are located on the unit disk. Here we selected a scaling such that the maximum length is slightly lower than one (for example 0.97). This implies that there is a known upper bound for the integrals involved.

Next ten illumination spectra with temperatures (in kelvins) between 2000 K and 9000 K (sampled in equal distances in the inverse temperature, mired, scale which is more closely related to properties of human color perception) were used which resulted in the 11×2782 chroma values z(k,T),k=1…2782,T=2000,…,9000,w. The images in Figure 1 illustrate the differences between the second image (2190 K) in the black body series and the image generated with a flat illumination spectrum.

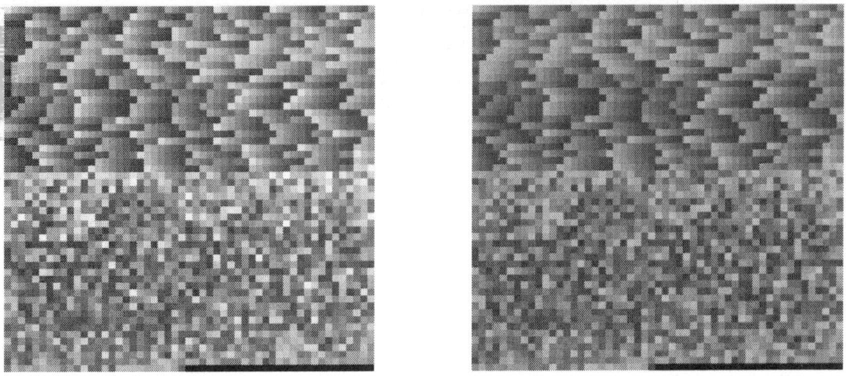

(a) White (b) 2190K

Figure 1. Simulated RGB Images. (**a**) White; (**b**) 2190 K.

The corresponding projections z(m,w) onto the unit disk (computed from the two images shown in Figure 1) are shown in Figure 2. They show that the points related to the coordinate system defining the PCA are located near the center of the disk whereas the shift of the illumination towards the red moves the corresponding points in the direction of the border of the disk.

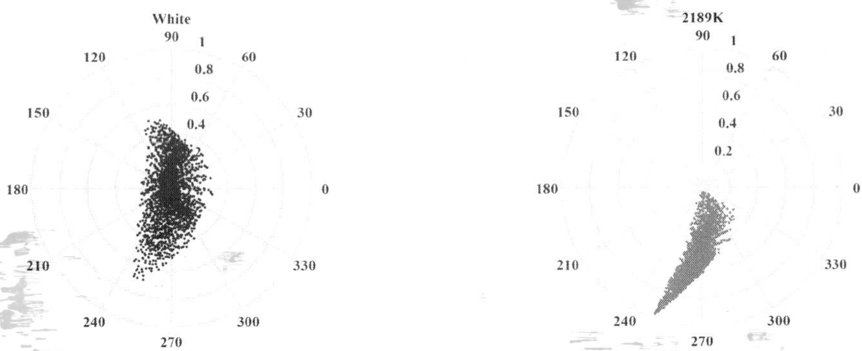

(a) White (b) 2190K

Figure 2. Chroma Points. (**a**) White; (**b**) 2190 K.

The next figures show how the type of the statistical distributions depends on the coordinate system used. Here the 'saturation' data from the illuminations with temperature 2000 K and constant spectrum ($l(\lambda)=1$, denoted by w earlier) are selected. Geometrically they are given by the radial position of the chroma points. The figures show the histograms of the two distributions and the fitted parametric distributions. In one case the radius ρ in the unit disk is used and in the other the group-theoretically motivated hyperbolic angle τ is used. Three different types of parametric distributions are used: normal distributions, generalized extreme value distributions (GEV) and the Beta distribution. Figure 3a shows the fitting of the distribution of the variables computed from the 2000 K blackbody radiator. It can be seen that the data is almost normally distributed with a slight upper tail which results in a better fit by the GEV. Figure 3b shows that the white illumination source produced a distribution mainly concentrated around the origin.

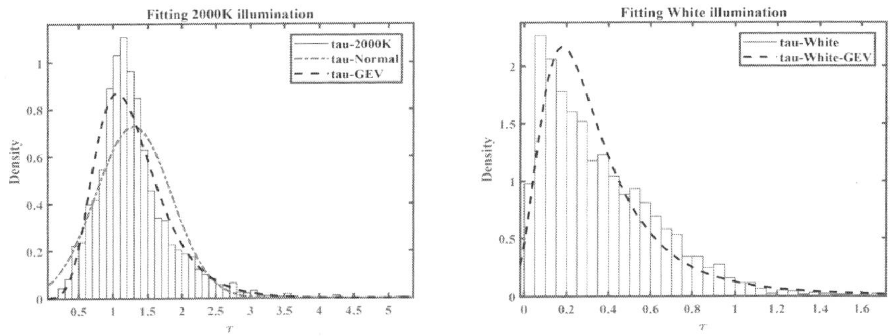

(a) 2000 K/τ-Variable/Parametric Fits (b) White/τ-Variable/Parametric Fits

Figure 3. τ-Parametrization. (**a**) 2000 K/τ-Variable/Parametric Fits; (**b**) White/ τ-Variable/Parametric Fits.

The general picture is the same for the data parametrized by the radius. This is illustrated in Figure 4. Again the 2000 K radiator generates a symmetrical distribution around the mean with good fittings of the normal and the GEV distributions. The white illumination has the same characteristic as in the hyperbolic parametrization. The difference here is that, by definition the output values are located in the unit interval. For such data the Beta-distribution is a popular statistical model and we see that this gives a good fitting for both cases.

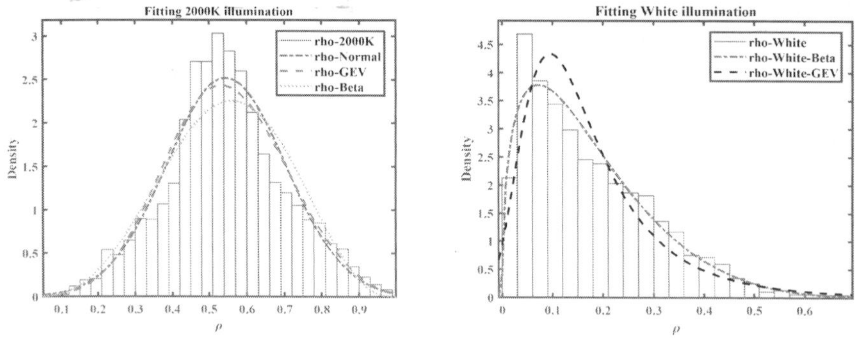

(a) 2000 K/ρ-Variable/Parametric Fits (b) White/ρ-Variable/Parametric Fits

Figure 4. ρ-Parametrization. (**a**) 2000 K/ρ-Variable/Parametric Fits; (**b**) White/ρ-Variable/Parametric Fits.

From Equation (9) it can be seen that in the case where the angle θ has the value zero (both points have the same angular value) the ALFs are zero for m≠0. This case will be illustrated next. From Equation (16) follows that $c_{\kappa l} = \gamma_{\kappa 0 l} w^{(k)}_{\kappa 0}$. The value of $\gamma_{\kappa 0 l}$ is given by the ALF and the values $w^{(k)}_{\kappa 0}$ are given by the MFT of the kernel. The definition of the ALFs given in Equation (9) is not suitable for numerical computations. Instead the connection between the conical functions and the Gauss hypergeometric function are often used (see [13] and the references mentioned there). In the following experiments the implementations of special functions in Mathematica were used and the numerical results exported to Matlab. In this process two problems have to be solved: (a) the integrals related to the coefficients $w^{(k)}_{\kappa m}$ from Equation (16) have to be calculated and (b) the discrete values in the κ-domain have to be selected.

For a few kernel functions *k* the MFT is known analytically but in the general case the integrals have to be computed numerically. In the experiments described here we used a two-stage procedure. First we selected a few relevant values in the κ-domain and for a given kernel *k* we computed the numerical values of the transform using the NIntegrate function in Mathematica using standard parameter settings. This is rather time-consuming and in the second stage of the process we therefore selected a grid on the relevant interval on the τ-axis and estimated the integral values by simple addition in Matlab. The grid points were chosen such that the Matlab and the Mathematica results were comparable. The result is a matrix whose entries are the values $w_{\kappa m}^{(k)}$. Note that this matrix only has to be computed once. An investigation of the numerical computation of these values (comparable to the work reported in [14]) lies outside the scope of this report. Also the κ-grid points where selected in an ad-hoc fashion and their importance was judged afterwards.

Figure 5 shows some of the ALFs in the relevant interval.

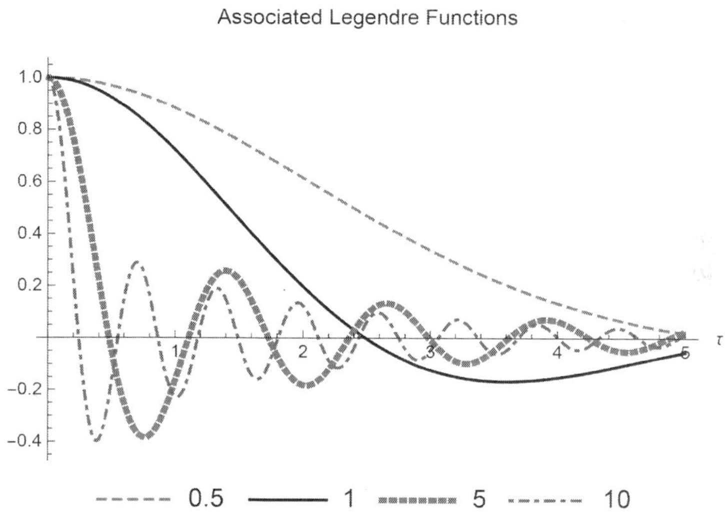

Figure 5. Associated Legendre Functions.

The values of the coefficients $w_{\kappa 0}^{(k)}$ are given by the MFT of the kernel. For every kernel they have to be computed only once. This can be done by some type of numerical integration (which has to take into account the oscillatory nature of the ALFs) or one can use a kernel for which an closed expression of the MFT is available. In the following illustrations kernels of the form k(coshτ)=(coshτ)−C are used.

The width of the kernel acts as a smoothing parameter. The effect of varying the exponent of the cosh-kernel is illustrated in Figure 6. The data used are the radial values (in the τ parametrization) computed from the simulated images under the 2000 K illuminant. It shows that for broad kernels some general characteristics can be obtained, whereas for very thin kernels the overall estimate starts to be sensitive to small local variations.

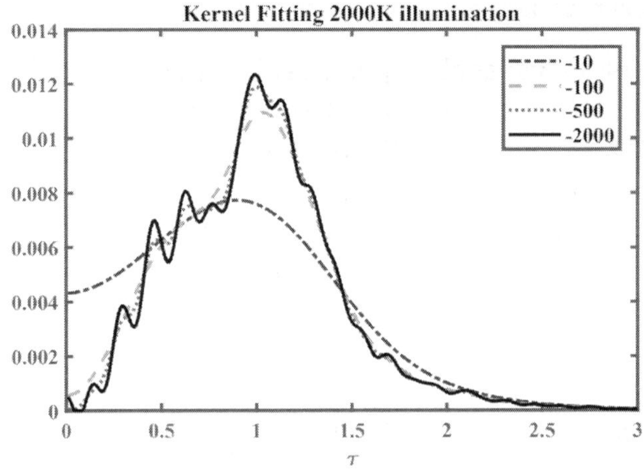

Figure 6. Kernels of varying sizes.

Figure 7 shows a section of the Mehler-Fock transform ($0.05 \leq \kappa \leq 12.5$) for three selected illuminants. This shows how the general shift to the border of the unit disk (for the red-shifted illuminants of lower temperature) changes the form of the Mehler-Fock transform.

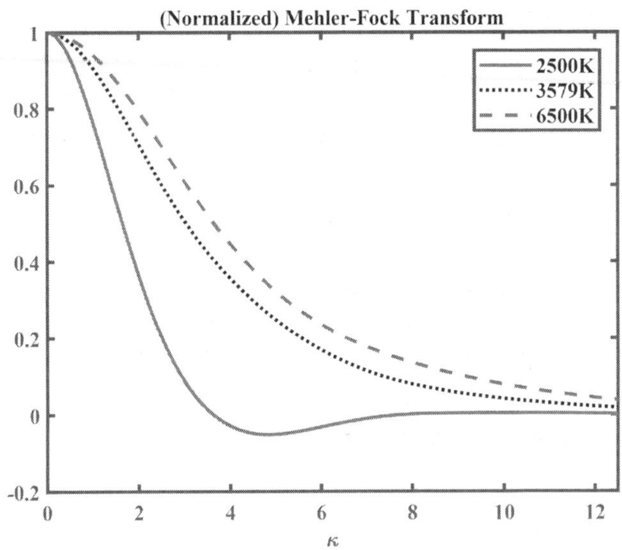

Figure 7. (Normalized) Mehler-Fock-Transform.

SUMMARY AND DISCUSSION

The first part of the paper shows how projection operators lead to the unit disk as a natural domain for signals relevant in applications. The simplest example was the decomposition of a three-parameter color space into intensity, saturation and hue descriptors where the saturation/hue part is located in the unit disk. Combining the Perron-Frobenius theorem with principal component analysis showed that similar parametrizations are natural for general spaces of positive signals. Next the group SU(1,1) was selected as a symmetry group of the unit disk. Decomposition of functions defined on domains that possess a symmetry group is studied in the theory of group representations. The group SU(1,1) is a non-commutative and non-compact group which makes the analysis rather complicated but in exchange the class of possible transformations is much larger than an analysis based on a geometry consisting of concentric rings. It is known that the most difficult part of the theory is the analysis of the radial part of functions on the unit disk. This leads to the Mehler-Fock transform based on the ALFs. Some properties of the Mehler Fock transform were illustrated by studying the illumination-induced chroma changes in the RGB images of chips in a color atlas. Also some basic problems in the numerical implementation of the transform were discussed.

The goal of this paper was thus to introduce the relatively unknown (in the signal processing community) Mehler Fock transform and to sketch some possible application areas. Due to limited length of the text a number of interesting topics have been mentioned very briefly or not at all. These should be studied in future investigations. Some of these topics are the following:

- Study of general functions defined on the whole disc and not only functions with no angular variation. The underlying theoretical tools are combination of the transform described here with expansions in Fourier series in the angular variable. The same holds for general functions on the group which requires the application of Fourier series in the third KAK coordinate. Generalizations to the case of spaces with higher-dimensional unit balls instead of the unit disc are also well-studied in the mathematical and theoretical literature. They involve the groups SU(1,N) and their representations.
- Efficient numerical implementations and the study of numerical

precisions. The transform is an integral transform and the effects of sampling and truncation decisions has to be investigated in detail.

- The separation of the data and the kernel properties was used as the main argument in favor of the transform. Not mentioned were other advantages of the combination of kernel density estimators and harmonic analysis. These are obtained from the fact that the MFT provides a decomposition of the probability density function into a superposition of ALFs. Operations, such as differentiations, on the pdf are therefore reduced to corresponding operations on the ALFS for which a large number of properties, like recursion formulas derivatives and integrals are known. Details can be found in [5,15].

ACKNOWLEDGMENTS

The support of the Swedish Research Council through a framework grant for the project Energy Minimization for Computational Cameras (2014-6227) is gratefully acknowledged.

REFERENCES

1. Lenz, R. Spectral Color Spaces: Their Structure and Transformations. *Adv. Imaging Electron Phys.* **2005**, *138*, 1–67.
2. Lenz, R. Lie methods for color robot vision. *Robotica* **2008**, *26*, 453–464.
3. Hastie, T.; Tibshirani, R.; Friedman, J. *The Elements of Statistical Learning: Data Mining, Inference, and Prediction*; Springer Series in Statistics; Springer: New York, NY, USA, 2013.
4. Said, S.; Bombrun, L.; Berthoumieu, Y.; Manton, J.H. Riemannian Gaussian distributions on the space of symmetric positive definite matrices. *IEEE Trans. Inf. Theory* **2017**, *63*, 2153–2170.
5. Vilenkin, N.; Klimyk, A. *Representation of Lie Groups and Special Functions: Volume 1: Simplest Lie Groups, Special Functions and Integral Transforms*; Springer: New York, NY, USA, 2012.
6. Sneddon, I.N. *The Use of Integral Transforms*; McGraw-Hill: New York, NY, USA, 1972.
7. Terras, A. *Harmonic Analysis on Symmetric Spaces and Applications I*; Springer: New York, NY, USA, 1985.
8. Lebedev, N.N. *Special Functions and Their Applications*; Dover: New York, NY, USA, 1972.
9. Bateman, H.; Erdélyi, A.; States, U. *Tables of Integral Transforms*; McGraw-Hill: New York, NY, USA, 1954; Volume 2.
10. Wyszecki, G.; Stiles, W.S. *Color Science*; Wiley: New York, NY, USA, 1982.
11. Parkkinen, J.P.; Hallikainen, J.; Jaaskelainen, T. Characteristic spectra of Munsell colors. *J. Opt. Soc. Am. A Opt. Image Sci. Vis.* **1989**, *6*, 318–322.
12. Darrodi, M.M.; Finlayson, G.; Goodman, T.; Mackiewicz, M. Reference data set for camera spectral sensitivity estimation. *J. Opt. Soc. Am. A Opt. Image Sci. Vis.* **2015**, *32*, 381–391.
13. Gil, A.; Segura, J.; Temme, N.M. Computing the Conical Function pμ−1/2+iτ(x). *SIAM J. Sci. Comput.* **2009**, *31*, 1716–1741.
14. Gautschi, W. Computing the Kontorovich-Lebedev integral transforms and their inverses. *BIT Numer. Math.* **2006**, *46*, 21–40.

15. NIST Digital Library of Mathematical Functions. *NIST Handbook of Mathematical Functions Hardback and CD-ROM*; Olver, F.W.J., Olde Daalhuis, A.B., Lozier, D.W., Schneider, B.I., Boisvert, R.F., Clark, C.W., Miller, B.R., Saunders, B.V., Eds.; Cambridge University Press: Cambridge, UK, 2010.

CHAPTER 9

LOCAL FRACTIONAL -TRANSFORMS WITH APPLICATIONS TO SIGNALS ON CANTOR SETS

Kai Liu,[1] Ren-Jie Hu,[1] Carlo Cattani,[2] Gong-Nan Xie,[3] Xiao-Jun Yang,[4] and Yang Zhao[5]

[1]School of Electrical Engineering, Southeast University, Nanjing 210096, China

[2]Department of Mathematics, University of Salerno, Via Ponte don Melillo, Fisciano, 84084 Salerno, Italy

[3]School of Mechanical Engineering, Northwestern Polytechnical University, Xi'an, Shaanxi 710048, China

[4]Department of Mathematics and Mechanics, China University of Mining and Technology, Xuzhou, Jiangsu 221008, China

[5]Department of Electronic and Information Technology, Jiangmen Polytechnic, Jiangmen 529090, China

CITATION (APA): Liu, K., Hu, R. J., Cattani, C., Xie, G. N., Yang, X. J., & Zhao, Y. (2014). Local fractional-transforms with applications to signals on cantor sets. In Abstract and Applied Analysis (Vol. 2014). Hindawi. (6 pages)

ABSTRACT

The Z-transform has played an important role in signal processing. In this paper the Z-transform has been generalized by the coupling of both the Z-transform and the local fractional complex calculus. In the literature the local fractional Z-transform is applied to analyze signals, in the following it will be used to analyze signals on Cantor sets. Some examples are also given to show the efficiency and accuracy for handling the signals on Cantor sets.

INTRODUCTION

Integral transforms [1, 2], such as Fourier, Laplace, Mellin, Hilbert, and Hankel transforms, play important roles in solving the mathematical problems arising in applied mathematics, mathematical physics, and engineering science. In recent years, fractional calculus [3–11] was developed and used to model also some anomalous behaviors of diffusion [12–21] and transport [22–27]. Fractional integral transforms are suitable generalizations of the classical ones and were recently proposed by some researchers. For example, the fractional Fourier transforms were considered in [28, 29]. In [30], the fractional Hilbert transform was presented. The fractional Mellin transform [31, 32] was proposed to be used in image encryption. The fractional wavelet transform was presented and some applications were investigated in [33–35]. In [36], the fractional Hankel transform was reported in order to research the charge-amplitude state representations.

The Z-transform method [1, 2, 37] was applied to handle the linear time-invariant discrete-time systems (LTI discretetime systems) and difference equations in Z-domain. However, the fractional derivative and integrals (the fractional PDIs) were used to transfer the fractional LTI discrete-time systems to Z-domain [38]. There appear signals defined on Cantor sets, which are the most striking properties of nondifferentiable functions. The classical Z-transform method and PDIs did not deal with them. In order to overcome them, local fractional calculus [39–43] may be applied to handle the function defined on Cantor sets shown in Figure 1. The local fractional integral transforms via local fractional calculus theory were proposed in [44–51]. For example, local fractional Fourier transforms reported in [40, 44] were used to find nondifferentiable solutions for local fractional ODEs and PDEs [45–47]. Laplace transforms via local fractional calculus [40] were generalized and reported in order to solve the local fractional ODEs and PDEs [48–50].

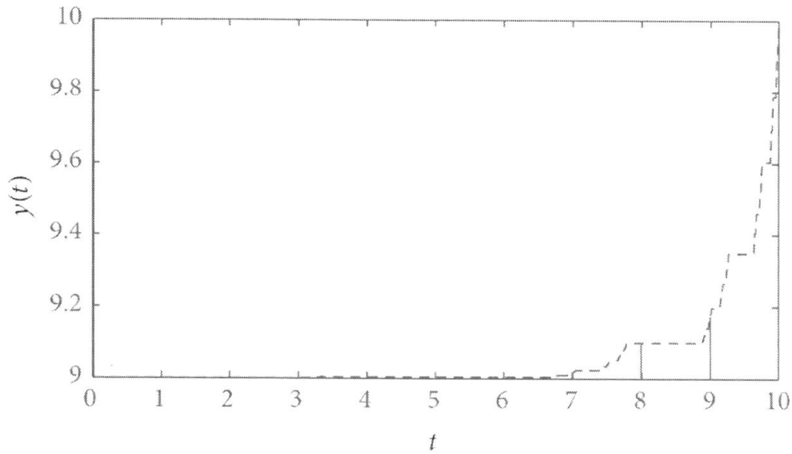

Figure 1: The chart of the signal y(t) defined on Cantor sets.

Fractal signal processing [51–59] is a hot topic for scientists and engineers. Very recently, the concept of the Z-transform method via local fractional calculus was considered only in [60]. However, there is no report on signal processing by using the local fractional Z-transforms. The main aim of this paper is to investigate the properties of local fractional Z-transforms and to present some examples for processing signals defined on Cantor sets.

The paper is organized as follows. In Section 2, the concepts of local fractional complex derivatives and integrals are given. In Section 3, the notions and properties of local fractional Z-transform method are presented. In Section 4, some examples and applications of this method are shown. Finally, Section 5 is the conclusions.

LOCAL FRACTIONAL DERIVATIVES AND INTEGRALS OF COMPLEX FUNCTIONS AND RECENT RESULTS

In this section, we introduce the concepts of local fraction derivative and integrals of complex functions. Let us first give the local fractional continuity of complex functions.

Definition 1 (see [40, 60]). The function (z) is said to be local fractional continuous at z_0 if there exists

$$\lim_{z \to z_0} f(z) = f(z_0). \tag{1}$$

There is the local fractional continuous relation in the following form:

$$f(z) \in C_\alpha(\mathfrak{R}), \tag{2}$$

where

$$\lim_{z \to z_0} f(z) = f(z_0), \quad z, z_0 \in \mathfrak{R}. \tag{3}$$

Definition 2 (see [40, 60]). The local fractional derivative of complex function f(z) of order is defined as

$$f^{(\alpha)}(z) = \frac{d^\alpha f(z)}{d^\alpha z} = \lim_{z \to z_0} \frac{\Delta^\alpha f(z)}{(z - z_0)^\alpha}, \quad \alpha \in (0, 1], \tag{4}$$

where

$$\Delta^\alpha f(z) \cong \Gamma(1+\alpha)(f(z) - f(z_0)). \tag{5}$$

If the limit of (4) exists for all z_0 in a region R, then the complex function (z) is said to be local fractional analytic in a region R.

The properties of the local fractional derivatives of some complex functions are presented as follows [40]:

$$\begin{aligned} \frac{d^\alpha z^{k\alpha}}{dz^\alpha} &= \frac{\Gamma(1+k\alpha)}{\Gamma(1+(k-1)\alpha)} z^{(k-1)\alpha}, \\ \frac{d^\alpha E_\alpha(z^\alpha)}{dz^\alpha} &= E_\alpha(z^\alpha), \\ \frac{d^\alpha \sin_\alpha z^\alpha}{dz^\alpha} &= \cos_\alpha z^\alpha, \\ \frac{d^\alpha \cos_\alpha z^\alpha}{dz^\alpha} &= -\sin_\alpha z^\alpha, \end{aligned} \tag{6}$$

where

$$\begin{aligned} E_\alpha(z^\alpha) &= \sum_{k=0}^{\infty} \frac{z^{\alpha k}}{\Gamma(1+k\alpha)}, \\ \sin_\alpha z^\alpha &= \sum_{k=0}^{\infty} \frac{(-1)^k z^{\alpha(2k+1)}}{\Gamma[1+\alpha(2k+1)]}, \\ \cos_\alpha z^\alpha &= \sum_{k=0}^{\infty} \frac{(-1)^k z^{2\alpha k}}{\Gamma(1+2\alpha k)}. \end{aligned} \tag{7}$$

Definition 3 (see [40, 46–50, 60]). The local fractional integral of complex function (z) of order α along the closed contour C is defined as

$$I_C^{\alpha} f(z) = \frac{1}{\Gamma(1+\alpha)} \oint_C f(z)(dz)^{\alpha}, \quad \alpha \in (0,1]. \tag{8}$$

The properties of the local fractional integrals of some complex functions are suggested as follows [40]:

$$\frac{1}{\Gamma(1+\alpha)} \int_C (f(z)+g(z))(dz)^{\alpha}$$
$$= \frac{1}{\Gamma(1+\alpha)} \int_C f(z)(dz)^{\alpha} + \frac{1}{\Gamma(1+\alpha)} \int_C g(z)(dz)^{\alpha},$$
$$\frac{1}{\Gamma(1+\alpha)} \int_{C_1+C_2} f(z)(dz)^{\alpha}$$
$$= \frac{1}{\Gamma(1+\alpha)} \int_{C_1} f(z)(dz)^{\alpha} + \frac{1}{\Gamma(1+\alpha)} \int_{C_2} f(z)(dz)^{\alpha},$$
$$\frac{1}{\Gamma(1+\alpha)} \int_{C_1} f(z)(dz)^{\alpha} = -\frac{1}{\Gamma(1+\alpha)} \int_{-C_1} f(z)(dz)^{\alpha}. \tag{9}$$

Theorem 4 (see [40]). If (z) is local fractional analytic within and on a simple closed contour C and a is any point interior to C, then we have

$$\frac{1}{(2\pi)^{\alpha} i^{\alpha}} \cdot \left\{ \frac{1}{\Gamma(1+\alpha)} \oint_C \frac{f(z)}{(z-z_0)^{\alpha}} (dz)^{\alpha} \right\} = \frac{f(z_0)}{\Gamma(1+\alpha)}. \tag{10}$$

Proof. See [40].

Definition 5 (see [40, 60]). If z_0 is an isolated singular point of (z), then we have a local fractional Laurent series of $f(z)$ at $C : |z - z_0|$

$$f(z) = \sum_{i=-\infty}^{\infty} u_k (z-z_0)^{k\alpha}. \tag{11}$$

The coefficient a_{-1} of $(z - z_0)^{-\alpha}$ is called the local fractional residue of (z) at $z=z_0$ and is frequently written as

$$\mathrm{Re}_{z=z_0} s\{f(z)\} = a_{-1}. \tag{12}$$

Theorem 6 (see [40]). If (z) is local fractional analytic within and on the boundary $C$of a region R except at a number of poles a within R, having a residue a−1, then

$$\frac{1}{(2\pi)^{\alpha} i^{\alpha}} \cdot \left\{ \frac{1}{\Gamma(1+\alpha)} \oint_C f(z)(dz)^{\alpha} \right\} = \mathrm{Re}_{z=z_0} s \{f(z)\}. \tag{13}$$

Proof. See [40].

Theorem 7 (see [40]). If (z) is local fractional analytic within and on the boundary$C$of a region R except at a number of poles a within R, having numbers of residues, then

$$\frac{1}{(2\pi)^{\alpha} i^{\alpha}} \cdot \left\{ \frac{1}{\Gamma(1+\alpha)} \oint_C f(z)(dz)^{\alpha} \right\} = \sum_{i=1}^{n} \mathrm{Re}_{z=z_k} s \{f(z)\}. \tag{14}$$

Proof. See [40].

LOCAL FRACTIONAL -TRANSFORMS AND THEIR PROPERTIES

In this section, we give the local fractional Z-transforms and their properties.

Definition 8 (see [60]). Local fractional Z-transform of (n) of order α is defined as

$$Z_{\alpha}\{f(n)\} = F_{\alpha}(z) = \sum_{n=\infty}^{\infty} f(n) z^{-n\alpha}, \tag{15}$$

where the above formula is convergent.

For a given sequence, the set R of values of z for which its local fractional Z-transform converges is called the region of convergence (ROC), namely

$$\sum_{n=\infty}^{\infty} \left| f(n) z^{-n\alpha} \right| < \infty. \tag{16}$$

The inverse formula of local fractional Z-transform of (n) of order α reads as follows (see [60]):

$$Z_{\alpha}^{-1}\{F_{\alpha}(z)\} = f(n) = \frac{1}{(2\pi i)^{\alpha} \Gamma(1+\alpha)} \oint_C F_{\alpha}(z) z^{(n-1)\alpha} (dz)^{\alpha}, \tag{17}$$

where C is a counterclockwise closed fractal path encircling the origin and entirely in the region of convergence.

Let $Z\{f(n)\} = F_{\alpha}(z)$ within the region of convergence R_1 and let $Z_{\alpha}\{g(n)\} = G_{\alpha}(z)$ within the region of convergence R_2.

Property 1 (linearity). We have

$$Z_\alpha\{f(n)+g(n)\}=F_\alpha(z)+G_\alpha(z) \tag{18}$$

within the region of convergence $R_1 \cap R_2$.

Proof. From (15) we have

$$\begin{aligned} Z_\alpha\{f(n)+g(n)\} &= \sum_{n=\infty}^{\infty}(f(n)+g(n))z^{-n\alpha} \\ &= Z_\alpha\{f(n)\}+Z_\alpha\{g(n)\} \end{aligned} \tag{19}$$

within the region of convergence $R_1 \cap R_2$.

Property 2 (time shifting). If the variable z has a useful interpretation in terms of time delay, then we have

$$Z_\alpha\{f(n-k)\}=z^{-z\alpha}F_\alpha(z). \tag{20}$$

Proof. From (15), we have

$$\begin{aligned} Z_\alpha\{f(n-k)\} &= \sum_{n=\infty}^{\infty} f(n-k)z^{-n\alpha} = \sum_{n=\infty}^{\infty} f(n)z^{-(n+k)\alpha} \\ &= z^{-k\alpha}\sum_{n=\infty}^{\infty} f(n)z^{-n\alpha} = z^{-k\alpha}Z_\alpha\{f(n)\}. \end{aligned} \tag{21}$$

Property 3 (frequency modulation). We have

$$Z_\alpha\{z_0^{n\alpha}f(n)\}=F_\alpha\left(\frac{z}{z_0}\right). \tag{22}$$

Proof. From (15), we have

$$\begin{aligned} Z_\alpha\{z_0^{n\alpha}f(n)\} &= \sum_{n=\infty}^{\infty} f(n)z_0^{n\alpha}z^{-n\alpha} \\ &= \sum_{n=\infty}^{\infty} f(n)\left(\frac{z}{z_0}\right)^{-n\alpha} = F_\alpha\left(\frac{z}{z_0}\right). \end{aligned} \tag{23}$$

SOME ILLUSTRATIVE EXAMPLES

In this section, we give some samples for nondifferentiable signals defined on Cantor sets.

Example 1. Let us consider the following signal in the form:

$$f(n) = \delta_\alpha(n) = \begin{cases} 1, & n = 0, \\ 0, & n \neq 0. \end{cases} \tag{24}$$

Taking local fractional Z-transform, we have

$$Z_\alpha\{f(n)\} = Z_\alpha\{\delta_\alpha(n)\} = \sum_{n=\infty}^{\infty} \delta_\alpha(n) z^{-n\alpha} = 1. \tag{25}$$

Example 2. We now suggest the following signal in the form:

$$f(n) = \delta_\alpha(n-k) = \begin{cases} 1, & n = k, \\ 0, & n \neq k. \end{cases} \tag{26}$$

Taking local fractional Z-transform, we obtain

$$Z_\alpha\{f(n)\} = Z_\alpha\{\delta_\alpha(n-k)\} = \sum_{n=\infty}^{\infty} \delta_\alpha(n-k) z^{-n\alpha} = z^{-k\alpha}. \tag{27}$$

When $z^\alpha = E(j^\alpha \omega^\alpha)$ with the imaginary unit j^α [40, 44–50], we get

$$\begin{aligned} \phi(\omega, k) &= Z_\alpha\{f(n)\} = E_\alpha\left(-j^\alpha \omega^\alpha k^\alpha\right) \\ &= \cos_\alpha\left(\omega^\alpha k^\alpha\right) - j^\alpha \sin_\alpha\left(\omega^\alpha k^\alpha\right). \end{aligned} \tag{28}$$

Hence, from (28), we get

$$\begin{aligned} \mathrm{Re}_\alpha \phi(\omega, k) &= \cos_\alpha\left(\omega^\alpha k^\alpha\right), \\ \mathrm{Im}_\alpha \phi(\omega, k) &= -\sin_\alpha\left(\omega^\alpha k^\alpha\right) \end{aligned} \tag{29}$$

with the real part graph in Figure 2 and imaginary part graph in Figure 3.

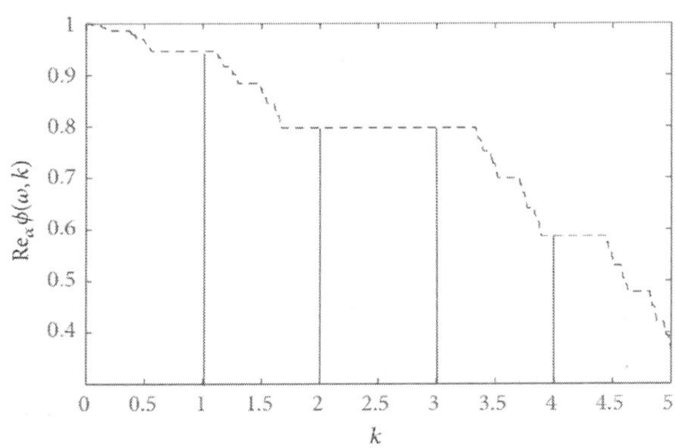

Figure 2: The graph of $\mathrm{Re}_\alpha(\omega, k)$ with premasters ω=1 and $\alpha = \ln 2/ \ln 3$.

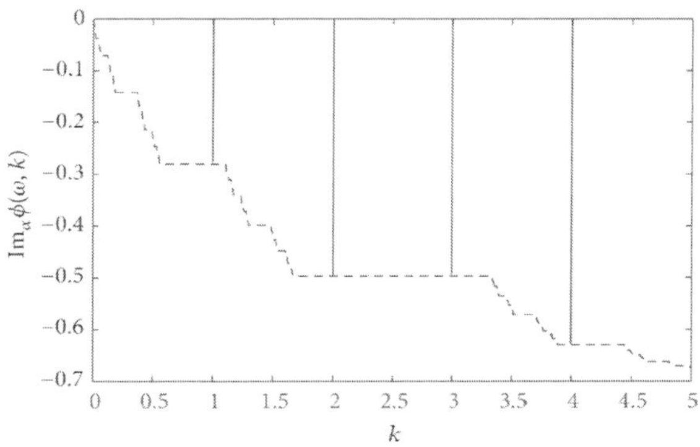

Figure 3: The plot of $\mathrm{Im}_\alpha(\omega, k)$ with premasters ω=1 and $\alpha = \ln 2/\ln 3$.

Example 3. There is the signal in the following form:

$$f(n) = \delta_\alpha(n-k) + \delta_\alpha(n+k) = \begin{cases} 1, & n = \pm k, \\ 0, & n \neq \pm k. \end{cases} \tag{30}$$

Taking local fractional -transform, we have

$$\begin{aligned} Z_\alpha\{f(n)\} &= Z_\alpha\{\delta_\alpha(n-k) + \delta_\alpha(n+k)\} \\ &= \sum_{n=\infty}^{\infty} (\delta_\alpha(n-k) + \delta_\alpha(n+k))\, z^{-n\alpha} = z^{-k\alpha} + z^{k\alpha}. \end{aligned} \tag{31}$$

When $z^\alpha = E(j^\alpha \omega^\alpha)$, we get

$$\phi(\omega, k) = Z_\alpha\{f(n)\} = E_\alpha(j^\alpha \omega^\alpha k^\alpha) = 2\cos_\alpha(\omega^\alpha k^\alpha) \tag{32}$$

with the graph of $\phi(\omega, k)$ shown in Figure 4.

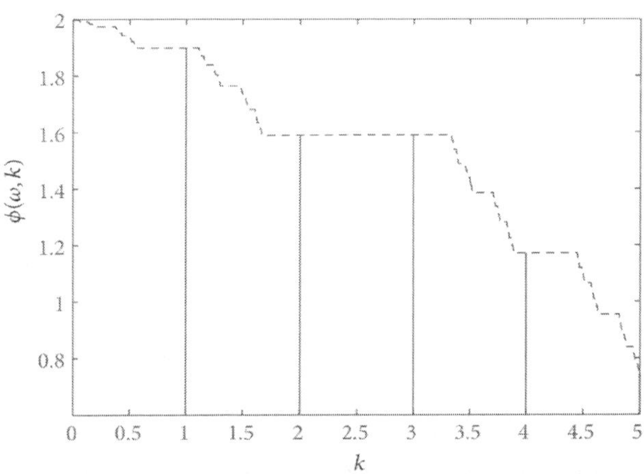

Figure 4: The map of (ω, k) with premasters ω=1 and $\alpha = \ln 2/\ln 3$.

Example 4. We have the following signal in the form:

$$f(n) = \begin{cases} a^{n\alpha}, & a \geq 0, \\ 0, & a < 0. \end{cases} \tag{33}$$

Local fractional -transform gives the following form:

$$Z_\alpha\{f(n)\} = Z_\alpha\{a^{n\alpha}\} = \sum_{n=0}^{\infty} a^{n\alpha} z^{-n\alpha} = \sum_{n=0}^{\infty} \left(\frac{z}{a}\right)^{-n\alpha} \tag{34}$$

with the region of convergence $|z| > |a|$.

Example 5. We consider the following signal in the form:

$$f(n) = \begin{cases} 0, & a \geq 0, \\ a^{n\alpha}, & a < 0. \end{cases} \tag{35}$$

Taking local fractional Z-transform, we arrive at the following form:

$$Z_\alpha\{f(n)\} = Z_\alpha\{a^{n\alpha}\} = \sum_{n=\infty}^{0} a^{n\alpha} z^{-n\alpha} = \sum_{n=\infty}^{0} \left(\frac{z}{a}\right)^{-n\alpha} \tag{36}$$

with the region of convergence $|z| < |a|$.

Example 6. We present the following signal in the form:

$$f(n) = \begin{cases} b^{n\alpha}, & a \geq 0, \\ a^{n\alpha}, & a < 0. \end{cases} \tag{37}$$

Local fractional Z-transform gives the following form:

$$Z_\alpha\{f(n)\} = Z_\alpha\{a^{n\alpha} + a^{n\alpha}\} = \sum_{n=0}^{\infty} b^{n\alpha} z^{-n\alpha} + \sum_{n=\infty}^{0} a^{n\alpha} z^{-n\alpha}$$

$$= \sum_{n=0}^{\infty} \left(\frac{z}{b}\right)^{-n\alpha} + \sum_{n=\infty}^{0} \left(\frac{z}{a}\right)^{-n\alpha} \tag{38}$$

with the region of convergence .

CONCLUSIONS

In this work, we investigated the local fractional -transforms based on the local fractional complex calculus and some properties are also obtained.

Some illustrative examples were also given. The obtained results show the accuracy and efficiency of the presented method.

ACKNOWLEDGMENT

This work was supported by Science and Technology Commission Planning Project of Jiangsu Province (no. BE2013737).

REFERENCES

1. B. Davies, Integral Transforms and Their Applications, Texts in Applied Mathematics, Springer, New York, NY, USA, 2002.
2. L. Debnath and D. Bhatta, Integral Transforms and Their Applications, CRC Press, New York, NY, USA, 2010.
3. V. E. Tarasov, Fractional Dynamics, Nonlinear Physical Science, Springer, New York, NY, USA, 2010.
4. R. Herrmann, Fractional Calculus: An Introduction for Physicists, World Scientific, Singapore, 2011.
5. A. B. Malinowska and D. F. M. Torres, Introduction to the Fractional Calculus of Variations, Imperial College Press, London, UK, 2012.
6. M. D. Ortigueira, Fractional Calculus for Scientists and Engineers, Lecture Notes in Electrical Engineering, Springer, New York, NY, USA, 2011.
7. F. Mainardi, Fractional Calculus and Waves in Linear Viscoelasticity, Imperial College Press, London, UK, 2010.
8. J. Sabatier, O. P. Agrawal, and J. A. T. Machado, Advances in Fractional Calculus: Theoretical Developments and Applications in Physics and Engineering, Springer, New York, NY, USA, 2007.
9. A. A. Kilbas, H. M. Srivastava, and J. J. Trujillo, Theory and Applications of Fractional Differential Equations, vol. 204 of North-Holland Mathematics Studies, Elsevier Science B.V., Amsterdam, The Netherlands, 2006.
10. I. Podlubny, Fractional Differential Equations, vol. 198 of Mathematics in Science and Engineering, Academic Press, San Diego, Calif, USA, 1999.
11. B. J. West, M. Bologna, and P. Grigolini, Physics of Fractal Operators, Institute for Nonlinear Science, Springer, New York, NY, USA, 2003.
12. R. Metzler and J. Klafter, "The random walk's guide to anomalous diffusion: a fractional dynamics approach," Physics Reports, vol. 339, no. 1, pp. 1–77, 2000.
13. L. Miller, "On the controllability of anomalous diffusions generated by the fractional Laplacian," Mathematics of Control, Signals, and Systems, vol. 18, no. 3, pp. 260–271, 2006.
14. B. J. West, P. Grigolini, R. Metzler, and T. F. Nonnenmacher, "Fractional

diffusion and Lévy stable processes," Physical Review E, vol. 55, no. 1, pp. 99–106, 1997.
15. V. V. Uchaikin, "Anomalous diffusion and fractional stable distributions," Journal of Experimental and Theoretical Physics, vol. 97, no. 4, pp. 810–825, 2003.
16. Y. Zhang, D. A. Benson, M. M. Meerschaert, E. M. LaBolle, and H. P. Scheffler, "Random walk approximation of fractional-order multiscaling anomalous diffusion," Physical Review E, vol. 74, no. 2, Article ID 026706, 2006.
17. Q. Yang, I. Turner, F. Liu, and M. Ilic, "Novel numerical methods for solving the time-space fractional diffusion equation in two dimensions," SIAM Journal on Scientific Computing, vol. 33, no. 3, pp. 1159–1180, 2011.
18. R. L. Magin, O. Abdullah, D. Baleanu, and X. J. Zhou, "Anomalous diffusion expressed through fractional order differential operators in the Bloch-Torrey equation," Journal of Magnetic Resonance, vol. 190, no. 2, pp. 255–270, 2008.
19. H. Sun, W. Chen, C. Li, and Y. Chen, "Fractional differential models for anomalous diffusion," Physica A, vol. 389, no. 14, pp. 2719–2724, 2010.
20. M. Xu and W. Tan, "Theoretical analysis of the velocity field, stress field and vortex sheet of generalized second order fluid with fractional anomalous diffusion," Science in China A, vol. 44, no. 11, pp. 1387–1399, 2001.
21. A. V. Chechkin, R. Gorenflo, and I. M. Sokolov, "Fractional diffusion in inhomogeneous media," Journal of Physics A, vol. 38, no. 42, pp. L679–L684, 2005.
22. G. M. Zaslavsky, "Chaos, fractional kinetics, and anomalous transport," Physics Reports, vol. 371, no. 6, pp. 461–580, 2002.
23. B. Berkowitz, J. Klafter, R. Metzler, and H. Scher, "Physical pictures of transport in heterogeneous media: advection-dispersion, random-walk, and fractional derivative formulations," Water Resources Research, vol. 38, no. 10, pp. 9-1–9-12, 2002.
24. J. D. Seymour, J. P. Gage, S. L. Codd, and R. Gerlach, "Anomalous fluid transport in porous media induced by biofilm growth," Physical Review Letters, vol. 93, no. 19, Article ID 198103, 4 pages, 2004.
25. Y. Luchko and A. Punzi, "Modeling anomalous heat transport in

geothermal reservoirs via fractional diffusion equations," GEM—International Journal on Geomathematics, vol. 1, no. 2, pp. 257–276, 2011.

26. J. Klafter, S. C. Lim, and R. Metzler, Fractional Dynamics, Recent Advances, World Scientific, Singapore, 2012.
27. P. Chun and J. Inoue, "Yet another possible mechanism for anomalous transport: theory, numerical method, and experiments," KSCE Journal of Civil Engineering, vol. 16, no. 1, pp. 45–53, 2012.
28. A. C. McBride and F. H. Kerr, "On Namias's fractional Fourier transforms," IMA Journal of Applied Mathematics, vol. 39, no. 2, pp. 159–175, 1987.
29. H. M. Ozaktas and O. Aytür, "Fractional Fourier domains," Signal Processing, vol. 46, no. 1, pp. 119–124, 1995.
30. A. W. Lohmann, D. Mendlovic, and Z. Zalevsky, "Fractional Hilbert transform," Optics Letters, vol. 21, no. 4, pp. 281–283, 1996.
31. N. Zhou, Y. Wang, and L. Gong, "Novel optical image encryption scheme based on fractional Mellin transform," Optics Communications, vol. 284, no. 13, pp. 3234–3242, 2011.
32. Y. Luchko and V. Kiryakova, "The Mellin integral transform in fractional calculus," Fractional Calculus and Applied Analysis, vol. 16, no. 2, pp. 405–430, 2013.
33. D. Mendlovic, Z. Zalevsky, D. Mas, J. García, and C. Ferreira, "Fractional wavelet transform," Applied Optics, vol. 36, no. 20, pp. 4801–4806, 1997.
34. J. Shi, N. Zhang, and X. Liu, "A novel fractional wavelet transform and its applications," Science China: Information Sciences, vol. 55, no. 6, pp. 1270–1279, 2012.
35. E. Dinç and D. Baleanu, "Fractional wavelet transform for the quantitative spectral resolution of the composite signals of the active compounds in a two-component mixture," Computers & Mathematics with Applications, vol. 59, no. 5, pp. 1701–1708, 2010.
36. H. Y. Fan, "Fractional Hankel transform studied by charge-amplitude state representations and complex fractional Fourier transformation," Optics Letters, vol. 28, no. 22, pp. 2177–2179, 2003.
37. E. I. Jury, Theory and Applications of the Z-Transform Method, John Wiley & Sons, New York, NY, USA, 1964.
38. J. A. T. Machado, "Analysis and design of fractional-order digital

control systems," Systems Analysis Modelling Simulation, vol. 27, no. 2-3, pp. 107–122, 1997.

39. X. J. Yang, Advanced Local Fractional Calculus and Its Applications, World Science, New York, NY, USA, 2012.
40. X. J. Yang, Local Fractional Functional Analysis and Its Applications, Asian Academic Publisher, Hong Kong, China, 2011.
41. X. J. Yang and D. Baleanu, "Fractal heat conduction problem solved by local fractional variation iteration method," Thermal Science, vol. 17, no. 2, pp. 625–628, 2013.
42. X. J. Yang, D. Baleanu, and W. P. Zhong, "Approximation solutions for diffusion equation on Cantor time-space," Proceeding of the Romanian Academy A, vol. 14, no. 2, pp. 127–133, 2013.
43. Y. Zhao, D. Baleanu, C. Cattani, D. F. Cheng, and X. J. Yang, "Maxwell's equations on Cantor sets: a local fractional approach," Advances in High Energy Physics, vol. 2013, Article ID 686371, 6 pages, 2013.
44. X. J. Yang, D. Baleanu, and J. A. T. Machado, "Mathematical aspects of the Heisenberg uncertainty principle within local fractional Fourier analysis," Boundary Value Problems, no. 1, article 131, 20132013.
45. F. Gao, W. P. Zhong, and X. M. Shen, "Applications of Yang-Fourier transform to local fractional equations with local fractional derivative and local fractional integral," Advanced Materials Research, vol. 461, pp. 306–310, 2012.
46. Y. Zhao, D. Baleanu, M. C. Baleanu, D. F. Cheng, and X. J. Yang, "Mappings for special functions on Cantor sets and special integral transforms via local fractional operators," Abstract and Applied Analysis, vol. 2013, Article ID 316978, 6 pages, 2013.
47. J.-H. He, "Asymptotic methods for solitary solutions and compactons," Abstract and Applied Analysis, vol. 2012, Article ID 916793, 130 pages, 2012.
48. C. F. Liu, S. S. Kong, and S. J. Yuan, "Reconstructive schemes for variational iteration method within Yang-Laplace transform with application to fractal heat conduction problem," Thermal Science, vol. 17, no. 3, pp. 715–721, 2013.
49. C. G. Zhao, A. M. Yang, H. Jafari, and A. Haghbin, "The Yang-Laplace transform for solving the IVPs with local fractional derivative," Abstract and Applied Analysis, vol. 2014, Article ID 386459, 5 pages, 2014.

50. S. Q. Wang, Y. J. Yang, and H. K. Jassim, "Local fractional function decomposition method for solving inhomogeneous wave equations with local fractional derivative," Abstract and Applied Analysis, vol. 2014, Article ID 176395, 7 pages, 2014.
51. M. Li and W. Zhao, "On bandlimitedness and lag-limitedness of fractional Gaussian noise," Physica A, vol. 392, no. 9, pp. 1955–1961, 2013.
52. C. Cattani, "Harmonic wavelet approximation of random, fractal and high frequency signals," Telecommunication Systems, vol. 43, no. 3-4, pp. 207–217, 2010.
53. J. Nagler and J. C. Claussen, "$1/f^{\alpha}$ spectra in elementary cellular automata and fractal signals," Physical Review E, vol. 71, no. 6, Article ID 067103, 4 pages, 2005.
54. A. Gupta and S. Joshi, "Variable step-size LMS algorithm for fractal signals," IEEE Transactions on Signal Processing, vol. 56, no. 4, pp. 1411–1420, 2008.
55. B. J. West, Fractal Physiology and Chaos in Medicine, World Scientific, Singapore, 2013.
56. M. Li, "Power spectrum of generalized fractional Gaussian noise," Advances in Mathematical Physics, vol. 2013, Article ID 315979, 3 pages, 2013.
57. M. Li, "Modeling autocorrelation functions of long-range dependent teletraffic series based on optimal approximation in Hilbert space—a further study," Applied Mathematical Modelling, vol. 31, no. 3, pp. 625–631, 2007.
58. M. Li and S. C. Lim, "Power spectrum of generalized Cauchy process," Telecommunication Systems, vol. 43, no. 3-4, pp. 219–222, 2010.
59. M. Li, "A class of negatively fractal dimensional Gaussian random functions," Mathematical Problems in Engineering, vol. 2011, Article ID 291028, 18 pages, 2011.
60. G. Yi, "Local fractional Z transform in fractal space," Advances in Digital Multimedia, vol. 1, no. 2, pp. 96–102, 2012.

CHAPTER 10

FAST TRANSFORMS IN IMAGE PROCESSING: COMPRESSION, RESTORATION, AND RESAMPLING

Leonid P. Yaroslavsky

Department of Physical Electronics, School of Electrical Engineering, Tel Aviv University, 69978 Tel Aviv, Israel

ABSTRACT

Transform image processing methods are methods that work in domains of image transforms, such as Discrete Fourier, Discrete Cosine, Wavelet, and alike. They proved to be very efficient in image compression, in image restoration, in image resampling, and in geometrical transformations and

CITATION (APA): Yaroslavsky, L. P. (2014). Fast transforms in image processing: compression, restoration, and resampling. Advances in Electrical Engineering, 2014. (23 pages)

can be traced back to early 1970s. The paper reviews these methods, with emphasis on their comparison and relationships, from the very first steps of transform image compression methods to adaptive and local adaptive filters for image restoration and up to "compressive sensing" methods that gained popularity in last few years. References are made to both first publications of the corresponding results and more recent and more easily available ones. The review has a tutorial character and purpose.

INTRODUCTION: WHY TRANSFORMS? WHICH TRANSFORMS?

It will not be an exaggeration to assert that digital image processing came into being with introduction, in 1965 by Cooley and Tukey, of the Fast Fourier Transform algorithm (FFT, [1]) for computing the Discrete Fourier Transform (DFT). This publication immediately resulted in impetuous growth of activity in all branches of digital signal and image processing and their applications.

The second wave in this process was inspired by the introduction into communication engineering and digital image processing, in the 1970s, of Walsh-Hadamard transform and Haar transform [2] and the development of a large family of fast transforms with FFT-type algorithms [3–5]. Whereas Walsh-Hadamard and Haar transforms have already been known in mathematics, other transforms, for instance, quite popular at the time Slant Transform [6], were being invented "from scratch." This development was mainly driven by the needs of data compression, though the usefulness of transform domain processing for image restoration and enhancement was also recognized very soon [3]. This period ended up with the introduction of the Discrete Cosine Transform (DCT, [7, 8]), which was soon widely recognized as the best choice among all available at the time transforms and resulted in JPEG and MPEG standards for image, audio, and video compression.

The third large wave of activities in transforms for signal and image processing was caused by the introduction, in the 1980s, of a family of transforms that was coined the name "wavelet transform" [9]. The main motivation was achieving a better local representation of signals and images in contrast to the "global" representation that is characteristic to Discrete Fourier, DCT, Walsh-Hadamard, and other fast transforms available at the time. During 1980s–1990s a large variety of discrete wavelet transforms were suggested [10] for solving various tasks in signal and in image processing.

Presently, fast transforms with FFT-type fast algorithms and wavelet transforms constitute the basic instrumentation tool of digital image processing.

The main distinctive feature of transforms that makes them so efficient in digital image processing is their energy compaction capability. In regular image representations, in form of sets of ordered pixels, some pixels, for instance, those that belong to object borders, are more important than the others and there are always some pixels in each particular image that are of no such importance and can be dropped out from image representation and restored from the remaining "important" pixels. But the problem is that one never knows in advance which pixels in the image are "important" and which are not.

The situation is totally different in image representation in transform domain. For orthogonal transforms that feature good energy compaction capability, a lion share of total image "energy" (sum of squared transform coefficients) is concentrated in a small fraction of transform coefficients, which indices are, as a rule, known in advance for the given type of images or can be easily detected. It is this feature of transforms that is called their energy compaction capability. It allows replacing images with their "band-limited," in terms of a specific transform, approximations, that is, approximations defined by a sufficiently small fraction of image transform coefficients.

The optimal transform with the best energy compaction capability can be defined for groups of images of the same type or ensembles of images that are subjects of processing. Customarily, the "band-limited" image approximation accuracy is evaluated in terms of the root mean square approximation error (RMSE) for the image ensemble. By virtue of this, the optimal transform that secures the least band limited approximation error is defined by the ensemble correlation function, that is, image autocorrelation function averaged over the image ensemble. For continuous (not sampled) signals, this transform is called the Karhunen-Loeve Transform[11, 12]. Its discrete analog for sampled signals is called the Hotelling transform [13] and the result of applying this transform to sampled signals is called signal principal component decomposition. Karhunen-Loeve and Hotelling transforms provide the theoretical lower bound for compact (in terms of the number of terms in signal decomposition) signal discrete representation for RMS criterion of signal approximation.

However, being optimal in terms of the energy compaction capability, Karhunen-Loeve and Hotelling transforms have, generally, high computational complexity: the per pixel number of operations required for their computation is of the order of image size. This is why for practical needs only fast transforms that feature fast transform algorithms with computational complexity of the order of the logarithm of image size are considered. A register of the most relevant fast transforms, their main characteristic features, and application areas is presented in Table 1. Their mathematical definitions are given in Table 2.

Table 1: Register of relevant fast transforms, their main characteristic features, and application areas

	Relevance to imaging optics	**Main characteristic features**	**Main application areas**
Discrete Fourier Transforms	Represents optical Fourier Transform	Cyclic shift invariance vulnerable to boundary effects	(i) Analysis of periodicities. (ii) Fast convolution and correlation. (iii) Fast and accurate image resampling. (iv) Image compression. (v) Image denoising and deblurring. (vi) Numerical reconstruction of holograms.
Discrete Cosine Transform	Represents optical Fourier Transform	Cyclic shift invariance (with double cycle); virtually not sensitive to boundary effects	
Discrete Fresnel Transforms	Represent optical Fresnel Transform	Computable through DFT/DCT	Numerical reconstruction of holograms
Walsh-Hadamard Transform	No direct relevance	Binary basis functions. Provides piece-wise constant separable image band-limited approximations	(i) Image compression (marginal). (ii) Coded aperture imaging.
Haar Transform and other Discrete Wavelet Transforms	Subband decomposition	Binary basis functions. The fastest algorithm. Multiresolution. Shift invariance in each particular scale.	(i) Signal/image wideband noise denoising. (ii) Image compression.

Table 2: Selected fast transforms most relevant for digital image processing and their computational complexity (in 1D denotations)

Name	Definition	Computational complexity (operations per signal sample) "Global": applied to entire signal of N samples	"Local": applied in mowing window of WSz pixels
Discrete Cosine Transform (DFT)	$\alpha_r = \frac{1}{\sqrt{N}} \sum_{k=0}^{N-1} a_k \cos\left(2\pi \frac{k+1/2}{N} r\right)$	Real number operations $C_{DCT}^{G} \log N$	Real number operations $C_{DCT}^{L}(<WSz)$
Discrete Fourier Transforms			
Canonic DFT	$\alpha_r = \frac{1}{\sqrt{N}} \sum_{k=0}^{N-1} a_k \exp\left(i2\pi \frac{kr}{N}\right)$	Complex number operations $C_{DFT}^{G} \log N$	Complex number operations $C_{DFT}^{L}(<WSz)$
General Shifted Scaled DFT	$\alpha_r = \frac{1}{\sqrt{N}} \sum_{k=0}^{N-1} a_k \exp\left[i2\pi \frac{(k+u)(r+v)}{\sigma N}\right]$; u, v—shift parameters that indicate shifts $u\Delta x$, $v\Delta f$ of sampling lattices in signal and spectrum domains; σ—sampling scale parameter ($\Delta x \Delta f = 1/\sigma N$)		
Discrete Fresnel Transforms (DFrT)			
Canonic DFrT	$\alpha_r = \frac{1}{\sqrt{N}} \sum_{k=0}^{N-1} a_k \exp\left[i2\pi\left(\frac{k\mu - r/\mu}{N}\right)\right]$; $\mu \geq 1$ $\mu = \sqrt{\lambda Z / N \Delta f^2}$	Complex number operations $C_{DFrT}^{G} \log N$	n/a
Convolutional DFrT	$\alpha_r = \frac{1}{\sqrt{N}} \sum_{k=0}^{N-1} a_k \exp\left[i2\pi \frac{(k-r)^2}{\mu^2 N}\right]$; $\mu \leq 1$		
Walsh-Hadamard Transform			
Hadamard Transform	$\alpha_r = \frac{1}{\sqrt{N}} \sum_{k=0}^{N-1} a_k \mathrm{had}_k(r)$; $\mathrm{had}_k(r) = (-1)^{\sum_{m=0}^{n-1} k_m r_m}$ $k = \sum_{m=0}^{n-1} k_m 2^m$; $r = \sum_{m=0}^{n-1} r_m 2^m$; $N = 2^n$	Addition operations only $C_{WHT}^{G} \log N$	n/a
Walsh Transform	$\alpha_r = \frac{1}{\sqrt{N}} \sum_{k=0}^{N-1} a_k \mathrm{wal}_k(r)$; $\mathrm{wal}_k(r) = (-1)^{\sum_{m=0}^{n-1} k_m r_{n-m-1}^{GrayCode}}$ $r_{n-m-1}^{GrayCode}$—binary digits of Gray code of r taken in bit reversal order		
Haar Transform (the simplest discrete wavelet transform)	$\alpha_r = \frac{1}{\sqrt{N}} \sum_{k=0}^{N-1} a_k \mathrm{har}_k(r)$; $\mathrm{har}_k(r) = 2^{\mathrm{msb}}(-1)^{k_{n-1-\mathrm{msb}}} \delta\left(r_{\mathrm{mod}\, 2^{\mathrm{msb}}} - \lfloor 2^{1-\mathrm{msb}} k \rfloor\right)$ Msb—index of the most significant nonzero bit of binary code of msb; $\lfloor \cdot \rfloor$—residual of $(\cdot)$	Addition operations only C_{Haar}^{G}	n/a

Different fast transforms have different energy compaction capability for different types of images. Figure 1illustrates the energy compaction capabilities of Discrete Fourier, Discrete Cosine, Walsh, and Haar transforms on a particular test image. It shows that for this image DCT demonstrates the best energy compaction capability: 95% of the image energy is contained in only 9.5% of all DCT spectral coefficients, whereas for DFT, Walsh and Haar transforms these fractions are 11%, 15%, and 9.8%, correspondingly. Note also that, though for Haar Transform the fraction of the "meaningful" transform coefficients (9.8%) is close to that for DCT, quality degradations of the band-limited approximation to the image are noticeably more severe.

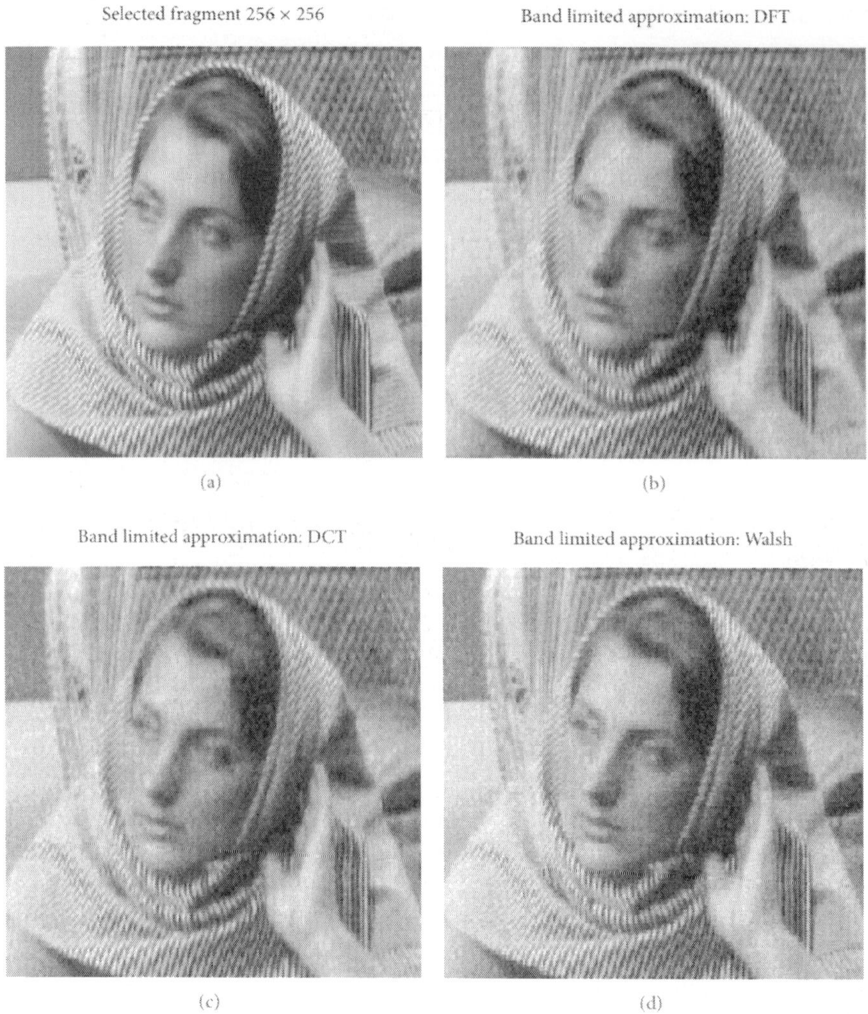
Selected fragment 256 × 256
Band limited approximation: DFT
(a)
(b)
Band limited approximation: DCT
Band limited approximation: Walsh
(c)
(d)

(e)

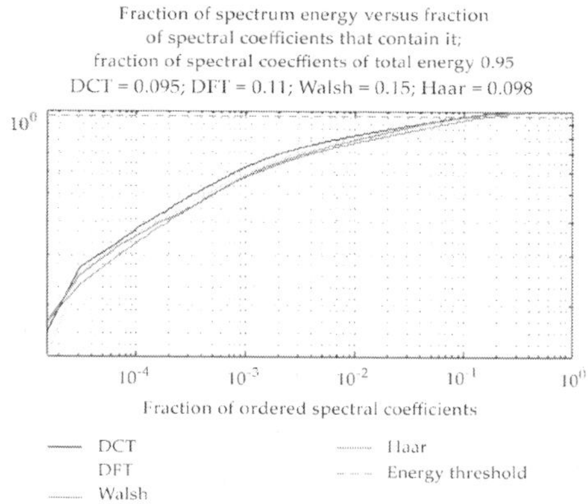

(f)

Figure 1: Illustration of the energy compaction capability of Discrete Fourier, Discrete Cosine, Walsh, and Haar transforms for a test image shown in the upper left corner. The rest of images show band-limited approximations of the test image in the domain of corresponding transforms for the approximation error variance equal to 5% of image variance. Graphs on the plot in the bottom right corner present energy contained in fractions of transform coefficients for different transforms.

Experience shows that DCT is in most applications advantageous with respect to other transforms in terms of the energy compaction capability. This property of DCT has a simple and intuitive explanation.

DCT of a discrete signal is essentially, to the accuracy of an unimportant exponential shift factor, DFT for the same signal extended outside its border to a double length by means of its mirrored, in the order of samples, copy [15, 16]. This way of signal extension has a profound implication on the speed of decaying to zero of signal DCT spectra and is a paramount and fundamental reason of the good energy compaction capability of DCT. DFT implies periodical extension of signals, which may cause severe discontinuities at signal borders and thus may require intensive high frequency components to reproduce them. DCT also implies periodical extension, however extension of not the signal itself but of its above-described extended copy. Thanks to such an "even" extension, the extended signal has no discontinuities at its borders, as well as at borders of the initial signal. Therefore such an extended signal is "smooth" and its DFT spectrum decays to zero faster than DFT spectrum of the initial, not extended signal.

It is noteworthy to mention that introducing DCT in [7] was motivated not by the above reasoning but by the finding that basis functions of DCT provide a good approximation to the eigenvectors of the class of Toeplitz matrices with entries $\{\mu_{k,} = \rho^{|k-l|}\}$; $\rho = 0.9$; $k, l = 0, 1, \ldots$; that is, DCT can be considered as a good approximation to the Karhunen-Loeve Transform for signals with the Toeplitz correlation matrix. In fact, the above-mentioned fundamental reason for good energy compaction capability of DCT is hidden in the even symmetry of the Toeplitz matrix.

In the family of orthogonal transforms, DFT and DCT occupy a special place. DFT and DCT are two discrete representations of the Integral Fourier Transform, which plays a fundamental role in signal and image processing as a mathematical model of signal transformations and of wave propagation in imaging systems [17, 18]. Thanks to this, DFT and DCT are applicable in much wider range of applications than the other fast transforms (see Table 1).

In conclusion of this section, come back to the common feature of fast transform, to the availability for all of these transform of fast algorithms of FFT type. Initially, fast transforms were developed only for signals with the number of samples equal to an integer power of two. These are the fastest algorithms. At present, this limitation is overcome, and fast transform algorithms exist for arbitrary number of signal samples. Note that 2D and 3D transforms for image and video applications are built as separable; that is, they work separably in each dimension. Note that the transform separability is the core idea for fast transform algorithms. All fast transform algorithms are based on representation of signal and transform spectrum sample indices as multidimensional, that is represented by multiple digits, numbers.

The data on computational complexity of the transform fast algorithms are collected in Table 2. These data are more or less commonly known. Not as widely known is the existence of so-called pruned algorithms for the cases, when transform input data contain substantial fraction of zero samples and/or one does not need to compute all transform coefficients [15–24]. These algorithms are useful, for instance, in numerical reconstruction of holograms and in interpolation of sampled data by means of transform zero padding [17, 25, 26].

In some applications it is advisable to apply transforms locally in running window rather than globally to entire image frames. For such application, Discrete Cosine and Discrete Fourier Transforms have an additional very useful feature. Computing these transforms in running window can be carried out recursively: signal transform coefficients at each window position can be found by quite simple modification of the coefficients found at the previous window position with per pixel computational complexity proportional to the window size rather than to the product of the window size and its logarithm, which is the computational complexity of the fast transforms [17, 26–28].

The rest of the paper is arranged as follows. In Section 2, applications of fast transforms for image data compression are reviewed. In Section 3, we proceed to transform methods for image restoration and enhancement. In Section 4, we show how DFT and DCT can be applied for image perfect, that is, error less, resampling. And finally in Section 5, we address the usage of fast transforms for solving a specific task of image resampling, the task of image recovery from sparse and irregularly taken samples, and, in particular, compressive sensing approach to this problem.

IMAGE COMPRESSION

Dilemma: Compressive Discretization or Compression

Ideally, image digitization, that is, converting continuous signals from image sensors into digital signals, should be carried out in such a way as to provide as compact image digital representation as possible provided that quality of image reproduction satisfies given requirements. Due to technical limitation, image digitization is most frequently performed in two steps: discretization, that is, converting image sensor signal into a set of real numbers, and scalar quantization, that is rounding off, of those numbers to a set of fixed quantized values [15, 18].

Image discretization can, in general, be treated as obtaining coefficients of image signal expansion over a set of discretization basis functions. In order to make the set of these representation coefficients as compact as possible, one should choose discretization basis functions that secure the least number of the signal expansion coefficients sufficient for image reconstruction with a given required quality. One can call this general method of signal discretization "Compressive discretization" because it secures the most compact discrete representation of signals, which cannot be further compressed. Note that this term should not be confused with terms "Compressive sensing" and "Compressive sampling," that gained large popularity in recent years [29–36]. We will discuss the "compressive sensing" approach later in Section 5.

There are at least two examples of practical implementation of the principle of general discretization by means of measuring transform domain image coefficients: Coded Aperture Imaging and Magnetic Resonance Imaging(MRI). In coded aperture imaging, images are sensed through binary masks that implement binary basis functions. In MRI imaging, coefficients of Fourier series expansion of sensor data are measured.

However, by virtue of historical reasons and of the technical tradition, image discretization is most frequently in imaging engineering is implemented as image sampling at nodes of a uniform rectangular sampling lattice using sampling basis functions, which are formed from one "mother" function by its shifts by multiple of fixed intervals called "sampling" intervals.

The theoretical foundation of image sampling is the sampling theorem. The traditional formulation of the sampling theorem states that signals with Fourier spectrum limited with bandwidth B can be perfectly restored from their samples taken with sampling interval $\Delta x = 1/B$, commonly called the Nyquist sampling interval [37–39].

In reality no continuous signal is band-limited, and the image sampling interval is defined not through specifying, in one or another way, of the image bandwidth, but directly by the requirement to reproduce smallest objects and borders of larger objects present in images sufficiently well. The selected in this way image sampling interval Δx specifies image base band $B = 1/\Delta x$.

Since small objcets and object borders usually occupy relatively small fraction of the image area, vast portions of images are oversampled, that is, sampled with redundantly small sampling intervals. Hence, substantial compression of image sampled representation is possible. It can be implemented by means of applying to the image primary and redundant sampled representation a discrete analog of the general compressive discretization, that is, by means of image expansion over a properly chosen set of discrete optimal bases functions and limitation of the amount of the expansion coefficients. This is exactly what is done in all transform methods of image compression.

Transform Methods of Image Compression

Needs of image compression were the primary motivations of digital image processing. Being started in 1950s from various predictive coding methods (a set of good representative publications can be found in [40]), by the beginning of 1970s image compression research began concentrating mostly on what we call now "transform coding" [41, 42], which literally implements the compressed discretization principle.

The principle of image transform coding is illustrated by the flow diagrams sketched in Figure 2.

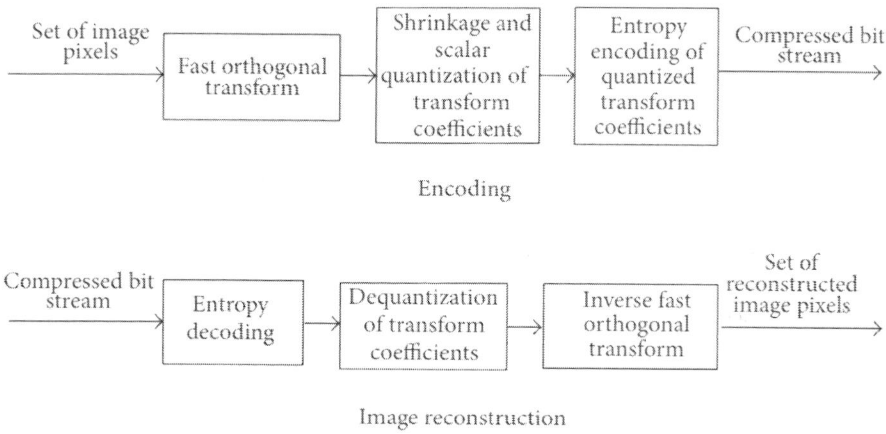

Figure 2: Flow diagrams of image transform coding and reconstruction.

According to these diagrams, set of image pixels is first subjected to a fast orthogonal transform. Then low intensity transform coefficients are discarded, which substantially reduces the volume of data. This is the main source of image compression. Note that discarding certain image transform coefficients means replacement of images by their band-limited, in terms of the selected transform, approximations. The remaining coefficients are subjected, one by one, to optimal nonuniform scalar quantization that minimizes the average number of transform coefficient quantization levels. Finally, quantized transform coefficients are entropy encoded to minimize the average number of bits per coefficient. For image reconstruction from the compressed bit stream, it must be subjected to corresponding inverse transformations.

In 1970s, main activities of researches were aimed at inventing, in addition to known at the time Discrete Fourier, Walsh-Hadamard, and Haar Transforms, new transforms that have guaranteed fast transform algorithms and improved energy compaction capability [3–5].

This transform invention activity gradually faded after introduction, in a short note, of the Discrete Cosine transform [7], which, as it was already mentioned, was finally recognized as the most appropriate transform for image compression due to its superior energy compaction capability and the availability of the fast algorithm. But the true breakthrough happened, when it was realized that much higher image compression efficiency can be achieved, if transform coding is applied not to entire image frames but blockwise [41, 42].

Images usually contain many objects, and their global spectra are a mixture of object spectra, whereas spectra of individual image fragments or blocks are much more specific and this enables much easier separation of "important" (most intense) from "unimportant" (least intense) spectral components. This is vividly illustrated in Figure 3, where global and block wise DCT power spectra of a test image are presented for comparison. Although, as one can see in the figure, spectral coefficients of DCT global spectrum are more sparse (spectrum sparsity of this image, i.e., fraction of spectral coefficients that contain 95% of the total spectrum energy, is 0.095) than, on average, those of image of 16 × 16 pixel fragments (average sparsity of spectra fragments is 0.14), spectra of individual fragments are certainly more specific and their components responsible for small objects and object borders have much higher energy than in the global spectrum and will not be lost when 95% of the most intense components are selected.

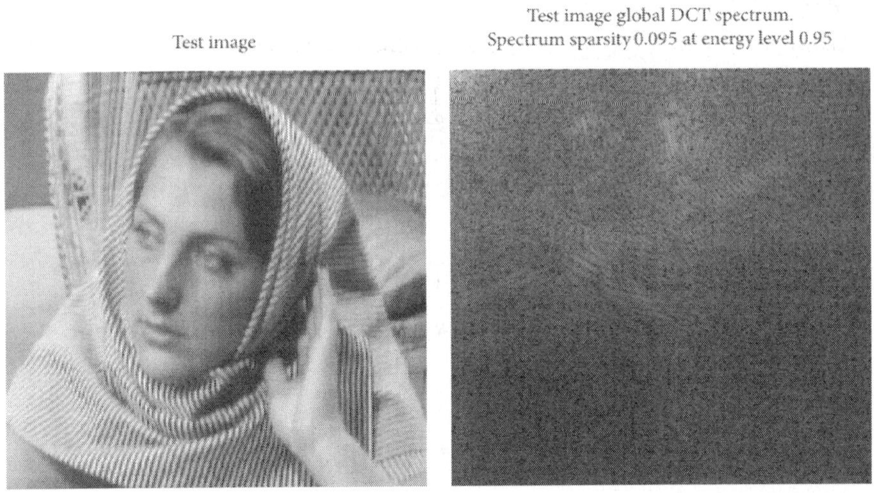

(a) (b)

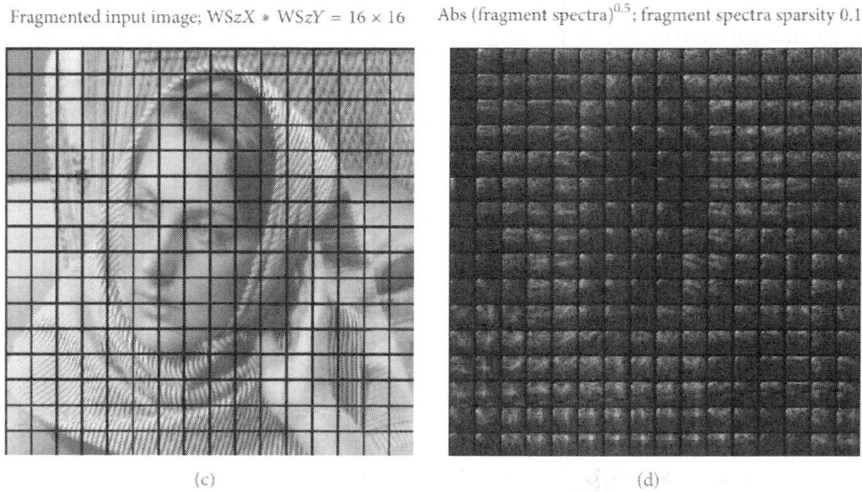

Figure 3: Comparison of global and local (blockwisc) image DCT spectra.

Blockwise image DCT transform compression proved to be so successful, that, by 1992, it was put in the base of the image and video coding standards "JPEG" and "MPEG" [43–46], which eventually resulted in the revolution in the industry of photographic and video cameras and led to the emergence of digital television as well [47].

Though the standard does not fix the size of blocks, blocks of 8×8 pixels are used in most cases. Obviously, the compression may, in principle, be more efficient, if the block size is adaptively selected in different areas of images; hence the use of variable size of blocks was discussed in number of publications (see, for instance, [48]).

TRANSFORM DOMAIN FILTERS FOR IMAGE RESTORATION AND ENHANCEMENT

Transform Domain Scalar Empirical Wiener Filters

Very soon after the image transform compression methods emerged, it was recognized that transforms represent a very useful tool for image restoration from distortions of image signals in imaging system and for image enhancement [49, 50]. The principle is very simple: for image improvement, image transform coefficients are modified in a certain way and then the image is reconstructed by the inverse transform of modified coefficients (Figure 4). This way of processing is called transform domain filtering.

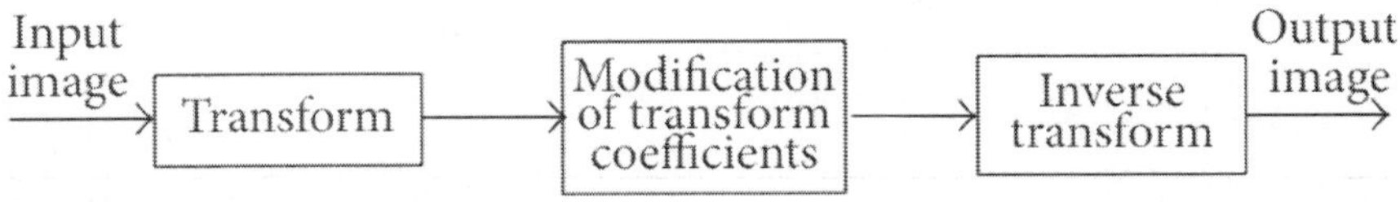

Figure 4: Flow diagram of transform domain filtering.

For the implementation of modifications of image transform coefficients, two options are usually considered.

- Modification of absolute values of transform coefficients by a nonlinear transfer function; usually it is a function that compresses the dynamic range of transform coefficients, which redistributes the coefficients' intensities in favor of less intensive coefficients and results in contrast enhancement of image small details and edges.
- Multiplication of transform coefficients by scalar weight coefficients; this processing is called transform domain "scalar filtering" [49].

For defining optimal scalar filter coefficients, a Wiener-Kolmogorov [51, 52] approach of minimization of mean squared filtering error is used and thus filters implement empirical Wiener filtering principles [17, 18, 26].

Three modifications of the filters based on this principle are (i) proper Empirical Wiener Filters, (ii) Signal Spectrum Preservation Filters, and (iii) Rejecting Filters [18]. For image denoising from additive signal independent noise and image deblurring, they modify input image Fourier spectra Sp_{inp} at each frequency component according to the equations correspondingly.

Empirical Wiener Filter

$$Sp_{\text{out}} = \frac{1}{\text{ISFR}} \max\left(0, \frac{\left|Sp_{\text{inp}}\right|^2 - Sp_{\text{noise}}}{\left|Sp_{\text{inp}}\right|^2}\right) Sp_{\text{inp}}; \tag{1}$$

Signal Spectrum Preservation Filter

$$Sp_{\text{out}} = \frac{1}{\text{ISFR}} \max\left(0, \frac{\left|Sp_{\text{inp}}\right|^2 - Sp_{\text{noise}}}{\left|Sp_{\text{inp}}\right|^2}\right)^{1/2} Sp_{\text{inp}}; \tag{2}$$

Rejecting Filter

$$Sp_{\text{out}} = \begin{cases} \dfrac{Sp_{\text{inp}}}{\text{ISFR}}, & \text{if } \left|Sp_{\text{inp}}\right|^2 > Sp_{\text{noise}}, \\ 0, & \text{otherwise}, \end{cases} \tag{3}$$

where Sp_{out} is Fourier spectrum of output image, *Sp*noise is power spectrum of additive noise, assumed to be known or to be empirically evaluated from the input noisy image [15–18], and ISFR is the imaging system frequency response assumed to be known, for instance, from the imaging system certificate.

As one can see in these equations, all these filters eliminate image spectral components that are less intensive than those of noise and the remaining components are corrected by the "inverse" filter with frequency response of 1/ISFR. Division of image spectra by ISFR compensates image blur in the imaging system due to imperfect ISFR; that is, it performs image deblurring. It is applicable in processing specifically in domains of DFT or DCT, which are, as it was already mentioned, two complementing discrete representations of the integral Fourier Transform.

In addition, empirical Wiener filter modifies image spectrum through deamplification of image spectral components according to the level of noise in them; spectrum preservation filter modifies magnitude of the image spectrum as well by making it equal to a square root of image power spectrum empirical estimate as a difference between power spectrum of noisy image and that of noise. Rejecting filter does not modify remaining, that is, not rejected, spectral components at all. In some publications, image denoising using the spectrum preservation filter is called "soft thresholding" and rejecting filtering is called "hard thresholding" [53, 54].

Versions of these filters are filters that combine image denoising-deblurring and image enhancement by means of P-law nonlinear modifications of the corrected signal spectral components

$$Sp_{\text{out}} = \begin{cases} \dfrac{1}{\text{ISFR}\left|Sp_{\text{inp}}\right|^{(1-P)}} Sp_{\text{inp}}, & \text{if } \left|Sp_{\text{inp}}\right|^2 > Sp_{\text{noise}}, \\ 0, & \text{otherwise}, \end{cases} \tag{4}$$

where $P \leq 1$ is a parameter that controls the degree of spectrum dynamic range compression and, by this, the degree of enhancement of low intensity image spectral components [18].

The described filters proved to be very efficient in denoising so-called narrow band noises, that is, noises, such as "moirè noise" or "banding noise," that contain only few nonzero components in its spectrum in the selected transform. Some illustrative examples are given in Figures 5 and 6.

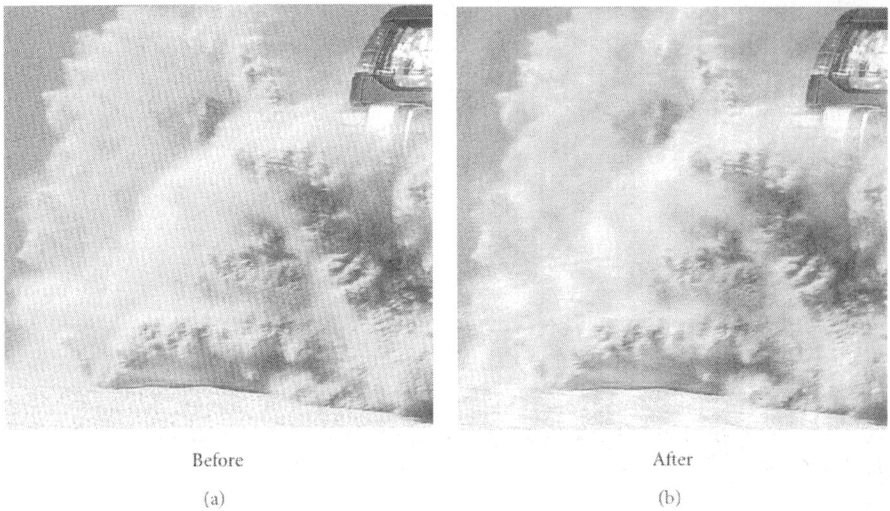

Figure 5: Image cleaning from moire noise.

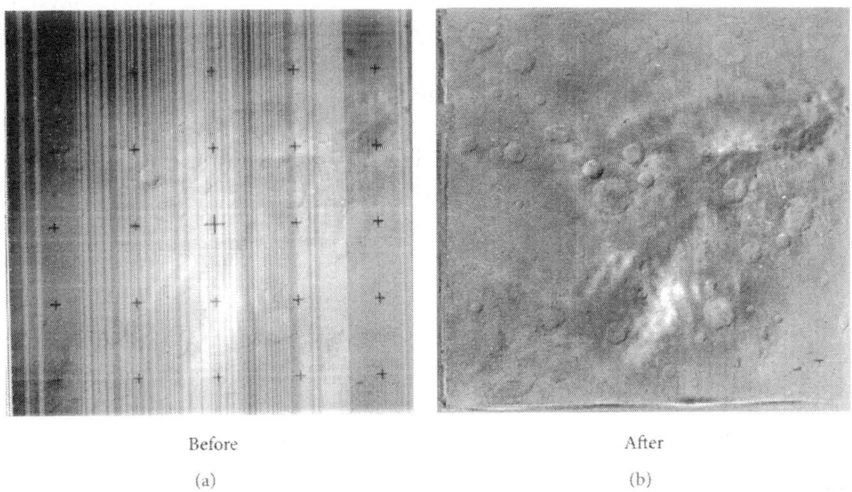

Figure 6: Cleaning a Mars satellite image from banding noise by means of separable (row-wise and column-wise) high pass rejecting filtering. Fiducial marks seen in the left image were extracted from the image before the filtering [14].

However for image cleaning from wideband or "white" noise, above transform domain filtering applied globally to entire image frames is not efficient. In fact it can even worsen the image visual quality. As one can see from an illustrative example shown in Figure 7, filtered image loses its sharpness and contains residual correlated noise, which is visually more annoying than the initial white noise.

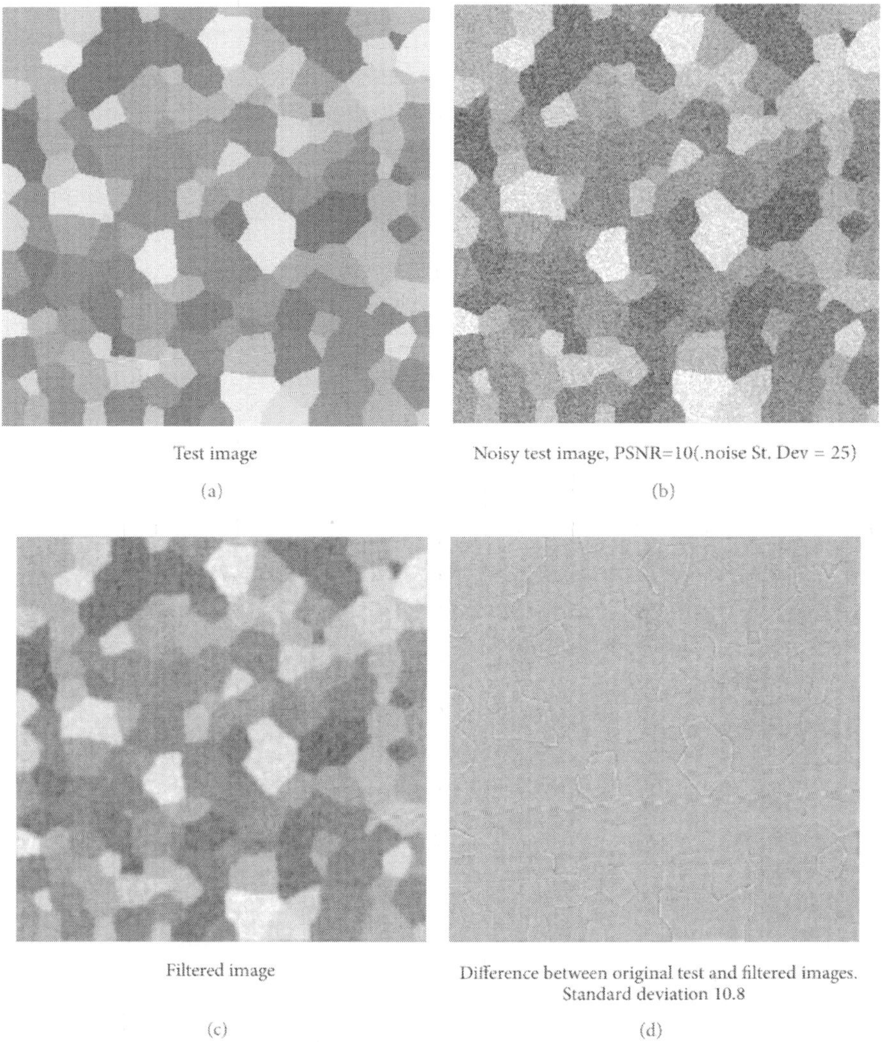

Test image (a)

Noisy test image, PSNR=10(.noise St. Dev = 25) (b)

Filtered image (c)

Difference between original test and filtered images. Standard deviation 10.8 (d)

Figure 7: Denoising of a piece-wise test image using empirical Wiener filter applied globally to the entire image.

The reason for this inefficiency of global transform domain filtering is the same as for the inefficiency of the global transform compression, which was discussed in Section 2.2: in global image spectra low intensity image components are hidden behind wideband noise spectral components. And the solution of this problem is the same: replace global filtering by local fragment wise filtering. This idea is implemented in local adaptive linear filters. Being applied locally, filters are becoming locally adaptive, as their frequency responses are determined, according to (1)–(4), by local spectra of image fragments within the filter window. Local adaptive linear filters date back to mid-1980s and were refined in subsequent years [16, 18, 26, 55]. As a theoretical base for the design of local adaptive filters, local criteria of processing quality were suggested [56].

It is instructive to note that this is not an accident that the evolution of human vision came up with a similar solution. It is well known that when viewing image, human eye's optical axis permanently hops chaotically over the field of view and that the human visual acuity is very nonuniform over the field of view. The field of view of a man is about 30°. Resolving power of man's vision is about 1'. However such a relatively high resolving power is concentrated only within a small fraction of the field of view that has size of about 2° (see, for instance, [57]); that is, the area of the acute vision is about 1/15th of the field of view. For images of 512 × 512 pixels this means window of roughly 35 × 35 pixels.

The most straightforward way to implement local filtering is to do it in a hopping window, just as human vision does, and this is exactly the way of processing implemented in the transform image coding methods. However "hopping window" processing, being very attractive from the computational complexity point of view, suffers from "blocking effects," artificial discontinuities that may appear, in the result of processing, at the edges of the hopping window. Blocking effects are artifacts characteristic for transform image coding. The desire to avoid them motivated the advancement of "Lapped" transforms [58], which are applied blockwise with the half of the block size overlap. Obviously, the ultimate solution of the "blocking effects" problem would be pixel by pixel processing in sliding window that scans image row-wise/column/wise.

Thus, local adaptive linear filters work in a transform domain in a sliding window and, at each position of the window, modify, according to the type of the filter defined by (1)–(4), transform coefficients of the image fragment within the window, and then compute an estimate of the window central pixel by means of the inverse transform of the modified transform coefficients. As an option, accumulation of estimates of the all window pixels overlapping in the process of image scanning by the filter window is possible as well [59]. As for the transform for local adaptive filtering, DCT proved to be the best choice in most cases. Figures 8 and 9 give examples of local adaptive filtering for image deblurring and enhancement.

Figure 8: Denoising and deblurring of a satellite image by means of local adaptive filtering. Top row: raw image and its fragments magnified for better viewing (marked by arrows); bottom row: resulting image and its corresponding magnified fragments.

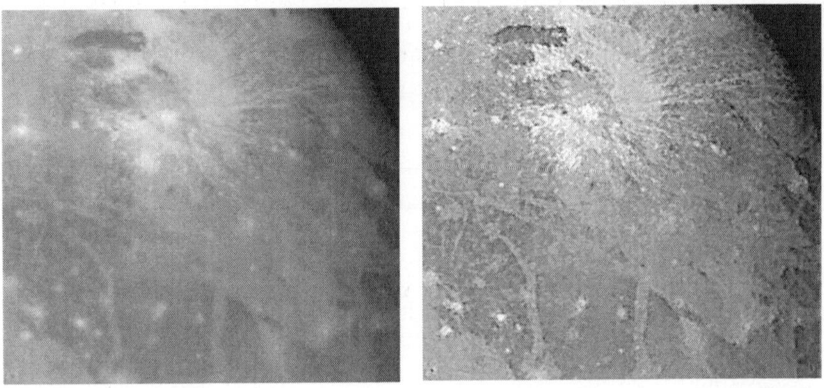

Before (a) After (b)

Figure 9: Enhancement of an astronomical image by means of Pth law modification of its spectrum.

Local adaptive filters can work with color and multicomponent images and videos as well. In this case, filtering is carried out in the corresponding multidimensional transform domains, for instance, domains of 3D (two spatial and one color component coordinates) spectra of color images or 3D spectra of a sequence of video frames (two spatial and time coordinates). In the latter case filter 3D window scans video frame sequence in both spatial and time coordinates. As one can see from Figures 10 and 11, the availability of the additional third dimension substantially adds to the filter efficiency [18, 55, 60].

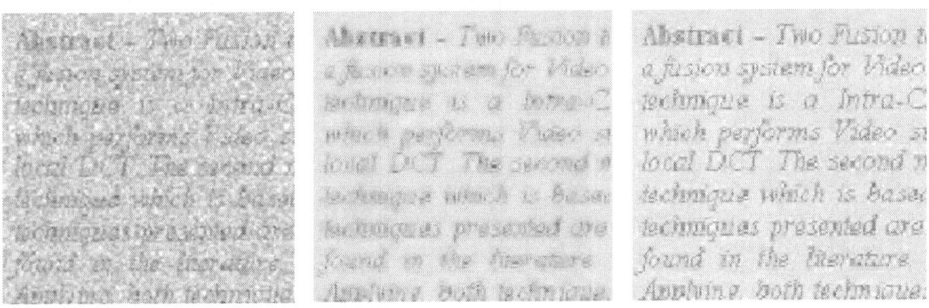

(a) (b) (c)

Figure 10: 2D and 3D local adaptive filtering of a simulated video sequence: (a) one of noisy frames of a test video sequence (image size 256 × 256 pixels); (b) a result of 2D local adaptive empirical Wiener filtering (filter window 5×5 pixels); (c) a result of 3D local adaptive empirical Wiener filtering (filter window 5×5 pixels × 5 frames).

Figure 11: 3D local adaptive empirical Wiener filtering for denoising and deblurring of a real thermal video sequence: (a) a frame of the initial video sequence; (b) a frame of a restored video sequence. Filter window is (5×5 pixels) × 5 frames; image size is 512 × 512 pixels.

Certain further improvement of image denoising capability can be achieved if, for each image fragment in a given position of the sliding window, similar, according to some criterion, image fragments over the whole image frame are found and included in the stack of fragments, to which the 3D transform domain filtering is applied. This approach in its simplest form was coined the name "nonlocal means" filtering [61–63]. In fact, similar method has been much earlier known as the correlational averaging (see, for instance, [64]). It assumes finding similar image fragments by their cross-correlation coefficient with a template object and averaging of similar fragments, that is, leaving only the dc component of the 3D spectra in this dimension. In [65] the simple averaging was replaced by full scale 3D transform domain filtering. One should, however, take into account that the correlational averaging is prone to produce artifacts due to false image fragments that may be erroneously taken to the stack just because of the presence of noise that has to be cleaned [66].

Obviously, image restoration efficiency of the sliding window transform domain filtering will be higher, if the window size is selected adaptively at each window position. To this goal, filtering can be, for instance, carried out in windows of multiple sizes and, at each window position, the best filtering result should be taken as the signal estimate at this position using methods of statistical tests, for instance, "intersection of confidence intervals method" described in [67]. Another option in multiple window processing is combining, in a certain way (say, by weighted summation or by taking a median or in some other way), filtering results obtained for different windows. All this, however, increases correspondingly the computational complexity of the filtering, which may become prohibitive, especially in video processing.

In conclusion of this section note that DFT and DCT spectra of image fragments in sliding window form the so called "time-frequency representation" of signals, which can be traced back to 1930s-1940s to Gabor and Wigner and works on "visible speech" [68–70]. For 1D signals, the dimensionality of the representation is 2D; for images, it is 3D. Such representation is very redundant and this opens an interesting option of applying to it image (for 1D signals) and video (for 2D signals) processing methods for signal denoising and, in general, signal separation.

Wavelet Shrinkage Filters for Image Denoising

In 1990s, a specific family of transform domain denoising filters, the so-called wavelet shrinkage filters, gained popularity after publications [53, 54, 71, 72]. The filters work in the domain of one of wavelet transforms and implement, for image denoising, the above-mentioned Signal Spectrum Preservation Filter (2) and Rejecting Filter (3), except that they do not include the "inverse filtering" component (1/ISFR) that corrects distortions of image spectra in imaging systems. As it was already mentioned, the filters were coined in these publications the names "soft thresholding" and "hard thresholding" filters, correspondingly.

Basis functions of wavelet transforms (wavelets) are formed by means of a combination of two methods of building transform basis functions from one "mother" function, shifting and scaling. Thanks to this, wavelets combine local sensitivity of shift basis functions and global properties of "scaled" basis functions and feature such attractive properties as multiresolution and good localization in both signal and transform domain. Thus, the wavelet shrinkage filters, being applied to entire image frames, possess both global adaptivity and local adaptivity in different scales and do not require for the latter specification of the window size, which is necessary for the sliding window filters.

These filters did show a good denoising capability. However, a comprehensive comparison of their denoising capability with that of sliding window DCT domain filters reported in [59] showed that even when for particular test images the best wavelet transform from a transform set [73] was selected, sliding window DCT domain filters demonstrated in most cases better image denoising results.

The capability of wavelets to represent images in different scales can be exploited for improving the image denoising performance of both families of filters and for overcoming the above-mentioned main drawback of sliding window DCT domain filters, the fixed window size that might not be optimal for different image fragments. This can be achieved by means of incorporating sliding window DCT domain (SWTD) filters into the wavelet filtering structure through replacing soft/hard thresholding of image wavelet decomposition components in different scales by their filtering with SWTD filters working in the window of the smallest size 3×3 pixels. This idea was implemented in hybrid wavelet/SWTD filters and proved to be fruitful in ([74], see also [55]).

Local Adaptive Filtering and Wavelet Shrinkage Filtering as Processing of Image Subband Decompositions

Obviously, sliding window transform domain and wavelet processing are just different implementation of the scalar linear filtering in transform domain. This motivates their unified interpretation in order to gain a deeper insight into their similarities and dissimilarities. The common base for such unified interpretation is the notion of signal subband decomposition [75]. One can show that both filter families can be treated as special cases of signal subband decomposition by band-pass filters with point spreads functions defined by the corresponding basis functions of the transform [26, 55]. From the point of view of signal subband decomposition, the main difference between sliding window transform domain and wavelet signal analysis is arrangement of bands in the signal frequency range. While for sliding window transform domain filtering subbands are uniformly arranged with the signal base band, subbands of the wavelet filters are arranged in a logarithmic scale. Hybrid wavelet/SWDCT filtering combines these two types of subband arrangements: "coarse" subband decomposition in logarithmic scale, provided by wavelet decomposition, is complemented with "fine" uniformly arranged sub-subbands within each wavelet subband, provided by the sliding window DCT filtering of the wavelet subbands.

It is curious to note that this "logarithmic coarse—uniform fine" subband arrangement resembles very much the arrangements of tones and semitones in music. In Bach's equal tempered scale, octaves are arranged in a logarithmic scale and 12 semitones are equally spaced within octaves [76].

IMAGE RESAMPLING AND BUILDING " CONTINUOUS" IMAGE MODELS

As it was indicated in the introductory section, Discrete Fourier and Discrete Cosine Transforms occupy the unique position among other orthogonal transforms. They are two versions of discrete representation of the integral Fourier Transform, the fundamental transform for describing the physical reality. Among applications specific for DFT and DCT are signal and image spectral analysis and analysis of periodicities, fast signal and image convolution and correlation, and image resampling and building "continuous" image models [17, 18, 26]. The latter is associated with the efficiency, with which DFT and DCT can be utilized for fast signal convolution.

Image resampling is a key operation in solving many digital image processing tasks. It assumes reconstruction of an approximation of the original nonsampled image by means of interpolation of available image samples to obtain samples "in-between" the available ones. The most feasible is interpolation by means of digital filtering implemented as digital convolution. A number of convolutional interpolation methods are known, beginning from the simplest and the least accurate nearest neighbor and linear (bilinear, for 2D case) interpolations to more accurate cubic (bicubic, for 2D case) and higher order spline methods [77, 78]. In some applications, for instance, in computer graphics and print art, the simplest nearest neighbor or linear (bilinear) interpolation methods can provide satisfactory results. In applications that are more demanding in terms of the interpolation accuracy, higher order spline interpolation methods gained popularity. The interpolation accuracy of spline interpolation methods grows with the spline order. However their computational complexity grows as well.

There exists a discrete signal interpolation method that is capable, given finite set of signal samples, to secure virtually error free interpolation of sampled signals, that is, the method that does not add to the signal any distortions additional to the distortions caused by the signal sampling. This method is the discrete sinc-interpolation [18, 25]. Interpolation kernel for the discrete sinc-interpolation of sampled signals of N samples is the discrete sinc-function $\text{sincd}x = \sin x/N \sin(x/N)$. This function replaces, for sampled signals with a finite number of samples, the continuous sinc-function $\text{sinc}\ x = \sin x/x$, which, according to the sampling theory, is the ideal interpolation kernel for reconstruction of continuous signals from their samples provided that the number of samples is infinitely large. Note that the discrete sinc-function should not be confused with a windowed or truncated sinc-function with a finite number of samples, which is very commonly believed to be a good replacement in numerical interpolation of the continuous sinc-function. The windowed sinc-function does not secure interpolation error free signal numerical interpolation, whereas the discrete sinc-interpolation does.

Fast Fourier transform and fast Discrete Cosine transform are the ideal computational means for implementing the convolutional interpolation by the discrete sinc-function. Of these two transforms, FFT has a substantial drawback: it implements the cyclic convolution with a period equal to the number N of signal samples, which is potentially prone to causing heavy oscillations at signal borders because of a discontinuity which is very probable between signal values at its opposite borders. This drawback is to a very high degree surmounted when signal convolution is implemented through DCT [18, 26]. As it was already indicated, convolution through DCT is also a cyclic convolution but with the period of $2N$ samples and of signals that are evenly extended to double length by their inversed, in the order of samples, copies. This completely removes the danger of appearance of heavy oscillations due to discontinuities at signal borders that are characteristic for discrete sinc-interpolation through processing in DFT domain.

Several methods of implementation of DFT/DCT based discrete sinc-interpolation. The most straightforward one is DFT/DCT spectrum zero padding. Interpolated signal is generated by applying inverse DFT (or, correspondingly, DCT) transform to its zero pad spectrum. Padding DFT/DCT spectra of signals of N samples with $L(N-1)$ zeroes produces LN samples of the discrete sinc interpolated initial signal with subsampling rate L, that is, with intersample distance $1/L$ of the initial sampling interval. The computational complexity of this method is (log LN) per output sample.

The same discrete sinc interpolated L times subsampled signal can be obtained more efficiently computation-wise by applying to the signal L times the signal perfect shifting filter [25, 79] with shifts by k/L at each kth application and by subsequent combining the shifted signal copies. The computational complexity of this method is (log N) per output sample.

The method is based on the property of Shifted DFT (SDFT, [18, 26, 80]): if one computes SDFT of a signal with some value of the shift parameter in signal domain and then computes inverse SDFT with another value of the shift parameter, the result will be a discrete sinc interpolated copy of the signal shifted by the difference between those values of the shift parameter. In order to avoid objectionable oscillations at signal borders, the DCT based version of this filter is recommended [81–83].

Image subsampling using the perfect shifting filter can be employed for creating "continuous" image models for subsequent image arbitrary resampling with any given accuracy [18, 84]. It can also be used for computing image correlations and image spectral analysis with a subpixel accuracy [18, 26]. Particular examples of using created in this way continuous signal models for converting image spectra from polar coordinate system to Cartesian coordinate system in the direct Fourier method for image reconstruction from projections and for converting (rebinning) of image fan beam projections into parallel projections one can find in [18].

Except for creating "continuous" image models, the perfect shifting filter is very well suited for image sheering in the three-step method for fast image rotation by means of three subsequent (horizontal/vertical/horizontal) image sheerings [85].

Perfect interpolation capability of the discrete sinc-interpolation was demonstrated in a comprehensive comparison of different interpolation methods in experiments with multiple 360° rotations reported in [25]. The results of experiments illustrated in Figure 12 clearly evidence that, in contrast to other methods, including high order spline interpolation ones [78, 86, 87], discrete sinc-interpolation does not introduce any appreciable distortions into the rotated image.

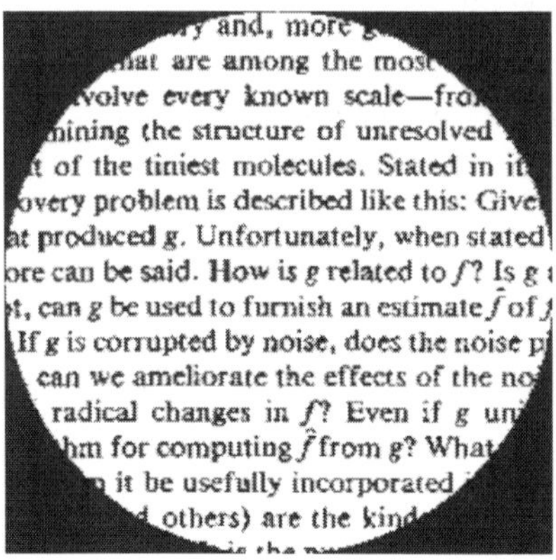

Test image

(a)

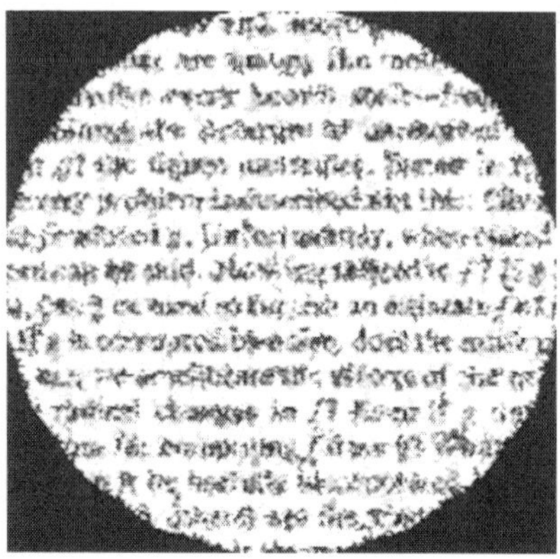

Nearest neighbor interpolation
(rotation by 1080° in 60 steps)

(b)

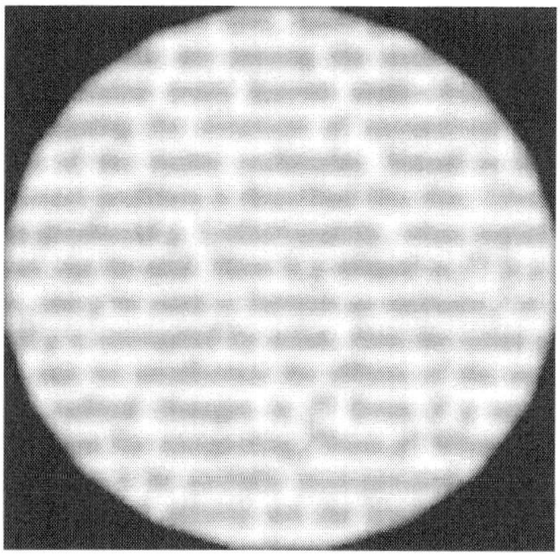

Bilinear interpolation
(rotation by 1080° in 60 steps)

(c)

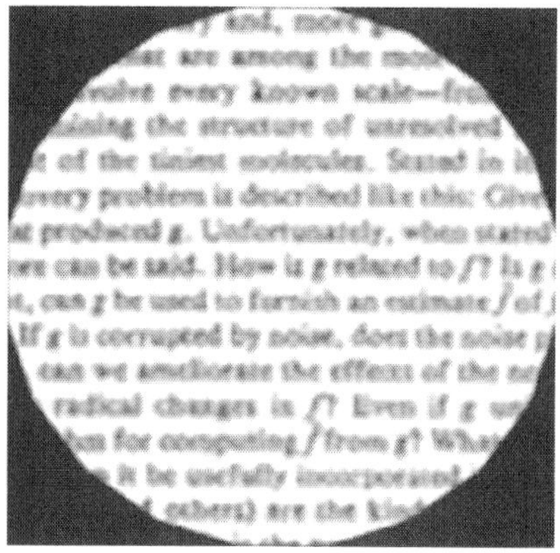

Bicubic interpolation
(rotation by 1080° in 60 steps)

(d)

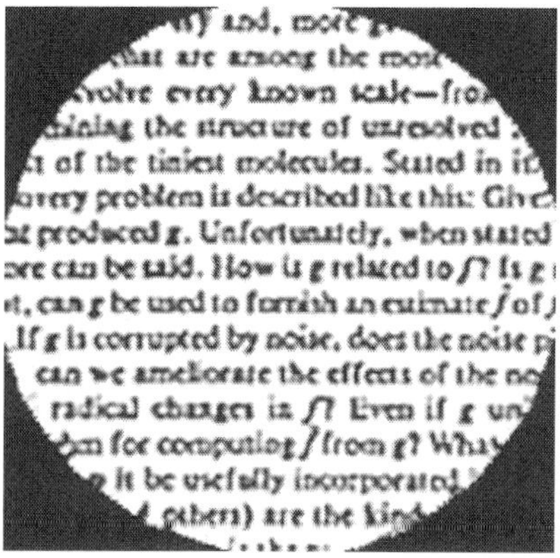

Spline interpolation Mems531 ([78], [87])
(rotation by 18000° in 1000 steps)

(e)

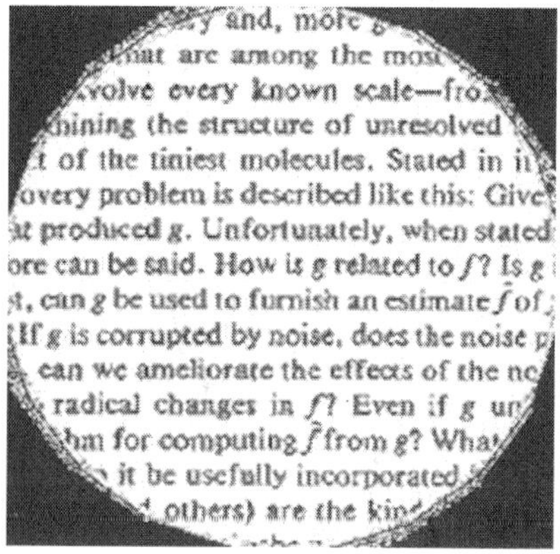

Discrete sinc-interpolation
(rotation by 18000° in 1000 steps)

(f)

Figure 12: Discrete sinc-interpolation versus other interpolation methods: results of multiple image rotations.

In some applications, "elastic" or space variant image resampling is required, when shifts of pixel positions are specified individually for each image pixel. In these cases, the perfect shifting filter can be applied to image fragments in sliding window for evaluating interpolated value of the window central pixel at each window position. Typical application examples are imitation of image retrieval through turbulent media (for illustration see [88]) and stabilization and superresolution of turbulent videos [89–91].

Being working in DCT transform domain, "elastic" resampling in sliding window can be easily combined with above-described local adaptive denoising and deblurring [25]. Yet another additional option is making resampling in sliding window adaptive to contents of individual image fragments by means of switching between the perfect shifting and another interpolation method better suited for specific image fragments. This might be useful in application to images that contain very heterogeneous fragments such as photographs and binary drawings and text [25].

A certain limitation of the perfect shifting filter in creating "continuous" image models is its capability of subsampling images only with a rate expressed by an integer or a rational number. In some cases, this might be inconvenient, for instance, when the required resampling rate is a value between one and two, say 1.1, 1.2, or alike. For such cases, there exists the third method of signal resampling with discrete sinc-interpolation. It is based on the general Shifted Scaled DFT, which includes arbitrary analog shift and scale parameters. Using Shifted Scaled (ShSc) DFT, one can apply to the image DFT spectrum inverse ShScDFT with the desired scale parameter and obtain a correspondingly scaled discrete sinc interpolated image [82, 83]. Being the discrete sinc-interpolation method, this method also does not cause any image blurring, which other interpolation methods are prone to do. Figure 13 illustrates this on an example of multiple iterative image zooming-in and zooming-out.

Figure 13: Discrete sinc-interpolation versus bilinear and bicubic interpolations in image iterative zoom-in/zoom-out with the scale parameter $\sqrt{2}$.

The Shifted Scaled DFT can be presented as a convolution and, as such, can be computed using Fast Fourier or Fast Cosine Transforms [18, 26]. Thus, the computational complexity of the method is (σ log N) per output signal sample, where σ is the scale parameter. As in all other cases, the use of DCT guaranties the absence of severe boundary effects. This method is especially well suited for numerical reconstruction of digitally recorded color holograms, when it is required to rescale images reconstructed from holograms recorded with different wave lengths of coherent illumination [82].

Among the image processing tasks, which involve "continuous" image models, are also signal differentiation and integration, the fundamental tasks of numerical mathematics that date back to such classics of mathematics as Newton and Leibnitz. One can find standard numerical differentiation and integration recipes in numerous reference books, for instance, [92, 93]. All of them are based on signal approximation through Taylor expansion. However, according to the sampling theory, approximation of sampled signal by Taylor expansion is not optimal, and the direct use of these methods for sampled signals may cause substantial errors in cases, when signals contain substantial high frequency components. In order to secure sufficiently accurate signal differentiation and integration by means of those standard algorithms, one must substantially oversample such signals.

The exact solution for the discrete representation of signal differentiation and integration provided by the sampling theory tells that, given the signal sampling interval and signal sampling and reconstruction devices, discrete frequency responses (in DFT domain) of digital filters for perfect differentiation and integrations should be, correspondingly, proportional and inversely proportional to the frequency index [18, 26]. This result directly leads to fast differentiation and integration algorithms that work in DFT or DCT domains using corresponding fast transforms with computational complexity (log N) operation per signal sample for signals of N samples. As in all cases, realization of the integration and, especially, differentiation filters in DCT domain is preferable because of much lower, for DCT, border effect artifacts associated with the finite number of signal samples.

The comprehensive comparison of the accuracy of standard numerical differentiation and integration algorithms with perfect DCT-based differentiation and integration reported in [25, 94] shows that the standard numerical algorithms in reality work not with original signals but with their blurred to a certain degree copies. This conclusion is illustrated in Figure 14 on an example of multiple alternative differentiations and integrations of a test rectangular impulse.

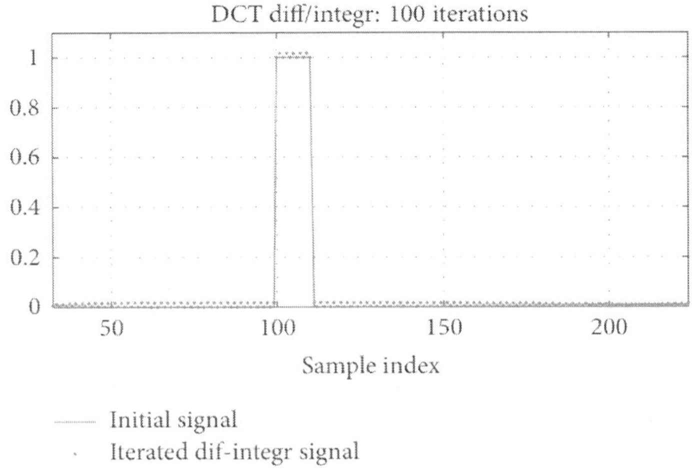

(a)

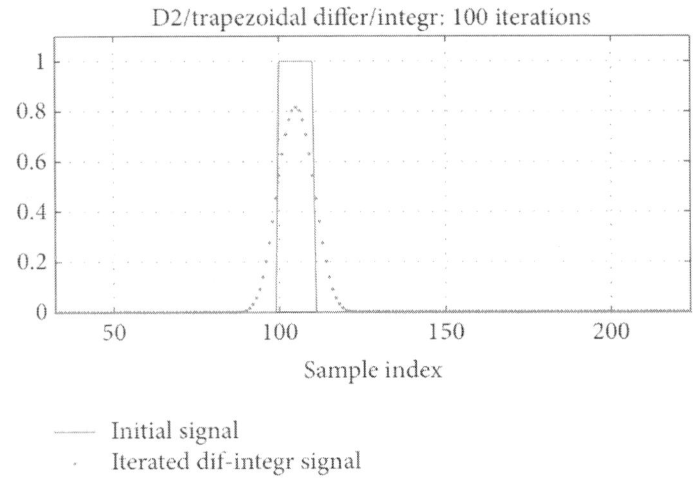

(b)

Figure 14: Comparison of results of iterative alternated 100 differentiations and integrations of a test rectangular signal using the DCT-based algorithms (left) and standard numerical algorithms (right, differentiation filter with point spread function [−0.5, 0, 0.5] and trapezoidal rule integration algorithm).

Computational efficiency of the DFT/DCT based interpolation error free discrete sinc-interpolation algorithms is rooted in the use of fast Fourier and Fast DCT transforms. Perhaps, the best concluding remark for this discussion of the discrete sinc-interpolation DFT/DCT domain methods would be mentioning that a version of what we call now Fast Fourier Transform algorithm was invented more than 200 hundred years ago by Gauss just for the purpose of facilitating numerical interpolation of sampled data of astronomical observation [95].

IMAGE RECOVERY FROM SPARSE AND IRREGULARLY SAMPLED DATA

Discrete Sampling Theorem Based Method

As we mentioned at the beginning of Section 2, image discretization is usually carried out by image sampling at nodes of a uniform rectangular sampling lattice. In this section we address using fast transforms for solving the problem closely associated with the general compressive discretization: how and with what accuracy can one recover an image from its sparse and irregularly taken samples.

There are many applications, where, contrary to the common practice of uniform sampling, sampled data are collected in irregular fashion. Because image display devices as well as computer software for processing sampling data assume using regular uniform rectangular sampling lattice, one needs in all these cases to convert irregularly sampled images to regularly sampled ones.

Generally, the corresponding regular sampling grid may contain more samples than it is available, because coordinates of positions of available samples might be known with a "subpixel" accuracy, that is, with the accuracy (in units of image size) better than $1/K$, where K is the number of available pixels. Therefore one can regard available K samples as being sparsely placed at nodes of a denser sampling lattice with the total amount of nodes $N>K$.

The general framework for recovery of discrete signals from a given set of their arbitrarily taken samples can be formulated as an approximation task in the assumption that continuous signals are represented in computers by their $K<N$ irregularly taken samples and it is believed that if all N samples in a regular sampling lattice were known, they would be sufficient for representing those continuous signals [18, 96]. The goal of the processing is generating, out of an incomplete set of K samples, a complete set of N signal samples in such a way as to secure the most accurate, in terms of the reconstruction mean square error (MSE), approximation of the discrete

signal, which would be obtained if the continuous signal it is intended to represent were densely sampled at all N positions.

Above-described discrete sinc-interpolation methods provide band-limited, in terms of signal Fourier spectra, approximation of regularly sampled signals. One can also think of signal band-limited approximation in terms of their spectra in other transforms. This approach is based on the Discrete Sampling Theorem [18, 96, 97].

The meaning of the Discrete Sampling Theorem is very simple. Given K samples of a signal one can, in principle, compute K certain coefficients of a certain signal transform and then reconstruct N samples of a band-limited, in this transform, approximation of the signal by means of inverse transform of those K coefficients supplemented with $N- K$ zero coefficients. If the transform has the highest, for this particular type of signals, energy compaction capability and selected were those nonzero transform coefficients that concentrate the highest, for this particular transform, signal energy, the obtained signal approximation will have the least mean square approximation error. If the signal is known to be band-limited in the selected transform and the computed nonzero coefficients corresponded to this band limitation, signal will be reconstructed precisely.

The discrete sampling theorem is applicable to signals of any dimensionality, though the formulation of the signal band-limitedness depends on the signal dimensionality. For 2D images and such transforms as Discrete Fourier, Discrete Cosine, and Walsh Transforms, the most simple is compact "low-pass" band-limitedness by a rectangle or by a circle sector.

It was shown in ([96], see also [18, 97]) that, for 1D signal and such transforms as DFT and DCT, signal recovery from sparse samples is possible for arbitrary positions of sparse samples. For images signals, the same is true, if band-limitation conditions are separable over the image dimensions. For nonseparable band-limitations, such as circle or circle sector, this may not be true and certain redundancy in the number of available samples might be required to secure exact recovery of band-limited images.

As it was indicated, the choice of the transform must be made on the base of the transform energy compaction capability for each particular class of images, to which the image to be recovered is believed to belong. The type of the band-limitation also must be based on a priori knowledge regarding the class of images at hand. The number of samples to be recovered is a matter of a priori belief of how many samples of a regular uniform sampling lattice would be sufficient to represent the images for the end user.

Implementation of signal recovery/approximation from sparse nonuniformly sampled data according to the discrete sampling theorem requires matrix inversion, which is, generally, a very computationally demanding procedure. In applications, in which one can be satisfied with image reconstruction with a certain limited accuracy, one can apply to the reconstruction a simple iterative reconstruction algorithm of the Gerchberg-Papoulis [98, 99] type, which, at each iteration step, alternatively applies band limitation in spectral domain and then restores available pixels in their given positions in the image obtained by the inverse transform.

Among reported applications, one can mention superresolution from multiple video frames of turbulent video and superresolution in computed tomography that makes use of redundancy of object slice images, in which usually a substantial part of their area is an empty space surrounding the object [96].

"Compressive Sensing": Promises and Limitations

The described discrete sampling theorem based methods for image recovery from sparse samples by means of their band-limited approximation in a certain transform domain require explicit formulation of the desired band limitation in the selected transform domain. While for 1D signal this formulation is quite simple and requires, for most frequently used low pass band limitation, specification of only one parameter, signal bandwidth, in 2D case formulation of the signal band limitation, requires specification of a 2D shape of signal band-limited spectrum. The simplest shapes, rectangle and circle sector ones, may only approximate, with a certain redundancy, real areas occupied by the most intensive image spectral coefficients for particular images. Figure 15 illustrates this on an example of spectral binary mask that corresponds to 9.5% of the DCT spectral coefficients of the test image shown in Figure 3 that contain 95% of the spectrum energy and a rectangular outline of this mask, which includes 47% of the spectral coefficients.

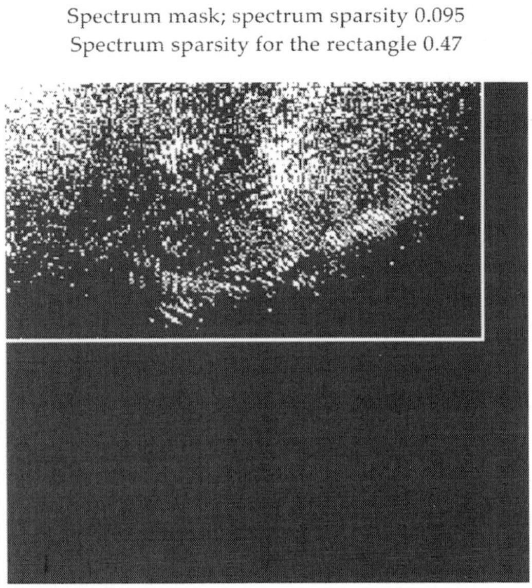

Figure 15: Spectral binary mask (shown white with the dc component in the left upper corner) that indicates components of DCT spectrum of the image in Figure 3, which contain 95% of the total spectrum energy and its rectangular outline.

In the cases, when exact character of spectrum band limitation is not known, image recovery from sparse samples can be achieved using the "compressive sensing" approach introduced in [29–32]. During last recent years, this approach to handling sparse spectral image representation has obtained considerable popularity [33–36].

The compressive sensing approach assumes obtaining a band-limited, in a certain selected transform domain, approximation of images as well, but it does not require explicit formulation of the image band-limitation and achieves image recovery from an incomplete set of samples by means of minimization of $L0$ norm of the image spectrum in the selected transform (i.e., of the number of signal nonzero transform coefficients), conditioned by preservation in the recovered image of its available pixels.

However, there is a price one should pay for the uncertainty regarding the band limitation: the number M of samples required for recovering N signal samples by this approach must be redundant with respect to the given number K nonzero spectral coefficients: $M=K \log N$ [30]. In real applications, this might be a serious limitation. For instance, for the test image of 256 × 256 pixels shown in Figure 3 spectrum sparsity (relative number of nonzero spectral components) on the energy level 95% is 0.095, whereas $\log N = \log(256 \times 256) = 16$, which means that the "compressive sensing" approach requires in this case more signal samples ($16 \cdot 9.5\% = 152\%$) than it is required to recover.

Discrete Signal Band-Limitedness and the Discrete Uncertainty Principle

Signal band-limitedness plays an important role in dealing with both continuous signals and discrete (sampled) signals representing them. It is well known that continuous signals cannot be both strictly band-limited and have strictly bounded support. In fact, continuous signals neither are band-limited nor have strictly bounded support. They can only be more or less densely concentrated in signal and spectral domains. This property is mathematically formulated in the form of the "uncertainty principle" [100]

$$X_{\varepsilon E} \times B_{\varepsilon E} > 1, \tag{5}$$

where $X_{\varepsilon E}$ and $B_{\varepsilon E}$ are intervals in signal and their Fourier spectral domains that contain a sufficiently large fraction εE of signal energy.

In distinction to that, sampled signals that represent continuous signals can be sharply bounded in both signal and spectral domains. This is quite obvious for some signal spectral presentation, such as Haar signal spectra: Haar basis functions are examples of sampled functions sharply bounded in both signal and Haar spectral domains. But it turns out that there exist signals that are sharply bounded in both signal and their spectral domains of DFT or DCT, which are discrete representations of the Fourier integral transform. An example of a space-limited/band-limited image is shown in Figure 16. Such images can be generated using the above-mentioned iterative Gershberg-Papoulis type algorithm which, at each iteration, applies to images alternatively space limitation in image domain and band limitation in transform domain. Such images are very useful as test images for testing image processing algorithms.

(a) (b)

Figure 16: Space-limited image "C. Shannon" (left) and its band-limited DFT spectrum (right; shown centered at the signal dc component).

In a similar way one can generate space-limited and band-limited analogs of the discrete sinc-function, i.e. functions, which form band-limited shift bases in the given space limits. In [97] such functions were coined a name "sinc-lets." Figure 17 shows an example of a sinc-let in its three positions within a limited interval and, for comparison, the discrete sinc-function for the same band limitation.

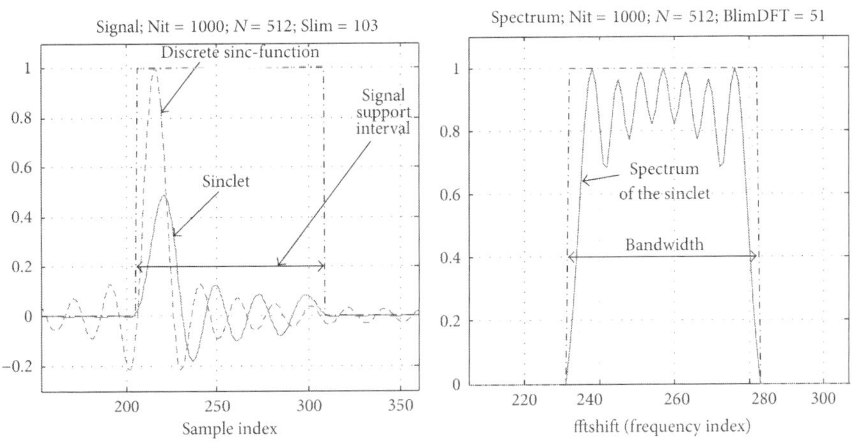

(a) (b)

Figure 17: Examples of a "sinc-let" (red plots) and, for comparison, of the discrete sinc-function for the same band limitation (blue plots) in their three different positions within the interval of 103 samples of 512 samples (boxes a–c). Plots of their DFT spectra are shown in boxes (d–f).

The following relationship between signal bounds in signal and DFT spectral domains can be derived:

$$N_{\text{sign}} \times N_{\text{spect}} < N, \qquad (6)$$

where N_{sign} is the number of signal nonzero samples, N_{spect} is the number of its nonzero spectral samples, and N is the number of samples in the signal sampling lattice. This is what one can call the discrete uncertainty principle.

CONCLUSION

Two fundamental features of fast transforms, the energy compaction capability and fast algorithms for transform computation, make them perfect tool for image compression, restoration, reconstruction, and resampling. Of all fast transforms, Discrete Fourier Transform and Discrete Cosine Transform are the most important as they are complementing each other discrete representations of the integral Fourier transform, one of the most fundamental mathematical models for describing physical reality, and, in addition, they enable fast digital convolution. From these two transforms, DCT is preferable in most applications thanks to its ability to very substantially reduce processing artifacts associated with image discontinuities at the borders.

Modern tendency in the imaging engineering is computational imaging. Computer processing of sensor data enables substantial price reduction and sometimes even complete removal of imaging optics and similar imaging hardware. It also gives birth to numerous new imaging techniques in astrophysics, biology, industrial engineering, remote sensing, and other applications. No doubts, this area promises many new achievements in the coming years. And it is certain that fast transforms will remain to be the indispensable tool in this process.

NOMENCLATURE

a_k:	Image samples
a_r:	Image spectral coefficients
$k = 0, \ldots, N-1$;	
$r = 0, \ldots, N-1$:	pixel and, correspondingly, spectra coefficient vertical and horizontal indices
$C_{(\cdot)}$:	Various constants

Δx:	Image sampling intervals
Δf:	Sampling interval in the transform domain
N:	Total number of image samples
λ:	Wavelength of coherent radiation
Z:	Distance between object plane and hologram.

REFERENCES

1. J. W. Cooley and J. W. Tukey, "An Algorithm for the Machine Calculation of Complex Fourier Series," Mathematics of Computation, vol. 19, no. 90, pp. 297–301, 1965.
2. H. F. Harmuth, Transmission of Information by Orthogonal Functions, Springer, Berlin, Germany, 1970.
3. H. C. Andrews, Computer Techniques in Image Processing, Academic Press, New York, NY, USA, 1970.
4. H. C. Andrews, "Two-dimensional Transforms," in Picture Processing and Digital Filtering, T. S. Huang, Ed., Springer, New, York, NY, USΛ, 1975.
5. N. Ahmed and K. R. Rao, Orthogonal Transforms for Digital Signal Processing, Springer, Berlin, Germany, 1975.
6. W. K. Pratt, W.-H. Chen, and L. R. Welch, "Slant Transform Image Coding," IEEE Transactions on Communications, vol. 22, no. 8, pp. 1075–1093, 1974.
7. N. Ahmed, N. Natarajan, and K. R. Rao, "Discrete cosine transform," IEEE Transactions on Computers, vol. 23, no. 1, pp. 90–93, 1974.
8. K. R. Rao and P. Yip, Discrete Cosine Transform: Algorithms, Advantages, Applications, Academic Press, Boston, Mass, USA, 1990.
9. I. Daubechies, "Where do wavelets come from? A personal point of view," Proceedings of the IEEE, vol. 84, no. 4, pp. 510–513, 1996.
10. S. Mallat, A Wavelet Tour of Signal Processing, Academic Press, Boston, Mass, USA, 3rd edition, 2008.
11. H.Karhunen,ÜberlincareMethodeninderWahrscheinlichkeitsrechnung, Series A.I, Annales Academiæ~Scientiarum Fennicæ, 1947.
12. M. Loeve, "Fonctions aleatoires de seconde ordre," in Processes Stochastiques et Movement Brownien, P. Levy, Ed., Hermann, Paris, France, 1948.
13. H. Hotelling, "Analysis of a complex of statistical variables into principal components," Journal of Educational Psychology, vol. 24, no. 6, pp. 417–441, 1933. ·
14. T. P. Belokova, M. A. Kronrod, P. A. Chochia, and L. P. Yaroslavakii, "Digital processing of martian surface photographs from mars 4 and mars 5," Space Research, vol. 13, no. 6, pp. 898–906, 1975.

15. L. P. Yaroslavsky, Digital Picture Processing: An Introduction, Springer, Berlin, Germany, 1985.

16. L. Yaroslavsky, Digital Signal Processing in Optics and Holography, Radio i Svyaz', Moscow, Russia, 1987 (Russian).

17. L. Yaroslavsky, Digital Holography and Digital Image Processing. Principles, Methods, Algorithms, Kluwer Academic Publishers, Boston, Mass, USA, 2004.

18. L. Yaroslavsky, Theoretical Foundations of Digital Imaging Using Matlab, Taylor & Francis, Boca Raton, Fla, USA, 2013.

19. L. Yaroslavsky, "Introduction to Digital Holography," in Digital Signal Processing in Experimental Research, L. Yaroslavsky and J. Astola, Eds., vol. 1 of Bentham Series of e-books, Bentham, 2009, http://www.benthamscience.com/ebooks/9781608050796/index.htm.

20. L. P. Yaroslavskii and N. S. Merzlyakov, Methods of Digital Holography, Consultance Bureau, New York, NY, USA, 1980.

21. L. P. Jaroslavski, "Comments on FFT algorithm for both input and output pruning," IEEE Transactions on Acoustics, Speech, and Signal Processing, vol. 29, no. 3, pp. 448–449, 1981.

22. J. Markel, "FFT pruning," IEEE Trans Audio Electroacoust, vol. 19, no. 4, pp. 305–311, 1971.

23. H. V. Sorensen and C. S. Burrus, "Efficient computation of the DFT with only a subset of input or output points," IEEE Transactions on Signal Processing, vol. 41, no. 3, pp. 1184–1200, 1993.

24. R. G. Alves, P. L. Osorio, and M. N. S. Swamy, "General FFT pruning algorithm," in Proceedings of the 43rd IEEE Midwest Symposium on Circuits and Systems, vol. 3, pp. 1192–1195, August 2000.

25. L. Yaroslavsky, "Fast discrete sinc-interpolation: a gold standard for image resampling," in Advances in Signal transforms: Theory and Applications, J. Astola and L. Yaroslavsky, Eds., vol. 7 of Eurasip Book Series on Signal Processing and Communications, pp. 337–405, 2007.

26. L. Yaroslavky, "Fast transforms in digital signal processing: theory, algorithms, applications," in Digital Signal Processing in Experimental Research, L. Yaroslavsky and J. Astola, Eds., vol. 2 of Bentham E-book Series, 2011, http://www.benthamscience.com/ebooks/9781608052301/index.htm.

27. R. Yu. Vitkus and L. P. Yaroslavsky, "Recursive Algorithms for Local Adaptive Linear Filtration," in Mathematical Research. Computer

Analysis of Images and Patterns, L. P. Yaroslavsky, A. Rosenfeld, and W. Wilhelmi, Eds., pp. 34–39, Academie Verlag, Berlin, Germany, 1987, Band 40.

28. J. Xi and J. F. Chicharo, "Computing running DCT's and DST's based on their second-order shift properties," IEEE Transactions on Circuits and Systems I: Fundamental Theory and Applications, vol. 47, no. 5, pp. 779–783, 2000.
29. E. Candès, "Compressive sampling," in Proceedings of the International Congress of Mathematicians, Madrid, Spain, 2006.
30. D. L. Donoho, "Compressed sensing," IEEE Transactions on Information Theory, vol. 52, no. 4, pp. 1289–1306, 2006.
31. E. J. Candès, J. K. Romberg, and T. Tao, "Stable signal recovery from incomplete and inaccurate measurements," Communications on Pure and Applied Mathematics, vol. 59, no. 8, pp. 1207–1223, 2006.
32. E. J. Candes and M. B. Wakin, "An introduction to compressive sampling: A sensing/sampling paradigm that goes against the common knowledge in data acquisition," IEEE Signal Processing Magazine, vol. 25, no. 2, pp. 21–30, 2008.
33. R. G. Baraniuk, "Compressive sensing," IEEE Signal Processing Magazine, vol. 24, no. 4, pp. 118–124, 2007.
34. M. Elad, Spatse and Redundant Representations: From Theory to Applications in Signal and Image Processing, Springer, New York, NY, USA, 2010.
35. R. M. Willet, R. F. Mareia, and J. M. Nichols, "Compressed sensing for practical optical imaging systems: a tutorial," Optical Engineering, vol. 50, no. 7, Article ID 072601, 2011.
36. "Compressive optical imaging: architectures and algorithms," in Optical and Digital Image Processing. Fundamentals and Applications, G. Cristobal, P. Schelkens, and H. Thienpont, Eds., Wiley-VCH, New York, NY, USA, 2011.
37. J. M. Whittaker, Interpolatory Function Theory, Cambridge University Press, Cambridge, UK, 1935.
38. V. A. Kotelnikov, "on the carrying capacity of the ether and wire in telecommunications," Material for the First All-Union Conference on Questions of Communication, Izd. Red. Upr. Svyazi RKKA, Moscow, Russia, 1933 (Russian).

39. C. Shannon, "A mathematical theory of communication," Bell System Technical Journal, vol. 27, no. 3, pp. 379–423, 623–656, 1948.
40. T. S. Huang and O. J. Tretiak, Picture Bandwidth Compression, Gordon and Breach, New York, NY, USA, 1972.
41. A. Habibi and P. Wintz, "Image coding by linear transformation and block quantization," IEEE Transactions on Communications, vol. 19, no. 1, pp. 50–62, 1971.
42. T. A. Wintz, "Transform picture coding," Proceedings of the IEEE, vol. 60, no. 7, pp. 809–823, 1972.
43. W. B. Pennebaker and J. L. Mitchell, JPEG Still Image Data Compression Standard, Springer, Berlin, Germany, 3rd edition, 1993.
44. I. E. G. Richardson, H.264 and MPEG-4 Video Compression, John Wiley & Sons, Chichester, UK, 2003.
45. T. Wiegand, G. J. Sullivan, G. Bjøntegaard, and A. Luthra, "Overview of the H.264/AVC video coding standard," IEEE Transactions on Circuits and Systems for Video Technology, vol. 13, no. 7, pp. 560–576, 2003.
46. P. Schelkens, A. Skodras, and T. Ebrahimi, The JPEG, 2000 Suite, John Wiley & Sons, Chichester, UK, 2009.
47. I. Pitas, Digital Video and Television, Ioannis Pitas, 2013.
48. D. Vaisey and A. Gersho, "Variable block-size image coding, acoustics, speech, and signal processing," in Proceedings of the IEEE International Conference on (ICASSP '87), vol. 12, pp. 1051–1054, 1987.
49. W. K. Pratt, "Generalized wiener filtering computation techniques," IEEE Transactions on Computers, vol. 21, pp. 636–641, 1972.
50. H. C. Andrwes, "Digital computers and image processing," Endeavour, vol. 31, no. 113, 1972.
51. N. Wiener, The Interpolation, Extrapolation and Smoothing of Stationary Time Series, Wiley, New York, NY, USA, 1949.
52. A. N. Kolmogorov, "Sur l'interpolation et extrapolation des suites stationnaires," Comptes Rendus de l'Académie des Sciences, vol. 208, pp. 2043–2045, 1939.
53. D. L. Donoho and J. M. Johnstone, "Ideal spatial adaptation by wavelet shrinkage," Biometrika, vol. 81, no. 3, pp. 425–455, 1994.
54. D. L. Donoho, "De-noising by soft-thresholding," IEEE Transactions on Information Theory, vol. 41, no. 3, pp. 613–627, 1995.

55. L. Yaroslavsky, "Space variant and adaptive transform domain image and video restoration methods," in Advances in Signal Transforms: Theory and Applications, J. Astola and L. Yaroslavsky, Eds., EURASIP Book Series on Signal Processing and Communications, 2007.
56. L. Yaroslavsky, "Local criteria: a unified approach to local adaptive linear and rank filterson," in Signal Recovery and Synthesis III, Technical Digest Series 15, 1989.
57. M. D. Levine, Vision in Man and Machine, McGraw-Hill, New York, NY, USA, 1985.
58. H. S. Malvar and D. H. Staelin, "LOT: transform coding without blocking effects," IEEE Transactions on Acoustics, Speech, and Signal Processing, vol. v, pp. 553–559, 1992.
59. R. Öktem, L. Yaroslavsky, K. Egiazarian, and J. Astola, "Transform domain approaches for image denoising," Journal of Electronic Imaging, vol. 11, no. 2, pp. 149–156, 2002.
60. L. P. Yaroslavsky, B. Fishbain, A. Shteinman, and S. Gepshtein, "Processing and fusion of thermal and video sequences for terrestrial long range observation systems," in Proceedings of the 7th International Conference on Information Fusion (FUSION '04), pp. 848–855, Stockholm, Sweden, July 2004.
61. A. Buades, B. Coll, and J. M. Morel, "A review of image denoising algorithms, with a new one," Multiscale Modeling and Simulation, vol. 4, no. 2, pp. 490–530, 2005. ·
62. A. Buades, B. Coll, and J.-M. Morel, "A non-local algorithm for image denoising," in Proceedings of the IEEE Computer Society Conference on Computer Vision and Pattern Recognition (CVPR '05), pp. 60–65, June 2005.
63. J. M. Morel and A. Buades Capo, Non Local Image Processing, http://www.farman.ens-cachan.fr/6_JeanMichel_Morel.pdf.
64. S. E. Hecht and J. J. Vidal, "generation of ECG prototype waveforms by piecewise correlational averaging," IEEE Transactions on Pattern Analysis and Machine Intelligence, vol. 2, no. 5, pp. 415–420, 1980.
65. K. Dabov, A. Foi, V. Katkovnik, and K. Egiazarian, "Image denoising by sparse 3-D transform-domain collaborative filtering," IEEE Transactions on Image Processing, vol. 16, no. 8, pp. 2080–2095, 2007.

66. L. Yaroslavsky and M. Eden, "Correlational accumulation as a method for signal restoration," Signal Processing, vol. 39, no. 1-2, pp. 89–106, 1994.

67. V. Katkovnik, K. Egiazarian, and J. Astola, "Adaptive varying window methods in signal and image processing," in Advances in Signal Transforms. Theory and Applications, J. Astola and L. Yaroslavsky, Eds., vol. 7 of Eurasip Book Series on Signal Processing and Communications, pp. 241–284, 2007.

68. D. Gabor, "Theory of Communication," Journal of IEEE, vol. 93, pp. 429–457, 1946.

69. E. Wigner, "On the quantum correction for thermodynamic equilibrium," Physical Review, vol. 40, no. 5, pp. 749–759, 1932.

70. R. K. Potter, G. Koppand, and H. C. Green, Visible Speech, Van Nostrand, New York, NY, USA, 1947.

71. D. L. Donoho, "Nonlinear wavelet methods for recovery of signals, densities, and spectra from indirect and noisy data," Proceedings of Symposia in Applied Mathematics, American Mathematical Society, vol. 47, pp. 173–205, 1993.

72. R. R. Coifman and D. L. Donoho, "Translation- Invariant De-Noising," in Wavelets and Statistics, A. Antoniadis, Ed., vol. 103 of Lecture Notes in Statistics, pp. 125–150, Springer, Berlin, Germany, 1995.

73. ftp://ftp.cis.upenn.edu/pub/eero/matlabPyrTools.tar.gz.

74. B. Z. Shaick, L. Ridel, and L. Yaroslavsky, "A hybrid transform method for image denoising," in Proceedings of the 10th European Signal Processing Conference (EUSIPCO '00), M. Gabbouj and P. Kuosmanen, Eds., Tampere, Finland, September 2000.

75. M. Vetterli and J. Kovacevic, Wavelets and Sub-Band Coding, Prentice Hall, Englewood Cliffs, NJ, USA, 1995.

76. J. Backus, The Acoustical Foundations of Music, Norton and Co., New York, NY, USA, 1969.

77. M. Unser, A. Aldroubi, and M. Eden, "Polynomial spline signal approximations: Filter design and asymptotic equivalence with Shannon's sampling theorem," IEEE Transactions on Information Theory, vol. 38, no. 1, pp. 95–103, 1992.

78. A. Gotchev, "Image interpolation by optimized spline-based kernels," in Advances in Signal transforms: Theory and Applications, J. Astola and L. Yaroslavsky, Eds., vol. 7 of Eurasip Book Series on Signal

Processing and Communications, pp. 285–335, 2007.

79. L. P. Yaroslavsky, "Efficient algorithm for discrete sine interpolation," Applied Optics, vol. 36, no. 2, pp. 460–463, 1997.
80. L. P. Yaroslavsky, "Shifted discrete fourier transforms," in Digital Signal Processing, V. Cappellini and A. G. Constantinides, Eds., pp. 69–74, Avademic Press, London, UK, 1980.
81. L. Yaroslavsky, "Boundary effect free and adaptive discrete signal sinc-interpolation algorithms for signal and image resampling," Applied Optics, vol. 42, no. 20, pp. 4166–4175, 2003.
82. L. Bilevich and L. Yaroslavsky, "Fast DCT-based image convolution algorithms and application to image resampling and hologram reconstruction," in Real-Time Image and Video Processing, Proceedings of SPIE, Brussels, Belgium, April 2010.
83. L. Bilevich and L. Yaroslavsky, "Fast DCT-based algorithm for signal and image accurate scaling," in Image Processing: Algorithms and Systems XI, vol. 8655 of Proceedings of SPIE, February 2013.
84. L. P. Yaroslavsky, "Linking analog and digital image processing," in Optical and Digital Image Processing. Fundamentals and Applications, G. Cristobal, P. Schelkens, and H. Thienpont, Eds., pp. 397–418, Wiley-VCH, New York, NY, USA, 2011.
85. M. Unser, P. Thevenaz, and L. Yaroslavsky, "Convolution-based interpolation for fast, high-quality rotation of images," IEEE Transactions on Image Processing, vol. 4, no. 10, pp. 1371–1381, 1995.
86. P. Thévenaz, EPFL/STI/IOA/LIB, Bldg. +BM-Ecublens 4.137, Station 17, CH-1015 Lausanne VD, Switzerland, http://bigwww.epfl.ch/thevenaz/.
87. A. Gotchev, "Spline and wavelet based techniques for signal and image processing," Tampere University of Technology, Publication 429, Tampere, Finland, TTY-Paino 2003.
88. http://www.eng.tau.ac.il/~yaro/Etudes/PDF/MonaLisa320x256CurvMirr.avi.
89. L. Yaroslavsky, B. Fishbain, G. Shabat, and I. Ideses, "Superresolution in turbulent videos: making profit from damage," Optics Letters, vol. 32, no. 21, pp. 3038–3040, 2007.
90. B. Fishbain, I. A. Ideses, G. Shabat, B. G. Salomon, and L. P. Yaroslavsky, "Superresolution in color videos acquired through turbulent media," Optics Letters, vol. 34, no. 5, pp. 587–589, 2009.

91. B. Fishbain, L. P. Yaroslavsky, and I. A. Ideses, "Real-time stabilization of long range observation system turbulent video," Journal of Real-Time Image Processing, vol. 2, no. 1, pp. 11–22, 2007.
92. W. H. Press, B. P. Flannery, S. A. Teukolsky, and W. T. Vetterling, Numerical Recipes. the Art of Scientific Computing, Cambridge University Press, Cambridge, UK, 1987.
93. J. H. Mathew and K. D. Fink, Numerical Methods Using MATLAB, Prentice-Hall, Englewood Cliffs, NJ, USA, 1999.
94. L. P. Yaroslavsky, A. Moreno, and J. Campos, "Frequency responses and resolving power of numerical integration of sampled data," Optics Express, vol. 13, no. 8, pp. 2892–2905, 2005.
95. C. F. Gauss, "Nachclass: theoria interpolationis methodo nova tractata," IEEE ASSP Magazine, vol. 1, no. 4, pp. 14–81, 1984.
96. L. F. Yaroslavsky, G. Shabat, B. G. Salomon, I. A. Ideses, and B. Fishbain, "Nonuniform sampling, image recovery from sparse data and the discrete sampling theorem," Journal of the Optical Society of America, vol. 26, no. 3, pp. 566–575, 2009.
97. L. P. Yaroslavsky, "Image recovery from sparse samples, discrete sampling theorem, and sharply bounded band-limited discrete signals," in Advances in Imaging and Electron Physics, P. Hawkes, Ed., vol. 167, chapter 5, pp. 295–331, Academic Press, 2011.
98. R. W. Gerchberg and W. O. Saxton, "Practical algorithm for the determination of phase from image and diffraction plane pictures," Optik, vol. 35, no. 2, pp. 237–250, 1972.
99. A. Papoulis, "New algorithm in spectral analysis and band-limited extrapolation," IEEE Trans Circuits Syst, vol. 22, no. 9, pp. 735–742, 1975.
100. D. Gabor, "Theory of communication," Journal of IEEE, vol. 93, pp. 429–457, 1946.

CHAPTER 11

MODELING OF THERMAL DISTRIBUTIONS AROUND A BARRIER AT THE INTERFACE OF COATING AND SUBSTRATE

Ali Sahin

Department of Mathematics, Fatih University, Buyukcekmece, 34500 Istanbul, Turkey

ABSTRACT

Due to constant heat flux, the thermal distribution around an insulated barrier at the interface of substrate and functionally graded material (FGM) which are essentially two-phase particulate composites is examined in such a way that the volume fractions of the constituents vary continuously in the thickness direction. Using integral transform method, two-dimensional steady-state diffusion equation with variable conductivity is turned into

CITATION (APA): Sahin, A. (2013). Modeling of Thermal Distributions around a Barrier at the Interface of Coating and Substrate. In Abstract and Applied Analysis (Vol. 2013). Hindawi. (8 pages)

constant coefficient differential equation. Reducing that equation to a singular integral equation with Cauchy type, the temperature distribution around the barrier is obtained by defining an unknown function, which is called density function, as a series expansion of orthogonal polynomials. Results are shown for different thickness and nonhomogeneity parameters of FGM.

INTRODUCTION

There are many engineering applications under severe thermal loading that require high temperature resistant materials in various forms of composites and bonded materials such as power generation, transportation, aerospace, and thermal barrier coatings. New developments in science and technology rely on the developments of new materials. Composites appear to provide the necessary flexibility in the design of these new materials, which are essentials that every part of the material in use exhibits uniform properties. In the mid-1980s, a new composite material, which was initially designed as a kind of thermal barrier coating used in aerospace structural applications and fusion reactors, was found by a group of scientists in Japan. Because of the material's structure, it is named as functionally graded material (FGM). FGMs were used in modern technologies as advanced structures where the composition or the microstructure is locally varied so that a certain variation of the local material properties is achieved [1]. FGMs are also developed for general use as structural components in extremely high-temperature environments. The concept is to make a composite material by varying the microstructure from one material to another material with a specific gradient. The transition between the two materials can usually be approximated by means of a power series or an exponential function [2–5].

The aircraft and aerospace industry and the computer circuit industry are very interested in the possibility of materials that can withstand very high thermal gradients. This is normally achieved by using a ceramic layer connected with a metallic layer. The composition profile which varies from 0% ceramic at the interface to 100% ceramic near the surface, in turn, is selected in such a way that the resulting nonhomogeneous material exhibits the desired thermomechanical properties. The concept of FGMs could provide great flexibility in material design by controlling both composition profile and microstructure [6].

A number of reviews dealing with various aspects of FGMs have been published in the past few decades. They show that most of early research studies in FGMs had focused more on thermal stress analysis and fracture mechanics. Fracture mechanics of FGMs have been studied analytically by Erdogan and coworkers [7–9]. Erdogan identified a number of typical problem areas relating to the fracture of FGMs by considering mainly the investigation of the nature of stress singularity near the tip of a crack in a different geometry [10]. Erdogan also investigated the nature of the crack-tip stress field in a nonhomogeneous medium having a shear modulus with a discontinuous derivative. The problem was considered for the simplest possible loading and geometry, namely, the antiplane shear loading of two bonded half spaces, in which the crack is perpendicular to the interface. It was shown that the square-root singularity of the crack-tip stress field is unaffected by the discontinuity in the derivative of the shear modulus [11]. Related to the fracture problems in composite materials, the solution of integral equations with strongly singular kernels is examined by Kaya and Erdogan [12, 13]. In an axisymmetric coordinate system, an embedded axisymmetric crack in a nonhomogeneous infinite medium was studied by Ozturk and Erdogan [14]. They showed the effect of the material nonhomogeneity on the stress intensity factors under constant Poisson's ratio.

Due to the material mismatch at the interface of substrate and coating, the thermal distributions and thermal stresses on the crack or insulated barrier at the interface are examined by researchers. A general analysis of one-dimensional steady-state thermal stresses in a hollow thick cylinder made of functionally graded material is developed by Jabbari et al. [15]. The material properties were assumed to be nonlinear with a power law distribution. The mechanical and thermal stresses were obtained through the direct method of solution of the Navier equation. The problem of general solution for the mechanical and thermal stresses in a short length functionally graded hollow cylinder due to the two-dimensional axisymmetric steady-state loads was solved using the Bessel functions by Jabbari et al. [16]. A standard method was used to solve a nonhomogeneous system of partial differential Navier equations with nonconstants coefficients, using Fourier series.

Jin and Noda [17] examined an internal crack problem in nonhomogeneous half-plane under thermal loading using the airy stress function method and Fourier transform. They reduced the problem to a system of singular integral equations and solved it by numerical methods. They used superposition method by defining the problem in two cases. One is the linear one-dimensional heat conduction problem under constant heat flux without crack, and the other one is the two-dimensional heat conduction problem with an insulated crack subject to constant heat flux which is acted in the opposite direction. It was also considered the problem of an axisymmetric penny-shaped crack embedded in an isotropic graded coating bonded to a semi-infinite homogeneous medium by Rekik et al. [18]. The coating's material gradient is parallel to the axisymmetric direction and is orthogonal to the crack plane. They used Hankel transform to convert the equations into coupled singular integral equations along with the density function, and they solved it numerically.

In this study, the Hankel integral transform method will be used to solve the heat equation in axisymmetric coordinate system. Problem will be examined as a one-dimensional and a two-dimensional heat conduction problem that is a mixed boundary value problem over the real line. Using mixed boundary conditions a Fredholm integral equation will be obtained with Cauchy type singularity and then it will be solved by using some known numerical techniques [19, 20].

DEFINITION OF THE PROBLEM

The thermal distribution around a penny-shaped barrier at the interface of graded composite coating and a substrate is given by the following steady-state heat equation in axisymmetric coordinate system:

$$\frac{1}{r}\frac{\partial}{\partial r}\left(rk(z)\frac{\partial T}{\partial r}\right)+\frac{\partial}{\partial z}\left(k(z)\frac{\partial T}{\partial z}\right)=0, \tag{1}$$

and the conductivities of the substrate and the graded composite coating are, respectively, given as

$$k(z)=k_0, \quad z\le 0,$$

$$k(z)=k_0e^{\delta z}, \quad z\ge 0, \tag{2}$$

where k_0 is a constant and δ is the nonhomogeneity parameter related to the graded coating. Note that the conductivity is continuous at the interface of substrate and the graded composite coating. As shown in Figure 1, it is considered a penny-shaped barrier with radius a, centered at the origin of the axisymmetric coordinate system. It is assumed that a uniform heat flux is applied over the stress free boundary, and the barrier faces remain insulated.

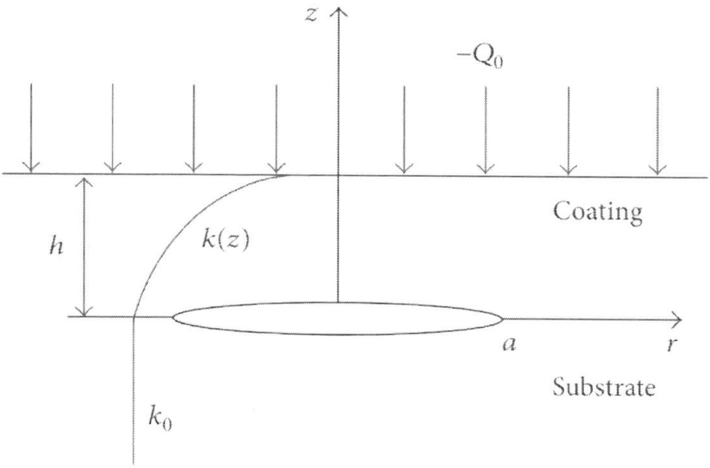

Figure 1: Geometry of the heat conduction problem.

The solution can be obtained using superposition method which is an addition of one- and two-dimensional heat conduction problems, $T_1(z)$ and $T_2(r, z)$, respectively, as shown in Figures 2(a) and 2(b). As in Figure 2(a), it will be assumed that there will be no barrier and in flux causes thermal distribution only z direction. On the other hand, in Figure 2(b), flux will be assumed in an opposite direction on the barrier that causes thermal distribution in (r, z) plane.

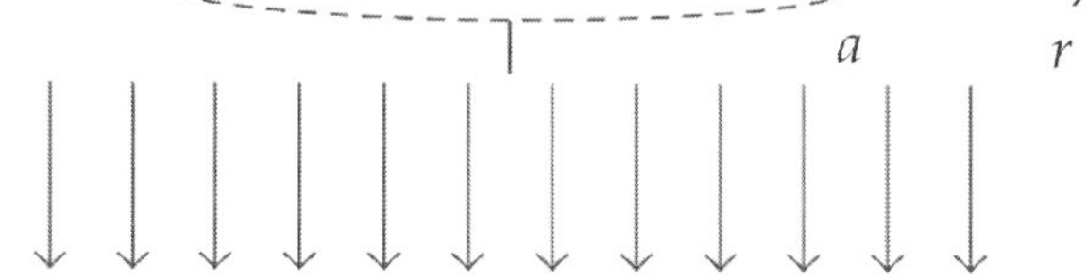

(a)

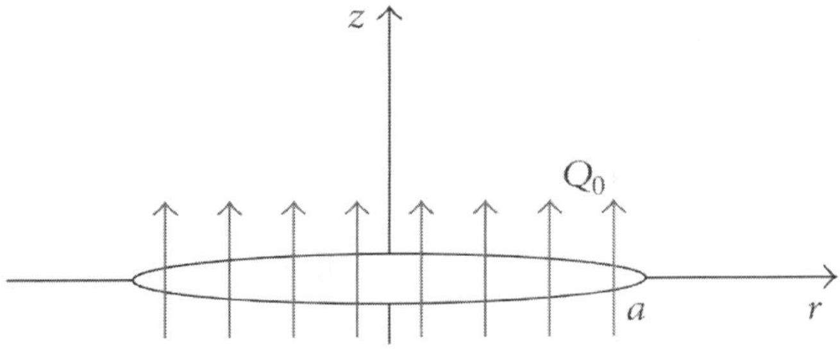

(b)

Figure 2: (a) One-dimensional heat conduction without an insulated barrier, (b) Two-dimensional heat conduction with an insulated barrier.

Rewriting (1) assuming that no changing in r direction:

$$\delta\frac{dT_1}{dz}+\frac{d^2T_1}{dz^2}=0,\quad 0\le z\le h;$$

$$\frac{d^2T_1}{dz^2}=0,\quad z\le 0, \tag{3}$$

along with suitable boundary conditions:

$$k(z)\frac{d}{dz}T_1(z)\longrightarrow \text{finite},\quad z\longrightarrow-\infty,$$

$$k(z)\frac{d}{dz}T_1(z)=-Q_0,\quad z=h,$$

$$T_1(0^+)=T_1(0^-), \tag{4}$$

the solution of the one-dimensional heat conduction is obtained straightforward as

$$T_1(z)=\begin{cases}\dfrac{Q_0}{k_0\delta}e^{-\delta z}, & 0\le z\le h,\\[2mm] \dfrac{Q_0}{k_0\delta}, & z\le 0,\end{cases} \tag{5}$$

where the continuity can be seen as $z\to 0$. For two-dimensional heat conduction problem, (1) can be simplified for 0

$$\frac{\partial^2}{\partial r^2}T_2(r,z) + \frac{1}{r}\frac{\partial}{\partial r}T_2(r,z) + \frac{\partial^2}{\partial z^2}T_2(r,z)$$
$$+\delta\frac{\partial}{\partial z}T_2(r,z) = 0, \tag{6}$$

and for $z<0$,

$$\frac{\partial^2}{\partial r^2}T_2(r,z) + \frac{1}{r}\frac{\partial}{\partial r}T_2(r,z) + \frac{\partial^2}{\partial z^2}T_2(r,z) = 0 \tag{7}$$

with standard boundary conditions:

$$k(z)\frac{\partial}{\partial z}T_2(r,z) \longrightarrow \text{finite}, \quad z \longrightarrow -\infty,$$

$$k(z)\frac{\partial}{\partial z}T_2(r,z) = 0, \quad z = h,\ 0 \le r < \infty,$$

$$\frac{\partial}{\partial z}T_2(r,0^+) = \frac{\partial}{\partial z}T_2(r,0^-), \quad a < r < \infty, \tag{8}$$

and mixed boundary conditions:

$$k(z)\frac{\partial}{\partial z}T_2(r,z) = Q_0, \quad z \longrightarrow 0,\ 0 \le r < a, \tag{9}$$

$$T_2(r,0^+) = T_2(r,0^-), \quad a < r < \infty. \tag{10}$$

Equation (6) will be solved using Hankel integral transform such that $T2(r, z)$ denote the Hankel transform of zero order and $\tau(\rho, z)$ denote inverse Hankel transform of zero order [21] shown as below, respectively,

$$T_2(r,z) = \int_0^\infty \tau(\rho,z) J_0(r\rho)\rho d\rho,$$

$$\tau(\rho,z) = \int_0^\infty T_2(r,z) J_0(r\rho) r dr. \tag{11}$$

Using Hankel transform, the solution of (6) along with boundary conditions (8) can be obtained as

$$\tau(\rho, z) = \begin{cases} A(\rho) e^{-(m_2+m_1)z} + A(\rho) \dfrac{m_2 + m_1}{m_2 - m_1} \\ \times e^{-2m_2 h + (m_2 - m_1)z}, \quad 0 < z \le h, \\ A(\rho) \dfrac{1}{\rho} \left(e^{-2m_2 h} - 1\right)(m_2 - m_1) \\ \times e^{\rho z}, \quad -\infty < z < 0, \end{cases} \tag{12}$$

where $m_1 = \delta/2$, $m_2 = \rho\sqrt{(\delta/2\rho)^2 + 1}$ and observing that $m_2 + m_1 > 0$ and $m_2 - m_1 > 0$.

EVALUATION OF INTEGRAL EQUATION

The unknown value (ρ) can be obtained by defining a new function, which is called density function [11], such as

$$\psi(r) = \frac{\partial}{\partial r}\left(T_2(r, 0^+) - T_2(r, 0^-)\right), \tag{13}$$

where $\psi(r)$ satisfies the following conditions:

$$\int_0^a \psi(r)\, dr = 0, \quad \psi(r) = 0, \quad \text{at } a < r < \infty,$$

$$\psi(r) = -\psi(-r). \tag{14}$$

Substituting (12) into (13) along with the conditions given in (14), the unknown function (ρ) can be obtained:

$$A(\rho) = \frac{1}{F(\rho)} \int_0^a \psi(s) J_1(s\rho)\, s ds, \tag{15}$$

where

$$F(\rho) = -\rho - \rho\left(\frac{m_2 + m_1}{m_2 - m_1}\right) e^{-2m_2 h} + \left(e^{-2m_2 h} - 1\right)(m_2 - m_1). \tag{16}$$

Using the boundary condition (9) in transformed domain as $z \to 0^-$, it can be obtained as

$$\int_0^\infty \left(e^{-2m_2 h} - 1\right)(m_2 - m_1) A(\rho) J_0(r\rho)\, \rho d\rho = \frac{Q_0}{k_0}, \tag{17}$$

and substituting the value of $A(\rho)$ into (17), the integral equation to be solved for unknown $\psi(s)$ can be obtained as

$$\int_0^a \left(\int_0^\infty \left(e^{-2m_2 h} - 1\right) \eta(\rho) J_0(r\rho) J_1(s\rho) \rho \, d\rho \right) \psi(s) s ds = \frac{Q_0}{k_0}, \tag{18}$$

where

$$\eta(\rho) = \frac{m_2 - m_1}{F(\rho)}. \tag{19}$$

Defining new normalized variables and parameters such as

$$s' = \frac{s}{a}, \qquad r' = \frac{r}{a}, \qquad \rho' = a\rho, \qquad h' = \frac{h}{a},$$
$$\delta' = a\delta, \qquad \psi(s) = \psi\left(as'\right) = \psi\left(s'\right), \tag{20}$$

the integral equation in (18) can be expressed in the form of

$$\int_0^1 \psi(s) s ds \int_0^\infty J_0(r\rho) J_1(s\rho) \rho d\rho + \int_0^1 \psi(s) s ds \int_0^\infty 2e^{-2m_2 h} \eta(\rho) J_0(r\rho) J_1(s\rho) \rho d\rho - \int_0^1 \psi(s) s ds \int_0^\infty (2\eta(\rho) + 1) J_0(r\rho) J_1(s\rho) \rho d\rho = q_0, \tag{21}$$

where $q_0 = 2Q_0/k_0$ and the prime sign is removed for the sake of simplicity. The first double integral in (21) can be expressed in terms of the first and second kinds of elliptic integrals, K and E, respectively. As $k \rightarrow 1$, the second kind of elliptic integral (k) has a finite value while the first kind of elliptic integral $K(k)$ has a logarithmic singularity such as

$$K(k) = \log\left(\frac{4}{\sqrt{1-k^2}}\right), \qquad E(k) = 1. \tag{22}$$

Defining a new function $M(s, r)$, given in the appendix, and using some algebraic manipulations, a Cauchy type singularity can obtained by using the properties of $\psi(s)$ in (14). Hence, the first integral of (21) becomes

$$\int_0^1 \psi(s)\, sds \int_0^\infty J_0(r\rho)\, J_1(s\rho)\, \rho d\rho$$

$$= \frac{1}{\pi}\int_{-1}^{1} \frac{\psi(s)}{s-r} ds$$

$$+ \frac{1}{\pi}\int_0^1 \left(\frac{M(s,r)-1}{s-r} + \frac{M(s,r)-1}{s+r}\right) \psi(s)\, ds. \quad (23)$$

The second double integral in (21) has an exponential integrand with negative exponent so that as $\rho \to \infty$, the integrand asymptotically approaches to zero. The asymptotic expansion of the integrand can be expressed as

$$e^{-2h\rho}\rho\left(-1 + \frac{h\delta^2 - \delta}{2^2\rho} - \frac{h^2\delta^4 - 2h\delta^3}{2^5\rho^2}\right.$$

$$+ \frac{(1/3)\, h^3\delta^6 - h^2\delta^5 - 2h\delta^4 + 2\delta^3}{2^7\rho^3}$$

$$- \frac{(1/3)\, h^4\delta^8 + (4/3)\, h^3\delta^7 - 8h^2\delta^6 - 40h\delta^5}{2^{11}\rho^4}$$

$$\left. + O\left(\rho^{-5}\right)\right). \quad (24)$$

Since there is no singularity and any discontinuity over the interval, and due to the asymptotic expansion as $\rho \to \infty$, the infinite integral can be approximated over the interval $[0, B]$. Depending on parameters h and δ, the value of B can be chosen large enough to obtain small enough value of the integral (less than 10^{-25}) over $[B,\infty)$. Hence, the second double integral in (21) may be expressed as

$$\int_0^1 \psi(s)\, sds \int_0^\infty 2e^{-2m_2 h}\eta(\rho)\, J_0(r\rho)\, J_1(s\rho)\, \rho d\rho$$

$$= \int_0^1 \psi(s)\, sds \int_0^{B(h,\delta)} 2e^{-2m_2 h}\eta(\rho)\, J_0(r\rho)\, J_1(s\rho)\, \rho d\rho. \quad (25)$$

Finally, the last integral in (21) can be evaluated by using a series

$$G_A(\rho) = -\frac{\delta}{2^2} + \frac{\delta^3}{2^6}\frac{1}{\rho^2} - \frac{\delta^5}{2^9}\frac{1}{\rho^4} + \frac{5\delta^7}{2^{14}}\frac{1}{\rho^6}$$

$$- \cdots + \frac{429\delta^{15}}{2^{30}}\frac{1}{\rho^{14}} - O\left(\rho^{-16}\right), \tag{26}$$

an asymptotic expansion of the integrand $G(\rho) = \rho(2\eta(\rho) + 1)$ as $\rho \rightarrow \infty$. Let us define a value C that depends on parameters h and δ and superposes the infinite integral as follows:

$$\int_0^\infty G(\rho) J_0(r\rho) J_1(s\rho)\, d\rho$$

$$= \int_0^{C(h,\delta)} G(\rho) J_0(r\rho) J_1(s\rho)\, d\rho$$

$$\times \int_{C(h,\delta)}^\infty G_A(\rho) J_0(r\rho) J_1(s\rho)\, d\rho$$

$$+ \int_{C(h,\delta)}^\infty [G(\rho) - G_A(\rho)] J_0(r\rho) J_1(s\rho)\, d\rho. \tag{27}$$

For a particular value of C,$(C) - G_A(C) \approx 0$ can be obtained; then the last infinite integral can be expressed as

$$\int_0^\infty G(\rho) J_0(r\rho) J_1(s\rho)\, d\rho$$

$$= \int_0^{C(h,\delta)} G(\rho) J_0(r\rho) J_1(s\rho)\, d\rho$$

$$+ \int_{C(h,\delta)}^\infty G_A(\rho) J_0(r\rho) J_1(s\rho)\, d\rho. \tag{28}$$

Since the function (ρ) is a smooth function over the interval$[0, C]$, it can be evaluated numerically using Gaussian quadrature along with Chebyshev orthogonal polynomials. On the other hand, the evaluation of the integral over the interval $[C,\infty)$ can be evaluated separately for each term in $G_A(\rho)$ like

$$\int_{C(h,\delta)}^{\infty} G_A(\rho)\, J_0(r\rho)\, J_1(s\rho)\, d\rho$$

$$= -\frac{\delta}{2^2}\int_{C(h,\delta)}^{\infty} J_0(r\rho)\, J_1(s\rho)\, d\rho$$

$$+ \int_{C(h,\delta)}^{\infty}\left(\frac{\delta^3}{64}\frac{1}{\rho^2} - \frac{\delta^5}{512}\frac{1}{\rho^4} + \frac{5\delta^7}{16384}\frac{1}{\rho^6} - \cdots\right)$$

$$\times J_0(r\rho)\, J_1(s\rho)\, d\rho, \tag{29}$$

where

$$H_0(r,s) = \int_{C(h,\delta)}^{\infty} J_0(r\rho)\, J_1(s\rho)\, d\rho$$

$$= \begin{cases} \dfrac{1}{s} - \displaystyle\int_0^{C(h,\delta)} J_0(r\rho)\, J_1(s\rho)\, d\rho, & s > r, \\ \dfrac{1}{2s} - \displaystyle\int_0^{C(h,\delta)} J_0(r\rho)\, J_1(s\rho)\, d\rho, & s = r, \\ -\displaystyle\int_0^{C(h,\delta)} J_0(r\rho)\, J_1(s\rho)\, d\rho, & s < r, \end{cases} \tag{30}$$

and the other integrals in (29) can be evaluated iteratively for $k = 1, 2, 3, \ldots$ as follows:

$$H_k(r,s) = \int_{C(h,\delta)}^{\infty} \frac{J_0(r\rho)\, J_1(s\rho)}{\rho^k}\, d\rho$$

$$= \frac{J_0(rC)\, J_1(sC)}{kC^{k-1}} + \frac{s}{k} L_{k-1}(r,s) + \frac{r}{k} N_{k-1}(r,s), \tag{31}$$

where

$$L_k(r,s) = \int_{C(h,\delta)}^{\infty} \frac{J_0(r\rho)\, J_0(s\rho)}{\rho^k}\, d\rho$$

$$= \frac{J_0(rC)\, J_0(sC)}{(k-1)\, C^{k-1}} - \frac{rH_{k-1}(s,r)}{k-1} - \frac{sH_{k-1}(r,s)}{1-k}, \tag{32}$$

$$N_k(r,s) = \int_{C(h,\delta)}^{\infty} \frac{J_1(r\rho)\,J_1(s\rho)}{\rho^k} d\rho$$

$$= \frac{J_1(rC)\,J_1(sC)}{(k+1)\,C^{k-1}} + \frac{rH_{k-1}(r,s)}{k+1} + \frac{sH_{k-1}(s,r)}{k+1}. \tag{33}$$

The initial values of the integral $H_1(r, s)$, $L_2(r, s)$, and $N_3(r, s)$ are shown in the appendix.

NUMERICAL EVALUATION OF INTEGRAL EQUATION

The integral equation in (21) can be solved using Gaussian quadrature method. Using the condition given in (14), the unknown function (s) can be defined in terms of the truncated Nth term series expansion of Chebyshev orthogonal polynomials of the first kind, $T_n(s)$, as

$$\psi(s) = \sum_{n=1}^{N} A_{2n-1} \frac{T_{2n-1}(s)}{\sqrt{1-s^2}}, \quad -1 < s < 1, \tag{34}$$

where A_{2n-1} are coefficients. Substituting the series expansion of the unknown function (s) into the first integral in (23), it can be found that

$$\frac{1}{\pi}\int_{-1}^{1} \frac{\psi(s)}{s-r} ds = \sum_{n=1}^{N} A_{2n-1} \frac{1}{\pi}\int_{-1}^{1} \frac{T_{2n-1}(s)}{(s-r)\sqrt{1-s^2}} ds$$

$$= \sum_{n=1}^{N} A_{2n-1} U_{2n-2}(r), \tag{35}$$

where $U_{2n-2}(r)$ is the Chebyshev polynomials of the second kind, and eliminating the logarithmic singularities, shown in appendix, the second integral in (23) can be written as

$$\frac{1}{\pi}\int_0^1\left(\frac{M(s,r)-1}{s-r}+\frac{M(s,r)-1}{s+r}\right)\psi(s)\,ds$$

$$=\sum_{n=1}^{N}A_{2n-1}\frac{1}{\pi}$$

$$\times\int_0^1\left(\frac{M(r,s)-1}{s-r}+\frac{M(r,s)-1}{s+r}+\frac{1}{2r}\log\left|\frac{s-r}{s+r}\right|\right)$$

$$\times\frac{T_{2n-1}(s)}{\sqrt{1-s^2}}ds. \tag{36}$$

Finally, substituting the truncated series representation of (s) into (25) and (29) system of algebraic equation can be obtained the to be solved for A_{2n-1},

$$\sum_{n=1}^{N}A_{2n-1}\left(U_{2n-2}(r_i)+\frac{1}{2r_i}\frac{T_{2n-1}(r_i)}{2n-1}+Z_1(r_i)\right.$$

$$\left.+Z_2(r_i)-Z_3(r_i)\right)=q_0, \tag{37}$$

where ri, $i = 1, 2, 3, \ldots$, are collocation points and

$$Z_1(r_i)=\frac{1}{\pi}\int_0^1\left(\frac{M(r_i,s)-1}{s-r_i}+\frac{M(r_i,s)-1}{s+r_i}\right.$$

$$\left.+\frac{1}{2r_i}\log\left|\frac{s-r_i}{s+r_i}\right|\right)\frac{T_{2n-1}(s)}{\sqrt{1-s^2}}ds,$$

$$Z_2(r_i)=\int_0^1\left(\int_0^{B(h,\delta)}2e^{-2m_2h}\eta(\rho)J_0(r_i\rho)J_1(s\rho)\rho d\rho\right)$$

$$\times\frac{sT_{2n-1}(s)}{\sqrt{1-s^2}}ds, \tag{38}$$

$$Z_3(r_i)=\int_0^1\left(\int_0^{C(h,\delta)}G(\rho)J_0(r_i\rho)J_1(s\rho)\,d\rho\right.$$

$$\left.+\int_{C(h,\delta)}^{\infty}G_A(\rho)J_0(r_i\rho)J_1(s\rho)\,d\rho\right)$$

$$\times\frac{sT_{2n-1}(s)}{\sqrt{1-s^2}}ds.$$

RESULTS

Because of the nature of the problem it is necessary to increase the density of the collocation points near the ends $r = \pm 1$. Thus, these points may be selected as follows:

$$T_n(r_i) = 0, \quad r_i = \cos\left(\frac{(2i-1)\pi}{2N}\right), \quad i = 1, 2, \ldots, N. \tag{39}$$

Then, we get a ($N \times N$) system of equations whose solution gives the coefficients A_{2n+1}. With known coefficient values, the temperature distribution around the insulated barrier may be obtained by integrating (13) such as

$$T_2(r, 0^+) - T_2(r, 0^-) = \sum_{n=1}^{N} A_{2n-1} \int_{-1}^{r/a} \frac{aT_{2n-1}(s)}{\sqrt{1-s^2}} ds. \tag{40}$$

Defining new variable like

$$s = \cos\theta, \quad \pi \le \theta \le \arccos\left(\frac{r}{a}\right), \tag{41}$$

the integral in (40) can be evaluated using the relation

$$U_n(t) = \frac{\sin\{(n+1)\arccos t\}}{\sin(\arccos t)}, \tag{42}$$

and the difference in temperature distribution on the plane of the insulated barrier can be obtained as

$$T^*(r) = \frac{T_2(r, 0^+) - T_2(r, 0^-)}{a} = -\sqrt{1-\left(\frac{r}{a}\right)^2} \sum_{n=1}^{N} A_{2n-1} \frac{U_{2n-2}(r/a)}{2n-1}. \tag{43}$$

APPENDIX

The function (s, r) in (23) can be defined [14] as

$$M(s,r) = \begin{cases} \dfrac{r}{s}E\left(\dfrac{s}{r}\right) + \dfrac{s^2 - r^2}{rs}K\left(\dfrac{s}{r}\right), & s < r, \\ E\left(\dfrac{r}{s}\right), & s > r, \end{cases} \tag{A.1}$$

in terms of the complete elliptic integrals of the first and second kinds, respectively,

$$K(k) = \int_0^{\pi/2} \frac{d\theta}{\sqrt{1 - k^2 \sin^2\theta}},$$

$$E(k) = \int_0^{\pi/2} \sqrt{1 - k^2 \sin^2\theta}\, d\theta. \tag{A.2}$$

For $k \geq 2$, the values of the integrals in (31), (32), and (33), respectively, can be obtained by solving the initial value of each integral for $k=1$ such that

$$\begin{aligned} H_1(r,s) &= \int_{C(h,\delta)}^{\infty} \frac{J_0(r\rho) J_1(s\rho)}{\rho} d\rho \\ &= \frac{1}{\pi} \int_0^{\pi} \frac{s - r\cos\phi}{R} \left(1 + J_1(RC(h,\delta)) \right. \\ &\qquad \left. - \int_0^{RC(h,\delta)} J_0(v)\, dv \right) d\phi, \\ L_1(r,s) &= \int_{C(h,\delta)}^{\infty} \frac{J_0(r\rho) J_0(s\rho)}{\rho} d\rho \\ &= \frac{1}{\pi} \int_0^{\pi} \left(-\gamma - \log\left(\frac{RC(h,\delta)}{2}\right) \right. \\ &\qquad \left. + \int_0^{RC(h,\delta)} \frac{1 - J_0(u)}{u} du \right) d\phi, \\ N_1(r,s) &= \int_{C(h,\delta)}^{\infty} \frac{J_1(r\rho) J_1(s\rho)}{\rho} d\rho \\ &= \frac{rs}{\pi} \int_0^{\pi} \frac{\sin^2\phi}{R} \left(\int_{C(h,\delta)}^{\infty} J_1(R\rho)\, d\rho \right) d\phi, \end{aligned} \tag{A.3}$$

where $R^2 = r^2 + s^2 - 2rs\cos\varphi$

Due to the logarithmic singularity as $s \to r$ in (23), the ratio $(((s,r) - 1)/(s - r)) \to (0/0)$ has undetermined limiting case. Using (22) along with

L'Hospital's rule, we have

$$\lim_{s\to r}\frac{M(s,r)-1}{s-r}=-\frac{1}{2r}\log|s-r|+\frac{1}{r}\left(\log\sqrt{8r}-1\right). \tag{A.4}$$

Now, by adding and subtracting the leading part of the logarithmic function,

$$\frac{1}{\pi}\int_0^1\left(\frac{M(r,s)-1}{s-r}+\frac{M(r,s)-1}{s+r}+\frac{1}{2r}\log|s-r|\right)\psi(s)\,ds$$
$$-\frac{1}{2\pi r}\int_0^1(\log|s-r|)\,\psi(s)\,ds, \tag{A.5}$$

and using the symmetry property of $\psi(s)$ like

$$-\int_0^1(\log|s-r|)\,\psi(s)\,ds=-\int_{-1}^1(\log|s-r|)\,\psi(s)\,ds$$
$$-\int_0^1(\log|s+r|)\,\psi(s)\,ds, \tag{A.6}$$

we have

$$\frac{1}{\pi}\int_0^1\left(\frac{M(r,s)-1}{s-r}+\frac{M(r,s)-1}{s+r}+\frac{1}{2r}\log\left|\frac{s-r}{s+r}\right|\right)\psi(s)\,ds$$
$$-\frac{1}{\pi}\int_{-1}^1\frac{1}{2r}(\log|s-r|)\,\psi(s)\,ds, \tag{A.7}$$

where

$$\frac{1}{\pi}\int_{-1}^1\log|s-r|\,\frac{T_{2n-1}(s)}{\sqrt{1-s^2}}\,ds=-\frac{T_{2n-1}(r)}{2n-1}, \tag{A.8}$$

using the series expansion of unknown function $\psi(s)$ in (34).

ACKNOWLEDGMENT

This work is supported by the Scientific Research Fund of Fatih University under Project number P50040801-2.

REFERENCES

1. H.-S. Shen, Functionally Graded Materials Nonlinear Analysis of Plates and Shells, CRC Press, New York, NY, USA, 2009.
2. S. Suresh and A. Mortensen, Fundamentals of Functionally Graded Materials, Institute of Materials Communications Limited, London, UK, 1998.
3. K. Trumble, Functionally Graded Materials 2000, Ceramic Transactions 114, The American Ceramics Society, Westerville, Ohio, USA, 2001.
4. W. Pan, Functionally Graded Materials VII, Trans Tech Publications LTD, Zürich, Switzerland, 2003.
5. O. van der Biest, FGM, 2004, Functionally Graded Materials VIII, Trans Tech. Publications LTD, Zürich, Switzerland, 2005.
6. I. Shiota and Y. Miyamoto, Functionally Graded Materials 1996, Elsevier, New York, NY, USA, 1997.
7. Y. F. Chen and F. Erdogan, "The interface crack problem for a nonhomogeneous coating bonded to a homogeneous substrate," Journal of the Mechanics and Physics of Solids, vol. 44, no. 5, pp. 771–787, 1996.
8. A. Şahin, "An interface crack between a graded coating and a homogeneous substrate," Turkish Journal of Engineering and Environmental Sciences, vol. 28, no. 2, pp. 135–148, 2004.
9. A. Sahin and F. Erdogan, "On debonding of graded thermal barrier coatings," International Journal of Fracture, vol. 129, no. 4, pp. 341–359, 2004.
10. F. Erdogan, "Fracture mechanics of functionally graded materials," Composites Engineering, vol. 5, no. 7, pp. 753–770, 1995.
11. F. Erdogan, "The crack problem for bonded nonhomogeneous materials under antiplane shear loading," ASME Journal of Applied Mechanics, vol. 52, pp. 823–828, 1995.
12. A. Kaya and F. Erdogan, "On the solution of integral equations with a strongly singular kernels," Quarterly of Applied Mathematics, vol. 45, pp. 105–122, 1987.
13. A. Kaya and F. Erdogan, "On the solution of integral equations with a generalized cauchy kernel," Quarterly of Applied Mathematics, vol. 45, pp. 455–469, 1987.

14. M. Ozturk and F. Erdogan, "Axisymmetric crack problem in a nonhomogeneous medium," Journal of Applied Mechanics, Transactions ASME, vol. 60, no. 2, pp. 406–413, 1993.
15. M. Jabbari, S. Sohrabpour, and M. R. Eslami, "Mechanical and thermal stresses in a functionally graded hollow cylinder due to radially symmetric loads," International Journal of Pressure Vessels and Piping, vol. 79, no. 7, pp. 493–497, 2002.
16. M. Jabbari, A. Bahtui, and M. R. Eslami, "Axisymmetric mechanical and thermal stresses in thick short length FGM cylinders," International Journal of Pressure Vessels and Piping, vol. 86, no. 5, pp. 296–306, 2009.
17. Z. H. Jin and N. Noda, "An internal crack parallel to the boundary of a nonhomogeneous half plane under thermal loading," International Journal of Engineering Science, vol. 31, no. 5, pp. 793–806, 1993.
18. M. Rekik, M. Neifar, and S. El-Borgi, "An axisymmetric problem of an embedded crack in a graded layer bonded to a homogeneous half-space," International Journal of Solids and Structures, vol. 47, no. 16, pp. 2043–2055, 2010.
19. F. Erdogan and G. D. Gupta, "Numerical solution of singular integral equations," in Methods of Analysis and Solution of Crack Problems, G. C. Sih, Ed., Martinus Nijhoff Publishers, Dordrecht, The Netherlands, 1973.
20. F. Erdogan, "Mixed Boundary Value Problems in Mechanics," in Mechanics Today, N. Nasser, Ed., chapter 1, Pergamon Press, Oxford, UK, 1975.
21. G. N. Watson, A Treatise on the Theory of Bessel Functions, Cambridge University Press, Cambridge, UK, 1966.

CHAPTER 12

DYNAMIC BEHAVIOR OF AXIALLY FUNCTIONALLY GRADED PIPES CONVEYING FLUID

Chen An[1] and Jian Su[2]

[1]Institute for Ocean Engineering, China University of Petroleum-Beijing, Beijing 102249, China

[2]Nuclear Engineering Program, COPPE, Universidade Federal do Rio de Janeiro, CP 68509, 21941-972 Rio de Janeiro, RJ, Brazil

ABSTRACT

Dynamic behavior of axially functionally graded (FG) pipes conveying fluid was investigated numerically by using the generalized integral transform technique (GITT). The transverse vibration equation was integral transformed into a coupled system of second-order differential equations in the temporal variable. The Mathematica's built-in function, NDSolve, was employed to numerically solve the resulting transformed ODE system. Excellent convergence of the proposed eigenfunction expansions was

CITATION (APA): An, C., & Su, J. (2017). Dynamic behavior of axially functionally graded pipes conveying fluid. Mathematical Problems in Engineering, 2017. (11 pages)

demonstrated for calculating the transverse displacement at various points of axially FG pipes conveying fluid. The proposed approach was verified by comparing the obtained results with the available solutions reported in the literature. Moreover, parametric studies were performed to analyze the effects of Young's modulus variation, material distribution, and flow velocity on the dynamic behavior of axially FG pipes conveying fluid.

INTRODUCTION

Pipelines conveying fluid exist widely in many application fields, particularly in nuclear power plants, chemical plants, aeronautic, oil transportation, water supply, heat exchanger devices, human circulation, and so forth. The high velocity internal flow may cause severe flow-induced vibration of piping systems, which may further result in leakages, fatigue failures, high noise, fire, and explosions of the pipes [1]. Extensive investigations have been carried out in the past decades to understand the dynamical behavior of pipes conveying fluid, as described by Païdoussis and Li [2] and Païdoussis [3, 4].

Similar to other structural dynamic problems, the earliest concern of fluid-conveying pipes was the free vibration response [5]. Research reveals that the boundary conditions can affect significantly the natural frequencies of the dynamic systems [6, 7]. Some numerical methods such as homotopy perturbation method [8] and precise integration method [9] were developed to analyze the effect of fluid flow velocity on the natural frequencies. Although there are many studies considering the flow velocity as constant, the flow velocity varies with time for the actual industrial problems. The unsteady flow is usually modeled by the superposition of the steady flow and a time-dependent harmonic component, which may induce the dynamic instability due to parametric resonances [10]. For instance, Jin and Song [11] investigated the effect of some physical parameters of the system, such as damping, mean flow velocity, mass ratio, tension, and gravity, on the three regions of parametric resonances of pipes with supported ends conveying pulsating fluid. Panda and Kar [12] analyzed the nonlinear planar vibration of a hinged-hinged pipe conveying fluid with harmonic flow velocity pulsation in the presence of internal resonance. With the development of material technology and application, new materials exhibiting viscoelastic behaviors such as polymer matrix composites are now widely used for pipes. Zhao et al. [13] investigated the dynamic behavior and stability of Maxwell viscoelastic pipes conveying fluid with simply supported ends. Zhang et al. [14] presented a viscoelastic finite element approach to the vibration analysis of viscoelastic Timoshenko pipes conveying fluid. Wang et al. [15] studied the vibration and stability of viscoelastic curved pipes conveying fluid using normalized power series method. Yang et al. [16] investigated the dynamic stability for the transverse vibrations of pipes conveying fluid using the method of multiple scales. To avoid the failure caused by resonance due to the excitation of external forces, many researchers considered the forced vibrations of pipes conveying fluid. Gulyayev and Tolbatov [17] carried out the numerical modelling of self-excited vibrations of tubes containing inner flows of nonhomogeneous boiling fluid. Seo et al. [18] presented the finite element method to predict the forced vibration response of a pipe conveying harmonically pulsating fluid. Liang and Wen [19] studied the forced responses with both an internal resonance and an external periodic excitation of the constant-fluid-conveying pipe by the multidimensional Lindstedt-Poincaré method. In practice, most of the load applied on the industrial pipes is random, therefore, the dynamic response of pipes conveying fluid subjected to random excitation was studied by Zhai et al. [20, 21]. Since the high-temperature environment should be confronted in some industries such as nuclear reactors, space planes, and chemical plants, the vibration behaviors of pipes conveying fluid under thermal loads have been studied in recent years [22–26].

Due to the advantages of being able to withstand severe high-temperature gradient while maintaining structural integrity, functionally graded materials (FGMs) have attracted great interest in a broad range of applications including biomechanical, automotive, aerospace, mechanical, civil, nuclear, and naval engineering [27, 28], as it is known that FGMs are a novel class of composite materials whose composition and/or function is designed to change continuously within the solid. The composites are usually made from a mixture of metals and ceramics to ensure the elastic and toughness properties gradually vary in space, which can prevent delamination and stress concentration in traditional multilayer, laminated composites. For pipes conveying fluid, Sheng and Wang [24] reported the result of an investigation into the coupled vibration characteristics of fluid-filled functionally graded cylindrical shells, while Hosseini and Fazelzadeh [26] investigated the thermomechanical stability of functionally graded thin-walled cantilever pipes conveying flow and loading by compressive axial force. Both of the above-mentioned investigations assumed that the material properties vary along the thickness direction of pipes; however, dynamic behaviors of axially functionally graded systems (structures with material graduation through the longitudinal directions) should be also concerned, as reported by [29–34]. To the authors' knowledge, the literature dealing with the dynamic behavior of fluid-conveying pipes made of axially FGMs is very limited, which forms the motivation for the current work.

In this study, the dynamic behavior of axially functionally graded (FG) pipes conveying fluid is analytically and numerically investigated on the basis of the generalized integral transform technique (GITT), which has been successfully applied in solving the dynamic response of axially moving beams [35], axially moving orthotropic plates [36], fluid-conveying pipes [37] and pipes conveying gas-liquid two-phase flow [38], the wind-induced vibration on overhead conductors [39], the vortex-induced vibration of long flexible cylinders [40], and the transverse vibrations of a cantilever beam with an eccentric tip mass in the axial direction [41]. From an engineering viewpoint, the novel contribution of this investigation is to generate the reliable reference data on the dynamic behavior of axially FG conveying fluid, while, from a mathematical viewpoint, this work provides a feasible numerical solution of the variable coefficient partial differential equations governing the phenomenon. The rest of paper is organized as follows. In the next section, the mathematical formulation of the transverse vibration problem of axially FG pipes conveying fluid is presented. In Section 3, the hybrid numerical-analytical solution is obtained by carrying out integral transform. Numerical results of proposed method including transverse displacements and their corresponding convergence behavior and verification are presented in Section 4. A parameter study is then performed to investigate the effects of material distributions and mass ratios on natural frequencies and vibration amplitude of pipes conveying fluid, respectively. Besides, the variation of the dimensionless frequencies with Young's modulus ratio, power exponent, and flow velocity are also presented. Finally, the paper ends in Section 5 with conclusions and perspectives.

MATHEMATICAL FORMULATION

We consider a fluid-conveying pipe made of axially functionally graded (FG) material based on Euler-Bernoulli beam theory, as illustrated in Figure 1. If gravity, internal damping, external imposed tension, and pressurization effects are either absent or neglected, the equation of motion of the FG pipe can be derived following the Newtonian derivation by means of decomposing an infinitesimal pipe-fluid element into the pipe element and the fluid element, according to the procedure given by Païdoussis [3]:

$$\frac{\partial^2}{\partial x^2}\left[E(x)\,I\frac{\partial^2 w}{\partial x^2}\right] + m_f v^2\frac{\partial^2 w}{\partial x^2} + 2m_f v\frac{\partial^2 w}{\partial x \partial t}$$

$$+\left[m_f + \rho(x)\,A\right]\frac{\partial^2 w}{\partial t^2} = 0, \tag{1a}$$

subjected to the clamped-clamped boundary conditions

$$w(0, t) = 0, \tag{1b}$$

$$\frac{\partial w(0,t)}{\partial x} = 0, \tag{1c}$$

$$w(L, t) = 0, \tag{1d}$$

$$\frac{\partial w(L,t)}{\partial x} = 0, \tag{1e}$$

where $w(x, t)$ is the transverse displacement, $E(x)I$ is the flexural rigidity of the pipe which depends upon both Young's modulus $E(x)$ and the inertial moment of cross-sectional area I, m_f is the mass of fluid per unit length, V is the steady flow velocity, $\rho(x)A$ is the mass of the pipe per unit length which depends upon mass density $\rho(x)$ and crosssectional area A, and L is the pipe length. Note that, for the axially FG pipe, (x) and $\rho(x)$ are functions of the axial coordinate x. In addition, we assume that the coefficient of thermal expansion and the thermal conductivity of the FG material are constant along the pipe; therefore, the influence of the thermal environment can be ignored. The following dimensionless variables are introduced:

$$x^* = \frac{x}{L}, \tag{2a}$$

$$w^* = \frac{w}{L}, \tag{2b}$$

$$t^* = \frac{t}{L^2}\sqrt{\frac{E_0 I}{m_f + \rho_0 A}}, \tag{2c}$$

$$v^* = vL\sqrt{\frac{m_f}{E_0 I}}, \tag{2d}$$

$$\alpha(x^*) = \frac{E(x^*)}{E_0}, \tag{2e}$$

$$\beta = \sqrt{\frac{m_f}{m_f + \rho_0 A}}, \tag{2f}$$

$$\gamma(x^*) = \frac{m_f + \rho(x^*) A}{m_f + \rho_0 A}, \tag{2g}$$

where E_0 and ρ_0 are the corresponding Young's modulus and mass density at the end $x^* = 0$. Substituting (2a)–(2g) into (1a)–(1e) gives the dimensionless equation (dropping the superposed asterisks for simplicity)

$$\frac{\partial^2}{\partial x^2}\left[\alpha(x)\frac{\partial^2 w}{\partial x^2}\right] + v^2\frac{\partial^2 w}{\partial x^2} + 2\beta v\frac{\partial^2 w}{\partial x \partial t} + \gamma(x)\frac{\partial^2 w}{\partial t^2}$$

$$= 0, \tag{3a}$$

together with the boundary conditions

$$w(0, t) = 0, \tag{3b}$$

$$\frac{\partial w(0,t)}{\partial x} = 0, \tag{3c}$$

$$w(1, t) = 0, \tag{3d}$$

$$\frac{\partial w(1,t)}{\partial x} = 0. \tag{3e}$$

The initial conditions are defined as follows:

$$w(x, 0) = 0, \tag{4a}$$

$$\dot{w}(x,0) = v_0 \sin(\pi x). \tag{4b}$$

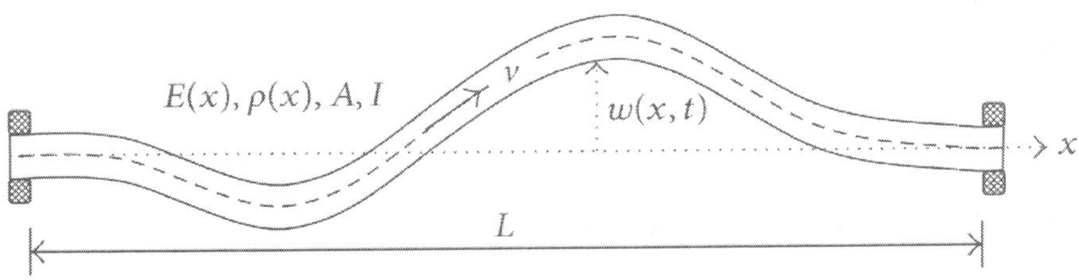

Figure 1: Illustration of an axially functionally graded pipe conveying fluid; the transverse displacement of which is described by (x, t).

INTEGRAL TRANSFORM SOLUTION

To conduct the integral transform technique, the additional eigenvalue problem should be introduced for the governing equation (3a) with the boundary conditions (3b)–(3e). The coordinate "x" is eliminated by using integral transformation, and the eigenvalue problem is employed for the transverse displacement as follows:

$$\frac{d^4 X_i(x)}{dx^4} = \mu_i^4 X_i(x), \quad 0 < x < 1, \tag{5a}$$

with the boundary conditions

$$X_i(0) = 0, \tag{5b}$$

$$\frac{dX_i(0)}{dx} = 0, \tag{5c}$$

$$X_i(1) = 0, \tag{5d}$$

$$\frac{dX_i(1)}{dx} = 0, \tag{5e}$$

where $X_i(x)$ and μ_i are the eigenfunctions and eigenvalues of problem (5a)–(5e), respectively. The orthogonality property should be satisfied by the eigenfunctions

$$\int_0^1 X_i(x) X_j(x)\, dx = \delta_{ij} N_i, \tag{6}$$

with $\delta_{ij} = 0$ when $i \neq j$ and $\delta_{ij} = 1$ when $i=j$. The norm is defined as

$$N_i = \int_0^1 X_i^2(x)\, dx. \tag{7}$$

Eigenvalue problem (5a)–(5e) can be solved analytically to generate

$$X_i(x) = \begin{cases} \dfrac{\cos[\mu_i(x-1/2)]}{\cos(\mu_i/2)} - \dfrac{\cosh[\mu_i(x-1/2)]}{\cosh(\mu_i/2)}, & \text{for } i \text{ odd,} \\ \dfrac{\sin[\mu_i(x-1/2)]}{\sin(\mu_i/2)} - \dfrac{\sinh[\mu_i(x-1/2)]}{\sinh(\mu_i/2)}, & \text{for } i \text{ even,} \end{cases} \tag{8}$$

where the eigenvalues are calculated through the transcendental equations:

$$\tanh\left(\frac{\mu_i}{2}\right) = \begin{cases} -\tan\left(\dfrac{\mu_i}{2}\right), & \text{for } i \text{ odd,} \\ \tan\left(\dfrac{\mu_i}{2}\right), & \text{for } i \text{ even,} \end{cases} \tag{9}$$

and the normalization of integral is

$$N_i = 1, \quad i = 1, 2, 3, \ldots. \tag{10}$$

Then, the normalized eigenfunction yields

$$\widetilde{X}_i(x) = \frac{X_i(x)}{N_i^{1/2}}. \tag{11}$$

For the transverse displacement, the integral transformation and the inversion equations are as follows:

$$\overline{w}_i(t) = \int_0^1 \widetilde{X}_i(x)\, w(x,t)\, dx, \quad \text{transform}, \tag{12a}$$

$$w(x,t) = \sum_{i=1}^{\infty} \widetilde{X}_i(x)\, \overline{w}_i(t), \quad \text{inverse}. \tag{12b}$$

Now, the integral transformation procedure is conducted by operation of (3a) with $\int_0^1 \widetilde{X}_i(x) dx$, to obtain the transformed transverse displacement equation system:

$$\sum_{j=1}^{\infty} \left[A_{ij} \frac{d^2 \overline{w}_j(t)}{dt^2} + 2\beta v B_{ij} \frac{d\overline{w}_j(t)}{dt} + \left(v^2 C_{ij} + D_{ij} + 2E_{ij} + \mu_j^4 F_{ij}\right) \overline{w}_j(t) \right] = 0,$$

$$i = 1, 2, 3, \ldots, \tag{13a}$$

where the coefficients are given by thc integrals below

$$A_{ij} = \int_0^1 \gamma(x)\, \widetilde{X}_i(x)\, \widetilde{X}_j(x)\, dx, \tag{13b}$$

$$B_{ij} = \int_0^1 \widetilde{X}_i(x)\, \widetilde{X}_j(x)'\, dx, \tag{13c}$$

$$C_{ij} = \int_0^1 \widetilde{X}_i(x)\, \widetilde{X}_j(x)''\, dx, \tag{13d}$$

$$D_{ij} = \int_0^1 \alpha(x)''\, \widetilde{X}_i(x)\, \widetilde{X}_j(x)''\, dx, \tag{13e}$$

$$E_{ij} = \int_0^1 \alpha(x)'\, \widetilde{X}_i(x)\, \widetilde{X}_j(x)'''\, dx, \tag{13f}$$

$$F_{ij} = \int_0^1 \alpha(x)\, \widetilde{X}_i(x)\, \widetilde{X}_j(x)\, dx. \tag{13g}$$

In a similar way, the initial conditions can be also transformed to eliminate "x" coordinate:

$$\overline{w}_i(0) = 0, \tag{14a}$$

$$\frac{d\bar{w}_i(0)}{dt} = v_0 \int_0^1 \tilde{X}_i(x)\sin(\pi x)\,dx,$$

$$i = 1, 2, 3, \ldots. \qquad (14b)$$

In the computational process, the expansion for the transverse displacement is truncated to finite orders *NW*. Equations (13a)–(13g), (14a), and (14b) in the truncated series are calculated using the NDSolve in Mathematica [43]. Once the values of $\overline{w}_i$ are determined, the inversion formulas equation (12b) is subsequently employed to yield explicit expression for the transverse displacement (x, t).

RESULTS AND DISCUSSION

Convergence Behavior of the Solution

We now present the convergence behavior of numerical results for the transverse displacement (x, t) of a functionally graded pipe conveying fluid calculated using the GITT approach. For the case examined, the geometrical parameters adopted by Zhai et al. [20] are taken in (1a)–(1e): $L = 1010$ mm, $D = 22.85$ mm, and $d = 19.65$ mm, where D and d are the outer and inner diameters of pipe cross section, respectively. In this study, we assume that the material properties of the pipe, such as Young's modulus E and mass density ρ, vary continuously as a power law through the pipe axis [34]: $P(x) = (P_L - P_R)(1 - x/L)^k + P_R$, where P_R and P_L are the corresponding material of the right and the left side of the pipe, and k is the nonnegative power-law exponent which dictates the material variation profile through the pipe axis. In the following calculations, aluminum and zirconia are chosen for the corresponding material of the left and the right sides of the pipe, respectively, the material properties of which are [29] $Ea = 70$ GPa, $\rho a = 2702$ kg/m^3 for aluminum, and $Ez = 200$ GPa, $\rho z = 5700$ kg/m^3 for zirconia. The fluid density conveying in the pipe is $\rho f = 1000$ kg/m^3 . The dimensionless variables can be obtained through (2a)–(2g). The solution of the system, (13a)–(13g), (14a), and (14b), is obtained with $NW \leq 16$ to analyze the convergence behavior.

The dimensionless transverse displacement $w(x, t)$ at different positions, $x = 0.1$, 0.3, 0.5, 0.7, and 0.9, of axially FG pipes conveying fluid is presented in Tables 1 and 2, respectively. The convergence behavior of the integral transform solution is examined for increasing truncation terms $NW = 4$, 8, 12, and 16 at$t = 10$, 20, and 30, respectively. For the dimensionless transverse displacement with V = 1.0 and k = 1.0, it can be observed that convergence is achieved essentially with a reasonably low truncation order ($N \leq 8$). For a full convergence to three significant digits, more terms (e.g., $N \leq 12$) are required. The results at t = 30 indicate that the excellent convergence behavior of the integral transform solution does not change with time, verifying the good longtime numerical stability of the scheme. For the dimensionless transverse displacement with V = 3.0 and k = 1.0, convergence to three significant digits is achieved with truncation order $N \leq 16$, which demonstrates that the increasing of V can make the solution with a relatively slow convergence. For the same cases, the profiles of the transverse displacement at t = 20 are illustrated in Figure 2 with different truncation orders, where it can be clearly seen that the convergence behavior of the integral transform solution for the case of V = 1.0 is better than the case of V = 3.0.

Table 1: Convergence behavior of the dimensionless transverse displacement $w(x, t)$ of FG pipes conveying fluid for V = 1.0 and k = 1.0

x	$NW = 4$	$NW = 8$	$NW = 12$	$NW = 16$
$t = 10$				
0.1	0.0044244	0.0037878	0.0039159	0.0039563
0.3	0.0191635	0.0206171	0.0206690	0.0207072
0.5	0.0207420	0.0244299	0.0245583	0.0246031
0.7	0.0127285	0.0124189	0.0123877	0.0124042
0.9	0.0021788	0.0012293	0.0012868	0.0013476
$t = 20$				
0.1	−0.0060519	−0.0062440	−0.0060664	−0.0060803
0.3	−0.0291388	−0.0296250	−0.0298559	−0.0298584
0.5	−0.0358924	−0.0382805	−0.0382784	−0.0382978
0.7	−0.0235454	−0.0256259	−0.0257693	−0.0257332
0.9	−0.0040988	−0.0044978	−0.0042350	−0.0042428
$t = 30$				
0.1	0.0064297	0.0066167	0.0064907	0.0064724
0.3	0.0325025	0.0313819	0.0315394	0.0315556
0.5	0.0405411	0.0390336	0.0384158	0.0383750
0.7	0.0251618	0.0234384	0.0238947	0.0239246
0.9	0.0040959	0.0038549	0.0037553	0.0037439

Table 2: Convergence behavior of the dimensionless transverse displacement $w(x, t)$ of FG pipes conveying fluid for V = 3.0 and $k = 1.0$

x	$NW = 4$	$NW = 8$	$NW = 12$	$NW = 16$
$t = 10$				
0.1	0.0030463	0.0024006	0.0026071	0.0026145
0.3	0.0083268	0.0100507	0.0099890	0.0099936
0.5	0.0025960	0.0057039	0.0059138	0.0059497
0.7	−0.0013099	−0.0021990	−0.0022931	−0.0023027
0.9	−0.0006775	−0.0011992	−0.0010060	−0.0010314
$t = 20$				
0.1	−0.0019432	−0.0013335	−0.0015234	−0.0015166
0.3	−0.0071319	−0.0088004	−0.0088014	−0.0088167
0.5	−0.0055658	−0.0118234	−0.0123907	−0.0124569
0.7	−0.0028723	−0.0073600	−0.0070684	−0.0071240
0.9	0.0005661	−0.0009220	−0.0011155	−0.0011267
$t = 30$				
0.1	0.0033760	0.0043476	0.0041930	0.0041604
0.3	0.0113849	0.0171511	0.0168645	0.0169404
0.5	0.0093804	0.0150589	0.0157554	0.0157545
0.7	0.0054698	0.0066738	0.0066753	0.0066903
0.9	0.0008500	0.0005556	0.0008388	0.0009427

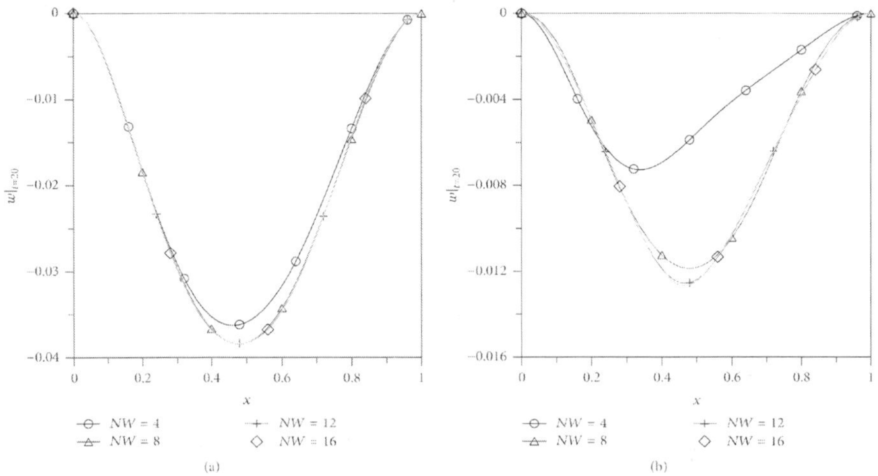

Figure 2: GITT solutions with different truncation orders *NW* for the dimensionless transverse displacement profiles: (a) $(x, t)|_{t=20}$ for V = 1.0 and (b) $w(x, t)|_{t=20}$ for V = 3.0 of FG pipes conveying fluid.

Verification of the Solution

The influence of fluid velocity on the first five dimensionless natural frequencies of axially FG pipes conveying fluid with the power-law exponent $k = 1.0$ is presented in Table 3. To obtain the natural circular frequencies for the transverse vibration of the system, the coupled ODEs, (13a)–(13g), can be represented in the matrix form as follows:

$$\mathbf{M}\ddot{\mathbf{w}}(t) + \mathbf{C}\dot{\mathbf{w}}(t) + \mathbf{K}\mathbf{w}(t) = \mathbf{F}(t). \qquad (15)$$

The fluid velocities of 0, 10, 20, 30, 40, and 50 m/s are considered, and, with the increasing of the velocity, all of the five natural frequencies of the system decrease. To demonstrate the validity and accuracy of the proposed GITT approach, GITT solution for dimensionless free vibration frequencies for clamped-clamped pipes with $m_f = 0$ and k=0 is calculated to compare with the results presented in the literature [42], where the excellent agreement between them can be found, as shown in Table 3.

Table 3: Influence of fluid velocity on the first five dimensionless natural frequencies of FG pipes conveying fluid ($k = 1.0$)

Fluid velocity (m/s)	ω_1	ω_2	ω_3	ω_4	ω_5
0	22.373*	61.673*	120.903*	199.859*	298.556*
0	22.373**	61.673**	120.903**	199.859**	298.556**
0	26.866	74.323	145.936	241.576	361.572
10	26.850	74.306	145.918	241.558	361.556
20	26.801	74.255	145.865	241.506	361.508
30	26.720	74.171	145.777	241.419	361.428
40	26.605	74.052	145.653	241.296	361.315
50	26.457	73.899	145.493	241.139	361.171

*: dimensionless free vibration frequencies for clamped-clamped beams [42].

**: GITT solution for dimensionless free vibration frequencies for clamped-clamped pipes ($mf = 0$ and k=0).

Parametric Study

In this section, transverse displacement of axially FG pipes conveying fluid with clampedclamped boundary conditions is analyzed to illustrate the applicability of the proposed approach. Different values of the mass ratio β, Young's modulus ratio E_{ratio}, the material distribution k, and the dimensionless flow velocity V are chosen to assess their effects on the dynamic behavior of the system. In the following analysis, we use a relative high truncation order, $NW = 16$, for a sufficient accuracy.

The Effect of Young's Modulus Variation

The first three dimensionless natural frequencies of the axially FG pipe conveying fluid for different Young's modulus ratios, powerlaw exponent, and mass ratios are tabulated in Tables 4–6, where the following parameters are adopted: $E_l = 70$ GPa, $E_{\text{ratio}} = E_L/E_R$, $\rho_f = 1000$ kg/m^3 , $\rho_{\text{ratio}} = \rho_L/\rho_R = 1.0$, and $\rho_L/\rho_f = 4.0$. The mass ratio is calculated by (2a)–(2g) with the specified value of pipe thickness t and $d = 19.65$ mm. The flow velocity of the fluid is 20 m/s. For the specified modulus ratio and power exponent, all of the first three dimensionless natural frequencies increase with the decrease of the mass ratio. The effect of the modulus ratios on the fundamental frequency of fluid-conveying pipe with the mass ratio of 0.796 is presented in Figure 3. It is observed that the fundamental frequency decreases significantly with increasing of modulus ratio especially for large power exponent. On the other hand, no significant changes can be seen in the fundamental frequency for different modulus ratios for lower value of power exponent. Figure 4 illustrates the variation of the fundamental frequency with the power exponent for fluidconveying pipe with the mass ratio of 0.796, which shows that the increase in power exponent causes the increase in frequency for Eratio < 1 and the decrease in frequency for $E_{ratio} > 1$ and no changes occur for $E_{ratio} = 1$.

Table 4: The variation of the first natural angular frequencies for different material distributions k and mass ratios β when flow velocity is 20 m/s, $E_{ratio} = E_l / E_r$, $\rho_{ratio} = \rho_l / \rho_r = 1.0$, and $\rho_l / \rho_f = 4.0$

t (mm)	b	E_{ratio}	$k = 0.0$	$k = 0.1$	$k = 0.2$	$k = 0.5$	$k = 1.0$	$k = 2.0$	$k = 5.0$	$k = 10.0$
1.0	0.796	0.25	22.1909	25.2702	27.1859	30.5936	33.4550	36.1690	39.0072	40.4687
		0.50	22.1909	23.3497	24.1498	25.6193	26.8588	28.0211	29.1967	29.7777
		1.00	22.1909	22.1909	22.1909	22.1909	22.1909	22.1909	22.1909	22.1909
		2.00	22.1909	21.5289	20.9829	19.8618	18.8849	17.9785	17.0936	16.6791
		4.00	22.1909	21.1690	20.2790	18.3220	16.5444	14.9059	13.3283	12.5955
2.0	0.672	0.25	22.3008	25.3695	27.2788	30.6755	33.5282	36.2347	39.0667	40.5260
		0.50	22.3008	23.4554	24.2525	25.7159	26.9498	28.1068	29.2778	29.8571
		1.00	22.3008	22.3008	22.3008	22.3008	22.3008	22.3008	22.3008	22.3008
		2.00	22.3008	21.6413	21.0977	19.9830	19.0138	18.1167	17.2418	16.8313
		4.00	22.3008	21.2828	20.3968	18.4519	16.6913	15.0763	13.5273	12.8067
3.0	0.587	0.25	22.3333	25.3989	27.3063	30.6998	33.5499	36.2543	39.0844	40.5430
		0.50	22.3333	23.4867	24.2829	25.7445	26.9767	28.1322	29.3018	29.8807
		1.00	22.3333	22.3333	22.3333	22.3333	22.3333	22.3333	22.3333	22.3333
		2.00	22.3333	21.6746	21.1317	20.0189	19.0520	18.1575	17.2856	16.8762
		4.00	22.3333	21.3165	20.4318	18.4904	16.7348	15.1266	13.5858	12.8688

Table 5: The variation of the second natural angular frequencies for different material distributions k and mass ratios β when flow velocity is 20 m/s, $E_{ratio} = E_l / E_r$, $\rho_{ratio} = \rho_l / \rho_r = 1.0$, and $\rho_l / \rho_f = 4.0$

t (mm)	b	E_{ratio}	$k = 0.0$	$k = 0.1$	$k = 0.2$	$k = 0.5$	$k = 1.0$	$k = 2.0$	$k = 5.0$	$k = 10.0$
1.0	0.796	0.25	61.5004	69.4713	74.6411	84.2515	92.8974	101.469	110.495	114.810
		0.50	61.5004	64.4750	66.6003	70.6875	74.4028	78.0533	81.7840	83.5138
		1.00	61.5004	61.5004	61.5004	61.5004	61.5004	61.5004	61.5004	61.5004
		2.00	61.5004	59.8213	58.4126	55.3901	52.5094	49.6838	46.9156	45.7453
		4.00	61.5004	58.9166	56.6441	51.4604	46.2746	41.1824	36.2901	34.2703

2.0	0.672	0.25	61.5959	69.5577	74.7221	84.3233	92.9617	101.527	110.547	114.860
		0.50	61.5959	64.5668	66.6895	70.7715	74.4821	78.1281	81.8547	83.5830
		1.00	61.5959	61.5959	61.5959	61.5959	61.5959	61.5959	61.5959	61.5959
		2.00	61.5959	59.9190	58.5124	55.4955	52.6216	49.8040	47.0441	45.8770
		4.00	61.5959	59.0156	56.7467	51.5737	46.4032	41.3315	36.4620	34.4515
3.0	0.587	0.25	61.6276	69.5864	74.7490	84.3471	92.9830	101.546	110.564	114.876
		0.50	61.6276	64.5973	66.7191	70.7995	74.5085	78.1529	81.8782	83.6060
		1.00	61.6276	61.6276	61.6276	61.6276	61.6276	61.6276	61.6276	61.6276
		2.00	61.6276	59.9515	58.5456	55.5305	52.6589	49.8439	47.0868	45.9207
		4.00	61.6276	59.0485	56.7809	51.6114	46.4459	41.3809	36.5192	34.5117

Table 6: The variation of the third natural angular frequencies for different material distributions k and mass ratios β when flow velocity is 20 m/s, $E_{\text{ratio}} = E_l / E_r$, $\rho_{\text{ratio}} = \rho_l / \rho_r = 1.0$, and $\rho_l / \rho_f = 4.0$

t (mm)	b	E_{ratio}	$k = 0.0$	$k = 0.1$	$k = 0.2$	$k = 0.5$	$k = 1.0$	$k = 2.0$	$k = 5.0$	$k = 10.0$
1.0	0.796	0.25	120.726	135.934	145.948	164.997	182.693	200.618	219.333	227.933
		0.50	120.726	126.374	130.466	138.522	146.073	153.635	161.360	164.876
		1.00	120.726	120.726	120.726	120.726	120.726	120.726	120.726	120.726
		2.00	120.726	117.564	114.895	109.017	103.185	97.3618	91.6997	89.3743
		4.00	120.726	115.872	111.598	101.632	91.1682	80.6355	70.6481	66.6398
2.0	0.672	0.25	120.822	136.020	146.028	165.068	182.757	200.677	219.386	227.984
		0.50	120.822	126.466	130.555	138.606	146.153	153.711	161.432	164.946
		1.00	120.822	120.822	120.822	120.822	120.822	120.822	120.822	120.822
		2.00	120.822	117.663	114.996	109.124	103.298	97.4820	91.8275	89.5047
		4.00	120.822	115.972	111.702	101.747	91.2969	80.7826	70.8185	66.8177
3.0	0.587	0.25	120.855	136.049	146.056	165.092	182.779	200.697	219.405	228.002
		0.50	120.855	126.498	130.586	138.635	146.180	153.737	161.457	164.970
		1.00	120.855	120.855	120.855	120.855	120.855	120.855	120.855	120.855
		2.00	120.855	117.697	115.031	109.160	103.336	97.5231	91.8713	89.5494
		4.00	120.855	116.006	111.738	101.786	91.3409	80.8330	70.8770	66.8789

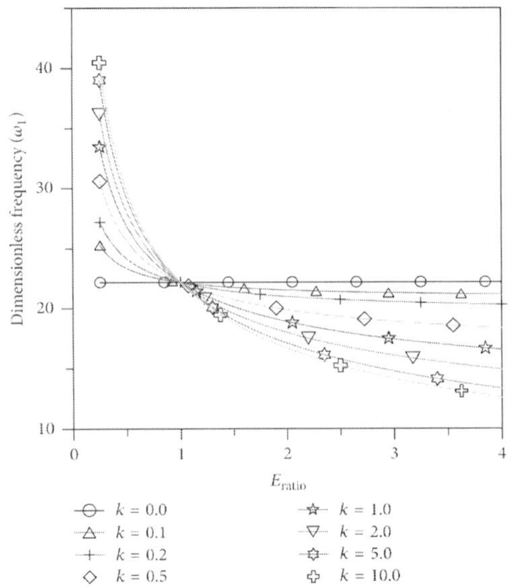

Figure 3: Variation of the dimensionless frequency (ω_1) with Young's modulus ratio.

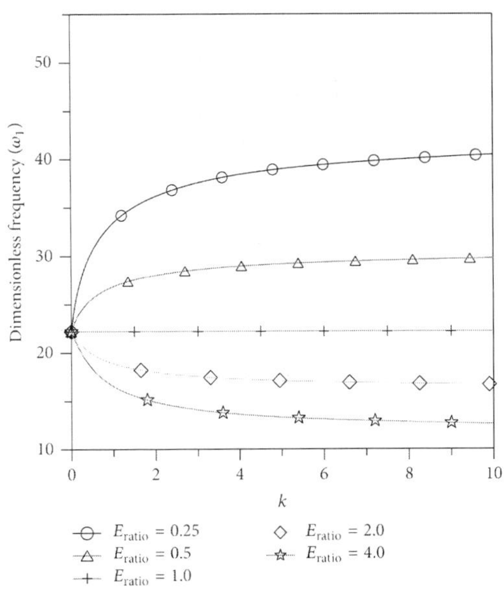

Figure 4: Variation of the dimensionless frequency (ω_1) with power exponent (k).

The Effect of Material Distribution

To examine the effect of material distribution on the frequencies of the axially FG pipe conveying fluid, the integral transform solutions are obtained based on the material properties given in Section 4.1 and the fluid velocity of 20m/s. The first five natural angular frequencies for different mass ratios and power exponents are reported in Table 7. All of the natural frequencies (ω_1, $\omega_{2,\ldots,5}$) increase with the power exponent for the specified mass ratio. Note that the natural frequencies increase with the decrease of the mass ratio for the case of k = 0.0, which means the pipe is made of singlecomponent, aluminum. However, when considering the FG material with the power exponents k = 0.1, 0.2, . . . , 10, the natural frequencies decrease with the decrease of the mass ratio.

Table 7: The variation of natural angular frequencies for different material distributions k and mass ratios β when flow velocity is 20 m/s

t (mm)	b	ω_i	k = 0.0	k = 0.1	k = 0.2	k = 0.5	k = 1.0	k = 2.0	k = 5.0	k = 10.0
1.0	0.796	i=1	22.1835	23.8868	24.9026	26.4715	27.4513	28.1186	28.8294	29.4111
		i=2	61.5036	65.7479	68.3906	72.7928	76.1318	79.0050	81.8223	83.2525
		i=3	120.732	128.668	133.700	142.428	149.573	156.161	162.514	165.285
		i=4	199.693	212.486	220.721	235.305	247.679	259.404	270.748	275.362
		i=5	298.392	317.258	329.420	351.361	370.905	389.776	407.194	413.686
2.0	0.672	i=1	22.2976	23.8315	24.6962	25.9020	26.4864	26.7497	27.1061	27.5738
		i=2	61.5972	65.3592	67.5855	71.0306	73.3695	75.2171	77.0180	78.0331
		i=3	120.825	127.785	131.993	138.872	144.125	148.770	153.121	155.007
		i=4	199.781	210.932	217.798	229.340	238.617	247.159	255.195	258.328
		i=5	298.477	314.839	324.951	342.322	357.127	371.143	383.736	388.144
3.0	0.587	i=1	22.3317	23.7692	24.5496	25.5605	25.9443	26.0152	26.2112	26.6270
		i=2	61.6283	65.1117	67.1047	70.0359	71.8650	73.2123	74.5377	75.3643
		i=3	120.856	127.252	131.000	136.890	141.179	144.870	148.272	149.752
		i=4	199.812	210.013	216.115	226.033	233.735	240.712	247.171	249.620
		i=5	298.508	313.428	322.400	337.340	349.740	361.370	371.651	375.089

The variation of vibration amplitudes for different material distributions and mass ratios is listed in Table 8. It can be seen that the vibration amplitude decreases with the increase of the power exponent and increases with the decrease of the mass ratio.

Table 8: The variation of vibration amplitudes for different material distributions k and mass ratios β when flow velocity is 20 m/s

t (mm)	b	$k = 0.0$	$k = 0.1$	$k = 0.2$	$k = 0.5$	$k = 1.0$	$k = 2.0$	$k = 5.0$	$k = 10.0$
1.0	0.796	0.05064	0.04742	0.04562	0.04265	0.04067	0.03966	0.03864	0.03787
2.0	0.672	0.05071	0.04779	0.04607	0.04365	0.04226	0.04176	0.04093	0.04042
3.0	0.587	0.05076	0.04794	0.04657	0.04431	0.04315	0.04276	0.04237	0.04192

In addition, the effect of material distribution on the critical velocity of fluid-conveying pipe with the material properties given in Section 4.1 is analyzed, as shown in Figure 5, which exhibits the variation of the fundamental frequency with the flow velocity of fluid for different power exponents. It can be clearly seen that the fundamental frequency decreases with the flow velocity for the specified power exponent, and the critical velocity ($\omega_1 = 0$) increases with the power exponent.

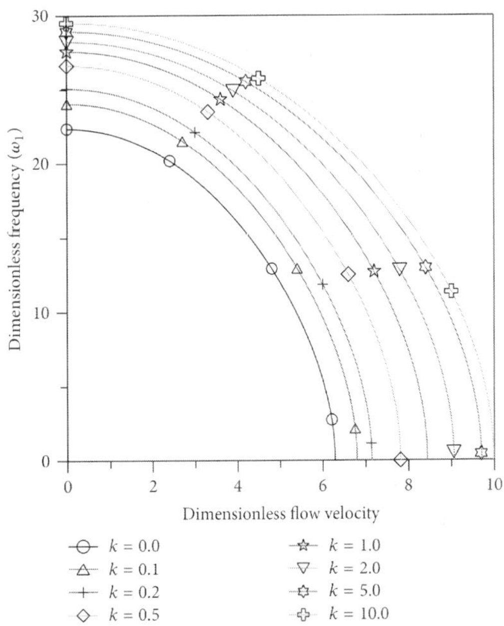

Figure 5: Variation of the dimensionless frequency (ω_1) with dimensionless flow velocity (V).

CONCLUSIONS

The generalized integral transform technique (GITT) has proved in this paper to be a good approach for the analysis of dynamic behavior of an axially FG pipe conveying fluid, providing an accurate numerical-analytical solution for the natural frequencies and transverse displacements. The investigation shows that the solutions converge to the values with three significant figures at a reasonable low truncation order $N \leq 12$ for V = 1.0, and the increasing of V can make the solution with a relatively slow convergence. The numerical results obtained are in good agreement with the ones presented in the literature. The parametric studies indicate that the fundamental frequency decreases significantly with increasing of modulus ratio especially for large power exponent, while no significant changes can be seen in the fundamental frequency for different modulus ratios for lower value of power exponent. The increase in power exponent causes the increase in frequency for $E_{ratio} < 1$ and the decrease in frequency for $E_{ratio} > 1$ and no changes occur for $E_{ratio} = 1$. The natural frequency increases with the power exponent, and the natural frequency decreases with the decrease of the mass ratio when considering the pipe is made of the FG material with the power exponents $k = 0.1, 0.2, \ldots, 10$. The vibration amplitude decreases with the increase of the power exponent and increases with the decrease of the mass ratio. The critical velocity of fluidconveying pipe increases with the power exponent. For future investigation, the proposed approach can be employed to predict the dynamic behavior of a transversally FG pipe conveying fluid and for more general boundary conditions.

ACKNOWLEDGMENTS

The work was supported by National Natural Science Foundation of China (Grant no. 51509258), Science Foundation of China University of Petroleum, Beijing (Grants nos. 2462013YJRC003 and C201602), National Key Research and Development Plan (Grant no. 2016YFC0303700), and CNPq (Grant no. 306618/2010-9) and CAPES and FAPERJ (Grant no. E-26/102.871/2012) of Brazil.

REFERENCES

1. M. H. Sadeghi and M. H. Karimi-Dona, "Dynamic behavior of a fluid conveying pipe subjected to a moving sprung mass—an FEM-state space approach," International Journal of Pressure Vessels and Piping, vol. 88, no. 4, pp. 123–131, 2011.
2. M. P. Païdoussis and G. X. Li, "Pipes conveying fluid: a model dynamical problem," Journal of Fluids and Structures, vol. 7, no. 2, pp. 137–204, 1993.
3. M. P. Païdoussis, Fluid-Structure Interactions: Slender Structures and Axial Flow, Academic Press, San Diego, Calif, USA, 1998.
4. M. P. Païdoussis, "The canonical problem of the fluid-conveying pipe and radiation of the knowledge gained to other dynamics problems across Applied Mechanics," Journal of Sound and Vibration, vol. 310, no. 3, pp. 462–492, 2008.
5. R. H. Long, "Experimental and theoretical study of transverse vibration of a tube containing flowing fluid," Journal of Applied Mechanics Transactions of the ASME, vol. 22, no. 1, pp. 65–68, 1955.
6. M. P. Païdoussis and N. T. Issid, "Dynamic stability of pipes conveying fluid," Journal of Sound and Vibration, vol. 33, no. 3, pp. 267–294, 1974.
7. M. P. Païdoussis, "Flutter of conservative systems of pipes conveying in compressible fluid," Journal of Mechanical Engineering Science, vol. 17, no. 1, pp. 19–25, 1975.
8. M.-R. Xu, S.-P. Xu, and H.-Y. Guo, "Determination of natural frequencies of fluid-conveying pipes using homotopy perturbation method," Computers & Mathematics with Applications, vol. 60, no. 3, pp. 520–527, 2010.
9. L. Liu and F. Xuan, "Flow-induced vibration analysis of supported pipes conveying pulsating fluid using precise integration method," Mathematical Problems in Engineering, vol. 2010, Article ID 806475, 15 pages, 2010.
10. J. H. Ginsberg, "The dynamic stability of a pipe conveying a pulsatile flow," International Journal of Engineering Science, vol. 11, no. 9, pp. 1013–1024, 1973. ·
11. J. D. Jin and Z. Y. Song, "Parametric resonances of supported pipes conveying pulsating fluid," Journal of Fluids and Structures, vol. 20, no. 6, pp. 763–783, 2005.

12. L. N. Panda and R. C. Kar, "Nonlinear dynamics of a pipe conveying pulsating fluid with combination, principal parametric and internal resonances," Journal of Sound and Vibration, vol. 309, no. 3–5, pp. 375–406, 2008.
13. F.-Q. Zhao, Z.-M. Wang, Z.-Y. Feng, and H.-Z. Liu, "Stability analysis of Maxwell viscoelastic pipes conveying fluid with both ends simply supported," Applied Mathematics and Mechanics, vol. 22, no. 12, pp. 1436–1445, 2001.
14. Y. L. Zhang, D. G. Gorman, and J. M. Reese, "A modal and damping analysis of viscoelastic Timoshenko tubes conveying fluid," International Journal for Numerical Methods in Engineering, vol. 50, no. 2, pp. 419–433, 2001.
15. Z.-M. Wang, Z.-W. Zhang, and F.-Q. Zhao, "Stability analysis of viscoelastic curved pipes conveying fluid," Applied Mathematics and Mechanics, vol. 26, no. 6, pp. 807–813, 2005.
16. X. Yang, T. Yang, and J. Jin, "Dynamic stability of a beam-model viscoelastic pipe for conveying pulsative fluid," Acta Mechanica Solida Sinica, vol. 20, no. 4, pp. 350–356, 2007.
17. V. I. Gulyayev and E. Y. Tolbatov, "Forced and self-excited vibrations of pipes containing mobile boiling fluid clots," Journal of Sound and Vibration, vol. 257, no. 3, pp. 425–437, 2002.
18. Y. S. Seo, W. B. Jeong, S. H. Jeong, J. S. Oh, and W. S. Yoo, "Finite element analysis of forced vibration for a pipe conveying harmonically pulsating fluid," JSME International Journal, Series C: Mechanical Systems, Machine Elements and Manufacturing, vol. 48, no. 4, pp. 688–694, 2006.
19. F. Liang and B. C. Wen, "Forced vibrations with internal resonance of a pipe conveying fluid under external periodic excitation," Acta Mechanica Solida Sinica, vol. 24, no. 6, pp. 477–483, 2011.
20. H.-B. Zhai, Z.-Y. Wu, Y.-S. Liu, and Z.-F. Yue, "Dynamic response of pipeline conveying fluid to random excitation," Nuclear Engineering and Design, vol. 241, no. 8, pp. 2744–2749, 2011.
21. H.-B. Zhai, Z.-Y. Wu, Y.-S. Liu, and Z.-F. Yue, "In-plane dynamic response analysis of curved pipe conveying fluid subjected to random excitation," Nuclear Engineering and Design, vol. 256, pp. 214–226, 2013.

22. R. Kadoli and N. Ganesan, "Parametric resonance of a composite cylindrical shell containing pulsatile flow of hot fluid," Composite Structures, vol. 65, no. 3-4, pp. 391–404, 2004.
23. N. Ganesan and R. Kadoli, "A study on the dynamic stability of a cylindrical shell conveying a pulsatile flow of hot fluid," Journal of Sound and Vibration, vol. 274, no. 3–5, pp. 953–984, 2004.
24. G. G. Sheng and X. Wang, "Thermomechanical vibration analysis of a functionally graded shell with flowing fluid," European Journal of Mechanics—A/Solids, vol. 27, no. 6, pp. 1075–1087, 2008.
25. Q. Qian, L. Wang, and Q. Ni, "Instability of simply supported pipes conveying fluid under thermal loads," Mechanics Research Communications, vol. 36, no. 3, pp. 413–417, 2009.
26. M. Hosseini and S. A. Fazelzadeh, "Thermomechanical stability analysis of functionally graded thin-walled cantilever pipe with flowing fluid subjected to axial load," International Journal of Structural Stability and Dynamics, vol. 11, no. 3, pp. 513–534, 2011.
27. H.-S. Shen, Functionally Graded Materials : Nonlinear Analysis of Plates and Shells, CRC Press, Boca Raton, Fla, USA, 2009.
28. Z. Zhong, L. Wu, and W. Chen, Mechanics of Functionally Graded Materials and Structures, Nova Science, New York, NY, USA, 2012.
29. Y. Huang and X.-F. Li, "A new approach for free vibration of axially functionally graded beams with non-uniform cross-section," Journal of Sound and Vibration, vol. 329, no. 11, pp. 2291–2303, 2010.
30. Y. Huang, L.-E. Yang, and Q.-Z. Luo, "Free vibration of axially functionally graded Timoshenko beams with non-uniform cross-section," Composites Part B: Engineering, vol. 45, no. 1, pp. 1493–1498, 2013.
31. A. Shahba, R. Attarnejad, M. T. Marvi, and S. Hajilar, "Free vibration and stability analysis of axially functionally graded tapered Timoshenko beams with classical and non-classical boundary conditions," Composites Part B: Engineering, vol. 42, no. 4, pp. 801–808, 2011.
32. A. Shahba and S. Rajasekaran, "Free vibration and stability of tapered Euler-Bernoulli beams made of axially functionally graded materials," Applied Mathematical Modelling, vol. 36, no. 7, pp. 3094–3111, 2012.
33. M. Şimşek, T. Kocatürk, and Ş. D. Akbaş, "Dynamic behavior of an axially functionally graded beam under action of a moving harmonic load," Composite Structures, vol. 94, no. 8, pp. 2358–2364, 2012.

34. A. E. Alshorbagy, M. A. Eltaher, and F. F. Mahmoud, "Free vibration characteristics of a functionally graded beam by finite element method," Applied Mathematical Modelling, vol. 35, no. 1, pp. 412–425, 2011.
35. C. An and J. Su, "Dynamic response of clamped axially moving beams: integral transform solution," Applied Mathematics and Computation, vol. 218, no. 2, pp. 249–259, 2011.
36. C. An and J. Su, "Dynamic analysis of axially moving orthotropic plates: integral transform solution," Applied Mathematics and Computation, vol. 228, pp. 489–507, 2014.
37. J. Gu, C. An, M. Duan, C. Levi, and J. Su, "Integral transform solutions of dynamic response of a clamped-clamped pipe conveying fluid," Nuclear Engineering and Design, vol. 254, pp. 237–245, 2015.
38. C. An and J. Su, "Dynamic behavior of pipes conveying gas-liquid two-phase flow," Nuclear Engineering and Design, vol. 292, pp. 204–212, 2015.
39. C. F. T. Matt, "On the application of generalized integral transform technique to wind-induced vibrations on overhead conductors," International Journal for Numerical Methods in Engineering, vol. 78, no. 8, pp. 901–930, 2009.
40. J.-J. Gu, C. An, C. Levi, and J. Su, "Prediction of vortex-induced vibration of long flexible cylinders modeled by a coupled nonlinear oscillator: integral transform solution," Journal of Hydrodynamics, vol. 24, no. 6, pp. 888–898, 2012.
41. C. F. Matt, "Simulation of the transverse vibrations of a cantilever beam with an eccentric tip mass in the axial direction using integral transforms," Applied Mathematical Modelling, vol. 37, no. 22, pp. 9338–9354, 2013.
42. A. W. Leissa and M. S. Qatu, Vibrations of Continuous Systems, McGrawHill, New York, NY, USA, 2011.
43. S. Wolfram, The Mathematica Book, Wolfram Media/Cambridge University Press, Champaign, Ill, USA, 5th edition, 2003.

INDEX

N

O

P

Q

R

S